Contemporary Mathematicians

Gian-Carlo Rota[†]
Joseph P.S. Kung

Editors

Gian-Carlo Rota on Analysis and Probability

Selected Papers and Commentaries

Jean Dhombres
Joseph P.S. Kung
Norton Starr
Editors

Birkhäuser
Boston • Basel • Berlin

Jean Dhombres
EHESS, Centre Koyre/CNRS
Histoire des Sciences et des Techniques
27 rue Damesme
Paris, 75013
France

Joseph P.S. Kung
University of North Texas
Department of Mathematics
General Academic Building
Denton, TX 76203-1430
U.S.A.

Norton Starr
Amherst College
Department of Mathematics
Mudd Building
Amherst, MA 01002
U.S.A.

Library of Congress Cataloging-in-Publication Data

Rota, Gian-Carlo, 1932
Gian-Carlo Rota on analysis and probability : selected papers and commentaries / Jean Dhombres, Joseph Kung, Norton Starr, editors.
p. cm.– (Contemporary mathematicians)
Includes bibliographical references.
ISBN 0-8176-4275-7 (alk. paper) — ISBN 3-7643-4275-7 (alk. paper)
1. Mathematical analysis. 2. Probabilities. I. Dhombres, Jean G. II. Kung, Joseph P. S. III. Starr, Norton. IV. Title. V. Series.

QA300.5 .R67 2002
515–dc21

2002035616
CIP

AMS Subject Classifications: 26-XX, 34-XX, 37-XX, 47-XX, 28-XX

Printed on acid-free paper.

Birkhäuser

ISBN 0-8176-4275-7 SPIN 10851306
ISBN 3-7643-4275-7

Printed in the United States of America.

9 8 7 6 5 4 3 2 1

Birkhäuser Boston • Basel • Berlin
A member of BertelsmannSpringer Science+Business Media GmbH

Gian-Carlo Rota

Photograph is reprinted with permission of the photographer Alfred Anderson, MIT Technique Yearbook

Contents

Chapter 3. Reynolds Operators

Chapter 4. Ergodic Theory

Chapter 5. Inequalities

Contributors

Jean Dhombres, L'École des Hautes Études en Sciences Sociales, Centre Alexandre Koyré, Centre National de la Recherche Scientifique, 27 rue Damesme, 75013 Paris, France.

Peter L. Duren, Department of Mathematics, University of Michigan, Ann Arbor, MI 48109, U. S. A.

W. Norrie Everitt, School of Mathematics and Statistics, University of Birmingham, Edgbaston, Birmingham B15 2TT, England.

Daniel A. Klain, Department of Mathematical Sciences, University of Massachusetts Lowell, Lowell, MA 01854.

Joseph P. S. Kung, Department of Mathematics, University of North Texas, Denton, TX 76203, U. S. A.

Arlan Ramsay, Department of Mathematics, University of Colorado, Boulder, CO 80309, U. S. A.

M. M. Rao, Department of Mathematics and Computer Science, University of California, Riverside, CA 92521, U. S. A.

James Rovnyak, Department of Mathematics, University of Virginia, Charlottesville, VA 22904, U. S. A.

Helmut H. Schaefer, Department of Mathematical Sciences, Florida Atlantic University, Boca Raton, FL 33431, U. S. A.

Bernd Schultze, FB6 - Mathematik und Informatik, Universität GH Essen, D 45117 Essen, Germany.

Jacob T. Schwartz, Courant Institute of Mathematical Sciences, New York University, New York, NY 10012, U. S. A.

Norton Starr, Department of Mathematics and Computer Science, Amherst College, Amherst, MA 01002, U. S. A.

Gilbert Strang, Department of Mathematics, Massachusetts Institute of Technology, Cambridge, MA 02139, U. S. A.

David C. Torney, Los Alamos National Laboratory, Los Alamos, NM 87545, U. S. A.

Radu Zaharopol, Department of Mathematics, Middle East Technical University, Ankara, Turkey, Department of Mathematical Sciences, Binghamton University (SUNY), Binghamton, NY 13902, U. S. A., and Mathematical Reviews, 416 Fourth Street, Ann Arbor, MI 48107, U. S. A.

Anton Zettl, Department of Mathematical Sciences, Northern Illinois University, DeKalb, IL 60115, U. S. A.

Xiao-Dong Zhang, Department of Mathematical Sciences, Florida Atlantic University, Boca Raton, FL 33431, U. S. A.

Editors' Preface

Gian-Carlo Rota was born in Vigevano, Italy, in 1932. He died in Cambridge, Massachusetts, in 1999. He had several careers, most notably as a mathematician, but also as a philosopher and a consultant to the United States government. His mathematical career was equally varied. His early mathematical studies were at Princeton (1950 to 1953) and Yale (1953 to 1956). In 1956, he completed his doctoral thesis under the direction of Jacob T. Schwartz. This thesis was published as the paper "Extension theory of differential operators I", the first paper reprinted in this volume.

Rota's early work was in analysis, more specifically, in operator theory, differential equations, ergodic theory, and probability theory. In the 1960's, Rota was motivated by problems in fluctuation theory to study some operator identities of Glen Baxter (see [7]). Together with other problems in probability theory, this led Rota to study combinatorics. His series of papers, "On the foundations of combinatorial theory", led to a fundamental re-evaluation of the subject. Later, in the 1990's, Rota returned to some of the problems in analysis and probability theory which motivated his work in combinatorics. This was his intention all along, and his early death robbed mathematics of his unique perspective on linkages between the discrete and the continuous. Glimpses of his new research programs can be found in [2,3,6,9,10].

This volume of *the Mathematical Papers of Gian-Carlo Rota* is devoted to his papers in analysis, convexity theory, and probability, or, roughly speaking, papers from the beginning and end of his career. The early papers show different aspects of Rota's genius. In addition to the paper derived from his doctoral thesis, the first chapter contains one of his favorite papers "On the spectra of singular boundary value problems." In this paper, he gave examples of naturally occurring singular operators (such as the Riemann differential operators) on L^p-spaces whose essential spectrums depend on p, and, even more unexpectedly, change from discrete to continuous as p changes. Another set of papers deals with Reynolds operators, which are abstractions of conditional expectation operators. These papers led Rota naturally to his work on Baxter algebras and his entry into combinatorics.

Together with the papers on Baxter algebras [7], the paper "An "alternierende Verfahren" for general positive operators marked a turning point in Rota's career. The theorem in this paper is a perfect illustration of Rota's *dictum* that the ideal way to overcome technical difficulties is to outflank them by the right idea. Rota was justifiably proud of this paper, but his pride was always tempered by a tincture of regret, since Burkholder and Chow independently discovered a weaker version of the theorem at about the same time. Rota played San Martín to Burkholder's Bolívar, ceding to Burkholder and his school further development of this area. Fortunately, unlike San Martín, Rota did not withdraw into inaction, but devoted his energies to the foundations of combinatorics.

Two other notable, yet contrasting, papers are "On models for linear operators" and "A note on the joint spectral radius", the latter written jointly with Gilbert Strang. The first was motivated by the invariant subspace conjecture for linear operators on Hilbert spaces. It has a grand design – like a sketch for a gigantic fresco by Tiepolo. This idea in this paper was to model linear operators as suboperators of shifts on ℓ_p-spaces. It initiated a very active area of research which continues to this day. It also formed the basis of an attempt to prove the invariant subspace conjecture in the 1980's. The second paper gives a definition of spectral radius for two matrices. It is smaller in scale, but displays perfect technique, rather like a Vermeer. As Strang describes in his article, this paper had to wait over thirty years before finding its "killer" application in the theory of wavelets.

An important contribution from Rota's early period is the book *Ordinary Differential Equations* which he wrote jointly with Garrett Birkhoff. Rota often asserted that the best things in it are in the exercises and that's where they worked the hardest. This book has gone into four editions [1] and gives a characteristically insightful view of a old subject.

In a famous (and perhaps notorious) foreword to [4], reprinted in [8], Rota proposed criteria for distinguishing the "three ages of a mathematical subject".

> There are the old subjects, loaded with distinctions and honors. As their problems are solved away and the applications reaped by engineers and other moneymen, ponderous treatises gather dust in library basements, awaiting the day when a generation as yet unborn will rediscover the lost paradise in awe.
>
> Then there are the middle-aged subjects. You can tell which they are by roaming the halls of Ivy League universities or the Institute for Advanced Study. Their high priests haughtily refuse fabulous offers from eager provincial universities while receiving special permission from the President of France to lecture in English at the Collège de France. Little do they know that the load of technicalities is already critical, about to crack and submerge their theorems in the dust of oblivion that once enveloped the dinosaurs.
>
> Finally, there are the young subjects – combinatorics for instance. Wild-eyed individuals gingerly pick from a mountain of intractable problems, childishly babbling the first words of what will soon be a new language. Childhood will end with the first *Séminaire Bourbaki*.

Rota had a unique ability to keep subjects young, by looking *beyond* technical details to the key foundational questions and concepts. One hopes that his papers in probability and convexity, written in his last decade, will inspire the same re-evaluation of these subjects as his middle-period papers did for combinatorics.

One function, perhaps the most important, of volumes of collected papers is to make papers of important mathematicians *humanly* accessible. We hope that this volume would not reside solely in institutional libraries (assuming optimistically that such libraries will still exist in our future), but be a living part of personal libraries, read both for instruction and pleasure.

In addition to reprinting published papers, we have also included three unpublished papers. The first two, "Dilations of positive operators" and "A method for the study of positive operators", are from around 1961 and can be found in Chapter 2. These papers contain results on positive operators which Rota had decided not to put into final form himself. The third, "An example of profinite combinatorics", is from 1997 and is in Chapter 6. This paper, intended for a general audience, is a non-technical exposition of results in one of his published papers. All three papers were lightly edited and corrected for publication here by the editors.

We thank Susan Kimball, Science Librarian at Amherst College, Dennis Porche at M.I.T., and our colleague Philip Hirschhorn at Wellesley College for their invaluable assistance during the preparation of this volume. The new material in this volume was typeset using TeX. The editors and publisher are particularly grateful to Teresa Rondon Rota, who kindly provided the photograph of Gian-Carlo Rota from his earlier years at M.I.T.

References

[1] G. Birkhoff and G.-C. Rota, *Ordinary Differential Equations.* First edition, Ginn, Boston, 1962; second edition, Blaisdell, Waltham MA, 1969; third edition, Wiley, New York, 1978; fourth edition, Wiley, New York, 1989.

[2] H. Crapo and D. Senato, eds., *Algebraic Combinatorics and Computer Science, a Tribute to Gian-Carlo Rota.* Springer-Verlag Italia, Milano, 2001.

[3] D. A. Klain and G.-C. Rota, *Introduction to Geometric Probability.* Lezioni Lincee, Cambridge Univ. Press, Cambridge, 1997.

[4] J. P. S. Kung, ed., *A Source Book in Matroid Theory.* Birkhäuser, Boston and Basel, 1986.

[5] J. P. S. Kung, ed., *Gian-Carlo Rota on Combinatorics.* Birkhäuser, Boston and Basel, 1995.

[6] J. P. S. Kung and C. H. Yan, Six problems of Gian-Carlo Rota in lattice theory and universal algebra. *Algebra Universalis,* Gian-Carlo Rota memorial issue, to appear.

[7] G.-C. Rota, Baxter algebras and combinatorial identities, I and II. *Bull. Amer. Math. Soc.* 75(1969), 325–329 and 330–334. Reprinted in [5], pp. 471–480.

[8] G.-C. Rota, *Indiscrete Thoughts.* Edited by F. Palombi, Birkhäuser, Boston and Basel, 1997.

[9] G.-C. Rota, Twelve problems in probability no one likes to bring up (The Fubini Lectures, Torino, Italy, June 3–5, 1998). In [2], pp. 57–93.

[10] G.-C. Rota, Ten mathematics problems I will never solve. *Mitt. Dtsch. Math.-Ver.,* **2**(1998), 45–52.

Papers Reprinted in this Volume in Chronological Order

[1] Extension theory of differential operators I. *Communications in Pure and Applied Mathematics,* **11**(1958), 23–65.

[2] Note on the invariant subspaces of linear operators. *Rendiconti del Circolo Matematico di Palermo, Serie II,* **8**(1959), 182–184.

[3] On the representation of averaging operators. *Rendiconti del Seminario Matematico dell'Università di Padova,* **30**(1960), 52–64.

[4] Spectral theory of smoothing operators. *Proceedings of the National Academy of Sciences, U.S.A.,* **46**(1960), 863–868.

[5] Endomorphismes de Reynolds et théorie ergodique. *Comptes Rendus des Séances de l'Académie des Sciences Paris,* **250**(1960), 2791-2793.

[6] Représentation des opérateurs de Reynolds. *Comptes Rendus des Séances de l'Académie des Sciences Paris,* **250**(1960), 2831–2833.

[7] On models for linear operators. *Communications in Pure and Applied Mathematics,* **13**(1960), 469–472.

[8] Une généralisation de l'espérance mathématique conditionnelle qui se présente dans la théorie statistique de la turbulence. *Comptes Rendus des Séances de l'Académie des Sciences Paris,* **251**(1960), 624–626.

[9] A note on the joint spectral radius. *Indagationes Mathematicae,* **22**(1960), 379–381, with W. Gilbert Strang.

[10] On Reynolds operators in finite-dimensional algebras. *Journal of Mathematics and Mechanics,* **9**(1960), 927–932, with Martin Billik.

[11] On the completeness of Sturm-Liouville expansions. *American Mathematical Monthly,* **67**(1960), 835–841, with Garrett Birkhoff.

[12] On the spectra of singular boundary value problems. *Journal of Mathematics and Mechanics,* **10**(1961), 83–90.

[13] Une théorie unifiée des martingales et des moyennes ergodiques. *Comptes Rendus des Séances de l'Académie des Sciences Paris,* **252**(1961), 2064–2066.

[14] On the passage from the Navier-Stokes equations to the Reynolds equations. In R. E. Langer, ed., *Partial Differential Equations and Continuum Mechanics, Proceedings of an International Conference (1960),* University of Wisconsin Press, Madison, WI, 1961, pp. 374–376

[15] On the eigenvalues of positive operators. *Bulletin of the American Mathematical Society,* **67**(1961) 556-558; Addendum, *ibid.,* **68** (1962) 49.

[16] Multiplicative extensions of positive linear operators. *Notices of the American Mathematical Society,* **8**(1961), Abstract 61T-125, 272–273.

[17] An "alternierende Verfahren" for general positive operators. *Bulletin of the American Mathematical Society,* **68**(1962), 95–102.

[18] On the classification of periodic flows. *Proceedings of the American Mathematical Society,* **13**(1962), 659–662.

[19] On the maximal ergodic theorem for Abel-limits. *Proceedings of the American Mathematical Society,* **14**(1963), 722–723.

[20] Reynolds operators. In *Stochastic Processes in Mathematical Physics and Engineering,* Proceedings of Symposia in Applied Mathematics, Vol. XVI, American Mathematical Society, Providence RI, (1964), pp. 70–83.

[21] A limit theorem for the time-dependent evolution equation. In *Notes of a seminar on Equazioni Differenziali Astratte at Varenna, May 31–June 8, 1963, Centro Internazionale Matematico Estivo,* Edizioni Cremonese, Roma, 1964, 15 pages.

[22] Review: *Linear Operators. Part II. Spectral Theory.* By Nelson Dunford and Jacob T. Schwartz. *Bulletin of the American Mathematical Society,* **71**(1965), 705–708.

[23] On the inequality $\|f'\|^2 \leq 4\|f\| \cdot \|f''\|$. In O. Shisha, ed., *Inequalities, II (Proceedings of the Second Symposium, U. S. Air Force Academy, Colorado, 1967),* Academic Press, New York, 1970, pp. 187–192, with Robert R. Kallman.

[24] Frobenius reciprocity in ergodic theory. *Bulletin of the American Mathematical Society,* **77**(1971), 713–718, with Kenneth Lange and Arlan Ramsay.

[25] A measure theoretic approach to logical quantification. *Rendiconti del Seminario Matematico dell'Università di Padova,* **59**(1978), 227–246, with David P. Ellerman.

[26] A stochastic interpretation of the Riemann zeta function. *Proceedings of the National Academy of Sciences, U. S. A.,* **90**(1993), 697–699, with Kenneth S. Alexander and Kenneth Baclawski.

[27] Totally invariant set functions of polynomial type. *Communications in Pure and Applied Mathematics,* **47**(1994), 187–197, with Beifang Chen.

[28] A continuous analogue of Sperner's theorem. *Communications in Pure and Applied Mathematics,* **50**(1997), 205–223, with Daniel A. Klain.

[29] Stochastic integrals: a combinatorial approach. *Annals of Probability,* **25**(1997), 1257–1283, with Timothy C. Wallstrom.

[30] Probability set functions. *Annals of Combinatorics,* **3**(1999), 13–25, with William J. Bruno and David C. Torney.

[31] On the combinatorics of cumulants. *Journal of Combinatorial Theory, Series A,* **91**(2000), 283–304, with Jianhong Shen.

Gian-Carlo Rota on His Work in Analysis

1. Linear operators.

In [2] and [7], the existence of a universal operator in an infinite dimensional Hilbert space H is established. Specifically, it is shown that there exists a linear operator T of norm one, such that, for every other linear operator S of norm at most one, there exists a closed invariant subspace H' of T such that the restriction of T to H' is similar to S. This result has been the point of departure for later work of de Branges and Rovnyak on the invariant subspace problem.

In [3], [4], [5], [6], [8], [10], [13], and [20], the theory of Reynolds operators is developed. A Reynolds operator R is defined as an operator on an algebra which satisfies the identity

$$R(fRg) + R(gRf) = RfRg + R(RfRg).$$

These operators can be viewed alternatively as generalizations of conditional expectation operators, or as resolvents of derivations on the space $L_\infty(S, \Sigma, \mu)$. A complete classification of these operators on $L_\infty(S, \Sigma, \mu)$ is given, subject to some mild continuity assumptions. They turn out to be mixtures of conditional expectation operators and resolvents of derivations. In particular, it is shown that a Reynolds operator satisfies the identity $R(fRg) = RfRg$ (an identity which characterizes conditional expectation operators) if and only if its range is closed. Using Reynolds operators, a unified treatment can be given of ergodic theorems and martingale convergence theorems.

Papers [1] and [12] deal with the spectral theory of non-selfadjoint ordinary differential operators. In [1], there is perhaps what is historically the first occurrence of an "invariance of index" theorem; in [12], an example is exhibited of an ordinary differential operator of the second order with polynomial coefficients whose essential spectrum in L_p varies with p, for $p \geq 1$.

2. Ergodic Theory.

In [15], it is shown that if λ is an eigenvalue of a doubly stochastic operator such that $|\lambda| = 1$, then λ^2 is also an eigenvalue. This result has led to an extensive theory, developed by Helmut Schaefer, E. B. Davies and others.

In [17], the following limit theorem is proved. Let $P_1, P_2, \ldots$ be an infinite sequence of doubly stochastic operators. For every function f in L_p for $p > 1$ over a probability space, the sequence

$$P_1 P_2 \ldots P_n P_n^* P_{n-1}^* \ldots P_2^* P_1^* f$$

converges boundedly almost everywhere. The proof uses a martingale argument. This theorem has been widely applied and generalized, but the original proof has not been notably simplified. A continuous-parameter generalization is given in [21].

In July, 1993, Rota wrote a brief account of his work entitled "Summary of the mathematical work of Gian-Carlo Rota, July 1993." This survey is extracted from that account. Paper $[n]$ is the nth paper in the chronological list of reprinted papers.

Paper [18] is (roughly) the ergodic analog of [7]. Using oversimplified language, it is shown that there exists a measure-preserving transformation U in a non-atomic probability space which is "universal" in the sense that given any other measure-preserving transformation U', there exists an invariant σ-subalgebra Σ' of U such that the restriction of U to Σ' is conjugate to U'. In [19], what is perhaps the simplest proof is given of the pointwise ergodic theorem for Abel limits (rather than the usual Cesàro limit). In [24], some ideas of Mackey on virtual groups are extended to a suitable categorical setting.

In the Baxter algebra papers,[1] the theory of Baxter algebras is developed. This theory is motivated by the problem of extending Spitzer's formula of probability theory to arbitrary group actions, a problem which is solved in these papers. The solution is achieved by giving an explicit decision procedure for simplifying expressions made up by iterations of the function $x^+ = \max(0, x)$ and ordinary addition of real numbers. Such a decision procedure is obtained by use of Baxter operators. A Baxter operator on an algebra is an operator B such that

$$B(fBg) + B(gBf) = BfBg + B(fg).$$

An explicit solution is given for the word problem for Baxter algebras, by showing that the free Baxter algebra "encodes" the ordinary algebra of symmetric functions. The main result of these papers, stating that the standard Baxter algebra is free, has been elegantly generalized by Cartier in a Bourbaki seminar.

3. Other work.

Paper [9] gives an estimate for the joint spectral radius of matrices which has been applied in numerical analysis. Paper [11] was obtained while writing the textbook *Ordinary Differential Equations* (first edition, Ginn, Boston, 1962), with Garrett Birkhoff. It is shown that if (x_n) and (y_n) are orthonormal sequences in Hilbert space, if (x_n) is complete and

$$\Sigma \|x_n - y_n\|^2 < \infty,$$

then (y_n) is also complete.

In [23], the inequality

$$\|f'\|^2 \leq 4\,\|f\|\|f''\|,$$

proved by Landau, Hardy and Littlewood for L_p spaces, is extended to "arbitrary" Banach spaces. A similar inequality is proved for arbitrary semigroup generators. Several generalizations and extensions of this inequality have since been published.

[1] Baxter algebras and combinatorial identities. I and II. *Bull. Amer. Math. Soc.* **75** (1969), 325–334; Fluctuation theory and Baxter algebras. In *Symposia Mathematica, Vol. IX (Convegno di Calcolo delle Probabilità, INDAM, Rome, 1971)*, Academic Press, London, 1972, pp. 179–201, with D. A. Smith. Reprinted in J. P. S. Kung, ed., *Gian-Carlo Rota on Combinatorics*, Birkhäuser, Boston and Basel, 1995.

In [27], it is shown that Hadwiger's classification of *Quermassintegrale* (or elementary mixed volumes) in terms of elementary symmetric functions can be extended to a correspondence between the algebra of all set functions on convex bodies (enjoying a stronger property than invariance under the Euclidean group), and the algebra of all (polynomial) symmetric functions.

Reminiscences of Gian-Carlo Rota

Peter Duren

I was Gian-Carlo's first Ph.D student. Our collaboration began even before he joined the Massachusetts Institute of Technology faculty. Let me set the scene with a few words of personal history.

I came to graduate school at M.I.T. in 1956, following an undergraduate degree from Harvard. M.I.T. was then very strong in hard analysis, with established stars (Norbert Wiener, Norman Levinson, Alberto Calderón, ...) and rising stars such as Jürgen Moser and Eli Stein. I passed the qualifying exams and arranged to work with Calderón, who started me reading the book of Riesz and Nagy to develop background in functional analysis and operator theory. I took his course on singular integrals. Then in the fall of 1958, a bombshell: Professor Calderón informed me that he was returning to Chicago! He suggested that I move with him, but I preferred to remain at M.I.T. At that point Calderón recommended that I work with a young man, then a Peirce Instructor at Harvard, who would be coming to M.I.T. the following year. This young man, whose name was Gian-Carlo Rota, had done his graduate work at Yale in the famous Dunford–Schwartz school of operator theory, and had in fact participated in the preparation of their now-classic treatise *Linear Operators.* I met Gian-Carlo in his office at Harvard and under his supervision began reading my way through *Part I* of that book, working the exercises he selected (my copy still carries his check-marks). My initial apprehension vanished after a few of those sessions, as it became clear I was in good hands. Over a period of time, Gian-Carlo taught me the basic theory of Banach spaces while opening my eyes to the diversity and power of functional analysis as a tool in classical problems. In the summer of 1959 he had me read the book of Akhieser and Glazman on linear operators in Hilbert space. He steered me to the seminal paper of Beurling on invariant subspaces of the shift operator, and my thesis became a generalization of that theory.

During his time at Harvard I think Gian-Carlo was primarily interested in the theory of differential operators, the topic of his thesis at Yale. He was writing his text with Garrett Birkhoff on differential equations, and he taught a graduate course at Harvard on advanced aspects of that subject. I attended the course and was struck both by the enthusiasm and clarity of Rota's lectures and by his broad knowledge of the classical theory. His strategy in learning the subject was to study the original sources, the classic papers of the 19th Century. His lectures did not work through the details of proofs, but they conveyed ideas. ("Read for ideas," he would advise me later. "Don't spend time checking details".) When my thesis was complete and a postdoctoral position was arranged at Stanford, I stayed at M.I.T. for the summer of 1960. Gian-Carlo proposed that we collaborate in writing an

American Mathematical Society *Memoir* on spectral theory of differential operators, expanding on themes in his previous work. That seemed more his project than mine, so after some consideration I declined the offer with thanks. Looking back, it is sad to see that the book was never written. I believe it would have included the projected continuation of his paper "Extension theory of differential operators *I* ". It would also have developed ideas in his subsequent paper "On the spectra of singular boundary value problems", which he gave me then as a preprint.

In fact, Gian-Carlo's mathematical interests were already shifting. He was getting into ergodic theory, and he did some basic work on Reynolds operators. Then he moved to combinatorics, wrote his famous paper on the Möbius inversion formula, and started the long series of papers "On the foundations of combinatorial theory". It turned out that he was actually creating a new field of mathematics, algebraic combinatorics. A few years later he told me that in combinatorics he felt he had "finally found the real mathematics".

We remained in contact, but because he had switched fields we did not see so much of each other in later years. At Stanford I met Max Schiffer and started working in geometric function theory. My move to Michigan brought me back into functional analysis, as I worked on Hardy space questions with Allen Shields and Harold Shapiro. Gian-Carlo encouraged my research efforts for some years after graduate school, despite the sharp divergence of mathematical paths, and we continued to exchange reprints. When I spent the year 1968–69 at the Institute for Advanced Study, he invited me back to Cambridge to address the Harvard – M.I.T. – Brandeis Colloquium. After the talk he organized a dinner in a special room (males only!) at Locke-Ober Restaurant in Boston. He was a great gourmet, a sophisticated lover of fine food and wine. In anticipation of the feast, he enthusiastically promoted the specialities of the house: Steak Tartare, Baked Alaska, etc. But when we actually convened for dinner he seemed a bit subdued. He had a slight stomach ache, he said, so he would just order a bowl of soup. The soup came, and he didn't touch it. He got up from the table and made a telephone call to his wife Teresa, then called a taxi. As he left he issued words of reassurance and exhorted us all once more to have the Baked Alaska for dessert. The next morning, after a sleepless night (due more to the Baked Alaska than to concern about Gian-Carlo), I called his apartment. "Oh, Gianco is in the hospital", Teresa told me. "He had his appendix taken out last night. It ruptured".

Fortunately, he made a good recovery ...

Many years later, when we were editing the A.M.S. Centennial volumes, *A Century of Mathematics in America*, we asked Gian-Carlo to contribute an article. He elected to write about his undergraduate days at Princeton and the mathematicians he had encountered there: Alonzo Church, William Feller, Emil Artin, Solomon Lefschetz. In typical style, he managed to raise some hackles with his frank recollections of those distinguished mathematicians. He always reveled in controversy, so when I reported that some preliminary readers had found his remarks inappropriate, he responded with mischievous glee.

In reading Gian-Carlo's writings, one becomes aware of his amazing command of languages. Of course his English was eloquent, but he was actually quadrilingual.

Growing up in Italy near the French border, he learned to speak both Italian and French. Then his family moved to Ecuador, where he attended high school, and he became fluent in Spanish. English was I think his fourth language. He also knew German quite well. However, I once heard him complain that although he spoke all of those languages fluently, he spoke no language without an accent, not even Italian.

In the spring of 1996, I returned to M.I.T. to attend the conference in honor of Gian-Carlo on his 64th birthday. It was a wonderful opportunity to reconnect with Gian-Carlo and to meet so many of my younger "brothers and sisters." Of course, most of the talks were on combinatorics. Gian-Carlo was at his preposterous best, reading a speech entitled "Ten lessons I wish I had been taught", which later appeared in the *Notices of the American Mathematical Society* [2]. Among other things, he offered strong advice on how to give a mathematical talk. "Relate to your audience, give them something to take home, never run overtime, ... " I then had the unenviable task of following immediately afterwards with a mathematical talk. With Gian-Carlo's dicta fresh in mind, the audience must have found my presentation highly deficient, despite all last-minute efforts at revision. To add to my difficulty, I had chosen to speak about recent work in Bergman spaces, which was far from the main theme of the conference. But it was actually a topic close to my thesis, and I was pleasantly surprised to discover that Gian-Carlo had retained an interest in the structure of invariant subspaces after all those years of work in combinatorics. In fact, his subsequent *Intelligencer* article about Yale [1] confirms that he never lost his enthusiasm for operator theory.

Even in the early days, Gian-Carlo was a man of broad interests. He wanted to understand the whole fabric of mathematics – the history, the philosophy, the great ideas, the interconnection of fields. When I was his student, he continually recommended books and articles well beyond the range of my thesis work. For instance, I recall that he urged me to read the book by François Le Lionnais, *Les Grands Courants de la Pensée Mathématique.* While others slaved in the trenches, Gian-Carlo stood high on a hill and surveyed the entire landscape. In my student days, he once told me there are two kinds of mathematicians, those who prove theorems and those who produce theories. Gian-Carlo Rota will be remembered as a mathematician who produced theories.

References

[1] G.-C. Rota, Light shadows: Remembrances of Yale in the early fifties. *Math. Intelligencer* **18** (1996), no. 3, 44–51.

[2] G.-C. Rota, Ten lessons I wish I had been taught. *Notices Amer. Math. Soc.* **44** (1997), no. 1, 22–25.

Gian-Carlo Rota (1932–1999)

Jacob T. Schwartz

In the recollections of Yale included in his *Indiscrete Thoughts,* Gian-Carlo mentions the 1954 functional analysis seminar at which we met. This seminar, in memory yet green, was organized by Nelson Dunford and addressed by an outstanding group of young researchers, including John Wermer, William Bade, Robert Bartle, and Henry Helson. It was a high point of early functional analysis at Yale: interesting new results were presented by their discoverers almost every week. Even though he was then in his first graduate year, Gian-Carlo's talents became apparent at once, and he was immediately recruited, along with Bartle, Bade, and me, as a junior member of the group then working on the "Dunford project" that subsequently became *Linear Operators.* In 1957 Dunford decided to take a year's sabbatical at NYU (New York University). Support from the ONR (Office of Naval Research) being available, Rota and I were able to tag along for this one-year visit, which turned into a three-year stay for Gianco and forty-two years for me. Though only a few years older than Gian-Carlo, I assumed for the first time the ponderous dignity of thesis advisor. This was a period of youthful friendship, punctuated by frequent risotto Milanese garnished by Asti Spumante at Gian-Carlo's bottom-price, sixth-floor walkup apartment in what was then something of a Mafia-dominated slum just south of NYU but which has since been gentrified.

The Dunford connection, the general prestige of functional analysis at Yale, the ONR contract, and the pattern of my own interests led Gian-Carlo to an initial specialization in functional analysis. His dissertation, "Extension theory of differential operators I", appeared in *Communications in Pure and Applied Mathematics* in 1958. (Not untypically for papers whose titles bear the fatal Roman digit "I", there never was a "II".) A series of other papers on operator theory followed: "Note on the invariant subspaces of linear operators", "On the spectra of singular boundary value problems", "On models for linear operators", "On the eigenvalues of positive operators" — all in the period 1958-61. But already in his two 1959-60 papers on Reynolds operators, Gianco had stepped onto the bridge of ergodic theory, often a highly combinatorial corner of analysis, that carried him into purer combinatorics. The first of his landmark papers in his new area, "On the foundations of combinatorial theory I", dates from 1964. (Here the "I" had many following integers.) To its own good fortune, combinatorics (and later philosophy) had captured him completely. His distinguished book with Garrett Birkhoff, *Ordinary Differential Equations,* was a farewell to analysis. Unlike many, he had the courage and curiosity to move on.

A slightly different version of this article first appeared in the obituary of Gian-Carlo Rota in the *Notices of the American Mathematical Society* **47**(2000), 203–204.

Our lifelong friendship made me not only an admirer of the depth, scholarship, and sheer energy of his mathematical work (and of his ceaseless activities as an editorial entrepreneur on behalf of mathematics) but one in awe of his status as the ultimate relaxed sophisticate. Gian-Carlo could always state with easy authority not only the current standing of all the top restaurants in Paris, Rome, Boston, and Milan but where to get the hottest and best chili in New Mexico and even what local hash house had the most unexpected culinary surprises. I shall miss him greatly.

Acknowledgements

The editors and publishers acknowledge gratefully permissions to reprint the following papers in the list of reprinted papers:

From *American Mathematical Monthly,* [11], ©1960, the Mathematical Association of America. Reprinted with the permission of the Mathematical Association of America.

From *Annals of Combinatorics,* [30], ©1999, Birkhäuser Verlag. Reprinted with the permission of Birkhäuser Verlag.

From *Annals of Probability,* [29], ©1997, Institute of Mathematical Statistics. Reprinted with the permission of the Institute of Mathematical Statistics.

From *Bulletin of the American Mathematical Society,* [15], ©1961 and 1962, [17], ©1962, [22], ©1965, [24], ©1971, American Mathematical Society. Reprinted with the permission of the American Mathematical Society.

From *Communications in Pure and Applied Mathematics,* [1], ©1958, [7], ©1960; [27], ©1994, [28], ©1997, John Wiley and Sons. Reprinted with the permission of John Wiley and Sons.

From *Comptes Rendus de l'Académie des Sciences Paris,* [5], [6], [8], ©1960, [13], ©1961, Gauthier-Villars, Paris. Reprinted with the permission of Éditions Elsevier, France.

From *Indagationes Mathematicae,* [9], ©1960, Koninklijke Nederlandse Akademie van Wetenschappen. Reprinted with the permission of the Koninklijke Nederlandse Akademie van Wetenschappen.

From *Inequalities. II (Proceedings of the Second Symposium, U. S. Air Force Academy, Colorado, 1967),* [23], ©1970, Academic Press. Reprinted with the permission of Academic Press.

From *Journal of Combinatorial Theory, Series A.* [31], ©2000, Academic Press. Reprinted with the permission of Academic Press.

From *Journal of Mathematics and Mechanics,* [10], ©1960, [12], ©1961, the Univeristy of Indiana. Reprinted with the permission of the Department of Mathematics, University of Indiana, Bloomington, Indiana.

From *Notes of a Seminar, Equazioni Differenziale Astratte, Varenna, May 31 - June 8, 1963, Centro Internazionale Matematico Estivo,* distributed by Istituto Matematico dell'Università Roma, 1963, [21]. Reprinted with the permission of the author.

From *Notices of the American Mathematical Society,* [16], ©1961, American Mathematical Society. Reprinted with the permission of the American Mathematical Society.

From *Partial Differential Equations and Continuum Mechanics, Proceedings of an International Conference (1960),* [14], ©1961, Regents of the University of Wisconsin. Reprinted with the permission of the Department of Mathematics, University of Wisconsin, Madison, Wisconsin.

From *Proceedings of the American Mathematical Society,* [18], ©1962, [19], ©1963, American Mathematical Society. Reprinted with the permission of the American Mathematical Society.

From *Proceedings of the National Academy of Sciences, U. S. A.*, [4], ©1960, Gian-Carlo Rota. Reprinted with the permission of the author.

From *Proceedings of the National Academy of Sciences, U. S. A.*, [26], ©1993, National Academy of Sciences, U. S. A. Reprinted with the permission of the National Academy of Sciences.

From *Rendiconti del Seminario Matematico dell'Università di Padova,* [3], ©1960, [25], ©1978, Università di Padova. Reprinted with the permission of the editors, *Rendiconti del Seminario Matematico dell'Università di Padova.*

From *Rendiconti del Circolo Matematico di Palermo*, [2], ©1959, Circolo Matematico di Palermo. Reprinted with the permission of the Circolo Matematico di Palermo.

From *Stochastic Processes in Mathematical Physics and Engineering,* Proceedings of Symposia in Applied Mathematics, Vol. XVI, [20], ©1964, American Mathematical Society. Reprinted with the permission of the American Mathematical Society.

CHAPTER 1

Differential Operators

COMMUNICATIONS ON PURE AND APPLIED MATHEMATICS, VOL. XI, 23–65 (1958)

Extension Theory of Differential Operators I*

G. C. ROTA

Harvard University

The problem of determining sufficient conditions on non-self-adjoint ordinary linear differential operators which will insure that the operators be "spectral" in the sense of Dunford [4, 5] is one of the outstanding tasks of spectral theory. While the theory is not yet sufficiently developed for a full scale attack of the general problem, the results obtained in many and important special cases [3, 7] are quite promising, and naturally lead to the study undertaken in the present paper. We shall be concerned with the following questions: (a) What is the nature of the spectra of singular non-self-adjoint differential operators? (b) How are boundary conditions to be imposed as a (singular) differential operator in order that the spectrum of the extended operator (in a given Banach space) will be of the type required by the general theory of spectral operators? A complete qualitative answer is provided in the following.

For reasons of space and time we have reserved for a later paper all questions of a "quantitative" nature, such as the rate of growth of the resolvent in the neighborhood of the essential spectrum, specific criteria on the coefficients of a differential operator to determine the essential spectrum and the comparison of essential spectra in various B-spaces. The following discussion is cast in a general L_p-space, $1 \leqq p \leqq \infty$, with a view to including the results required for the spaces L_1 and L_∞ by Hille's studies on semigroups and Levitan's generalized translations.

1. Closed Operators in Non-Reflexive Banach Spaces

We collect some results on Banach spaces which do not appear in the accessible literature and which are needed for later work. The notation is the standard one (see [2] and [3]). All Banach spaces will be tacitly assumed to be defined over the complex field, and all "functions" on an interval of the real line to be complex-valued. A star "*" indicates conjugate spaces and adjoints. The letters $f, g, \cdots$ have been preferred to indicate elements of Banach spaces which are function spaces over an interval of the real line. Greek letters will indicate complex numbers. Points on the real line will be

*This paper which embodies some of the results of a Doctoral Dissertation at Yale University was partially written while the author was a Temporary Member of the Institute of Mathematical Sciences, New York University, during the academic year 1956-1957. This Temporary Membership plan is supported by the National Science Foundation.

indicated by the letters $s, t, \cdots$, and linear operators by the letters $S, T, \cdots$. The null space of a linear operator T in a Banach space X, namely the set of all elements x in X which are mapped into zero by the operator T, will be indicated by $N(T)$. The range of an operator T will be indicated by $R(T)$. If A is a subspace of a Banach space X, the symbol $A^{\perp}$ will designate the subspace of all continuous linear functionals on X which vanish on the subspace A. If A is a closed subspace of the conjugate space X^*, the symbol $A^{\perp}$ will designate all elements of the space X upon which the linear functionals belonging to A vanish. The direct sum of two Banach spaces X and Y, designated by $X+Y$, will be understood in the usual topological as well as algebraic sense. The symbol for quotient spaces X/Y will be defined only for a closed subspace Y of a Banach space X. The customary notation will be used for closed operators T in a Banach space X. For an element x in $D(T)$ we set

$$|x|_T = |x| + |Tx|,$$

whenever T is a closed operator. We call this the T-norm. Thus the subspace $D(T)$ will be considered a Banach space with the norm $|x|_T$. The notation $S \supset T$ will indicate that the operator S is an extension of the operator T. If there exists a closed operator S which is an extension of an operator T, then the closure $\bar{T}$ is defined as the minimal closed operator which is an extension of T. If A is a subspace of the Banach space X, and if B is any subset of X, the elements of B will be said to be linearly independent modulo A if no linear combination of the elements of B lies in the subspace A. If x^* is a continuous linear functional on the Banach space X, the notation $(x^*, x) = x^*(x)$ will often be used.

LEMMA 1.1. *Let S be a closed subspace of a Banach space X such that the dimension of the space X/S is finite. Then*:

(a) *There exists a finite-dimensional subspace S_1 of X such that*

$$X = S + S_1. \tag{1.1}$$

(b) *If S_1 is any subspace of X such that X is the direct sum of S and S_1, then S_1 is finite-dimensional and*

$$\dim S_1 = \dim X/S.$$

(c) *Under the hypothesis of* (b), *there exists a bounded projection E of X onto the subspace S_1 which vanishes on all elements of S.*

The proof is left to the reader.

LEMMA 1.2. *Suppose that a Banach space X is the direct sum of the subspaces S and S_1, where S_1 is finite-dimensional. Let T be a bounded linear operator from X to a Banach space Y. Then the range of T is closed if and only if the subspace TS of Y is closed.*

Proof: If the subspace TS is closed, then $TX = TS + TS_1$, and, since

TS_1 is finite-dimensional, TX is closed. To prove the converse, let N be the null space of T. Clearly N is a closed subspace of X. Let ϕ be the canonical mapping of X onto the Banach space X/N. Suppose the subspace $S+N$ is at most the sum of S and a finite-dimensional subspace of S_1, the subspace $\phi(S) = \phi(S+N)$ of X/N is closed. Furthermore, the space X/N is the direct sum of $\phi(S)$ and a finite-dimensional subspace, and the operator T induces a continuous linear operator T' of X/N into Y whose range coincides with the range of T. Now if the range of T' is closed, then by the closed graph theorem T' maps closed sets onto closed sets and thus $T'\phi(S) = TS$ is closed.

LEMMA 1.3. *Let T be a closed operator in a Banach space X, and suppose that for some complex λ the inverse operator $(\lambda I-T)^{-1}$ is bounded and everywhere defined. Then the linear operator $(\mu I-T)^{-1}$, considered as an operator of X into the Banach space $D(T)$ (with the T-norm), is defined and continuous, and, considered as a function of μ, analytic in μ, for μ in a neighborhood of λ.*

Proof: The first assertion follows at once from the closed graph theorem and from the analyticity of the resolvent as a linear operator of X into itself. To prove the second statement it suffices to show that for every element x of X the vector-valued function of μ

$$F(\mu) = (\mu I-T)^{-1}x$$

is analytic in the space $D(T)$. This follows from a theorem of Dunford asserting that "weak analyticity" is equivalent to "strong analyticity".

Let N be the domain of analyticity of $F(\mu)$ when considered as a function with values in the space X. Then we have

$$F'(\mu) = -(\mu I-T)^{-2}x, \qquad \mu \in N, \tag{1.3}$$

the derivative being taken in the space X. From the identity

$$T(\mu I-T)^{-1}x = x-\mu(\mu I-T)^{-1}x$$

we now infer that the function

$$G(\mu) = T(\mu I-T)^{-1}x$$

is analytic in N as a function with values in the space X, and its derivative is

$$G'(\mu) = -(\mu I-T)^{-1}x+\mu(\mu I-T)^{-2}x = T(\mu I-T)^{-2}x.$$

We shall prove that the derivative of $F(\mu)$, considered as a function with values in the space $D(T)$, is also given by (1.3). Let λ_n be a sequence of points in N converging to a point λ_0 in N. Then

$$\begin{aligned}
\lim_n |(\lambda_n-\lambda_0)^{-1}[F(\lambda_n)-F(\lambda_0)]-F'(\lambda_0)|_T \\
= \lim_n |(\lambda_n-\lambda_0)^{-1}[F(\lambda_n)-F(\lambda_0)]-F'(\lambda_0)| \\
+ \lim_n |T(\lambda_n-\lambda_0)^{-1}[F(\lambda_n)-F(\lambda_0)]-TF'(\lambda_0)| \\
= \lim_n |(\lambda_n-\lambda_0)^{-1}[F(\lambda_n)-F(\lambda_0)]-F'(\lambda_0)| \\
+ \lim_n |(\lambda_n-\lambda_0)^{-1}[G(\lambda_n)-G(\lambda_0)]-G'(\lambda_0)| = 0.
\end{aligned}$$

The following notions on closed operators in non-reflexive Banach spaces will be needed for the study of differential operators on the space $L_1(I)$ for an interval I of the real line.

DEFINITION 1.1. Let X be a (not necessarily reflexive) Banach space, and let S be a linear operator from the conjugate space X^* into itself whose domain is total on X. We define the adjoint S^* of S in X as follows: Let x be an element of X to which we can assign an element x' in X in such a way that for all x^* in $D(S)$ we have

$$(Sx^*, x) = (x^*, x').$$

We shall then say that x is in the domain $D(S^*)$ and that $S^*x = x'$.

The operator S^* is well defined, for, if the element x' were not uniquely determined, then $(x^*, x'') = 0$ for some x'' in X and for all x^* in $D(S)$. Hence $x'' = 0$.

If, in particular, an operator T is given in X the domain of whose adjoint is X-total, then the above definition yields a second adjoint T^{**} in X.

The operator S^* is closed: Indeed, let x_n be a sequence in $D(S^*)$ such that

$$x_n \to x, \qquad S^*x_n \to y.$$

Then

$$\lim_n (x^*, S^*x_n) = (x^*, y) = \lim_n (Sx^*, x_n) = (Sx^*, x), \qquad x^* \in D(S).$$

Hence x is in $D(S^*)$ and $S^*x = y$.

The following properties of the second adjoint operator will be needed in the ensuing discussion.

LEMMA 1.4. (a) *If T has a closure $\bar{T}$, then $\bar{T}^* = T^*$.*

(b) *The domain $D(T^*)$ is X-total if and only if the operator T has a closure.*

(c) *If the operator T has a closure $\bar{T}$, then $\bar{T} = T^{**}$.*

Proof: (a) Clearly $\bar{T}^* \subset T^*$ from the definition. Conversely, let x^* be in $D(T^*)$, and let x lie in $D(\bar{T})$. Then there exists a sequence $\{x_n\}$ in $D(T)$ such that

$$x_n \to x, \qquad Tx_n \to \bar{T}_x .$$

Hence

$$(T^*x^*, x) = \lim_n (T^*x^*, x_n) = \lim_n (x^*, Tx_n) = (x^*, \bar{T}_x).$$

Therefore, x^* is in $D(\bar{T}^*)$ and $\bar{T}^*x^* = T^*x^*$.

(b) If $D(T^*)$ is X-total, then T^{**} exists, and from the definition we have $T^{**} \supset T$. Then T has a closure. To show the converse, we can assume, by (a), that T is a closed operator. Von Neumann's method of graphs for Hilbert space can be successfully generalized. There exists a natural imbedding of $D(T)$ in the space $X \oplus X$ given by

$$x \to [x, Tx].$$

Thus $D(T)$ is a closed subspace of $X \oplus X$. The subspace S of all linear functionals in $(X \oplus X)^* = X^* \oplus X^*$ which vanish on $D(T)$ consists of all elements of the form

$$[T^*x^*, -x^*], \qquad x^* \in D(T^*).$$

Now the domain of the operator T^{**} can be represented as the subspace D of all elements of $X \oplus X$ which vanish on all elements of the form $[T^*x^*, -x^*]$, $x^* \in D(T^*)$. Since the operator T is assumed to be closed, we have $D = D(T)$. Now suppose that the domain $D(T^*)$ is not X-total. Then there is a non-zero element x in X such that $x^*(x) = 0$ for all x^* in $D(T^*)$, and thus the element $[0, x]$ of $X \oplus X$ vanishes on all elements $[T^*x^*, -x^*]$. Therefore $[0, x] \in D(T)$, $T0 = x$, and hence $x = 0$.

(c) The domain $D(\bar{T})$ of the closure of T can be imbedded in $X \oplus X$ as the closure in $X \oplus X$ of the domain of T. Thus the statement is evident from the above discussion.

A word of caution is necessary at this point. If T is a closed operator in X^* whose domain is X-total, and if the adjoint operator T^* in X has dense domains, it is not true in general that $T^{**} = T$. In general, the operator T^{**} is a proper extension of T even if T is closed.

Of course, if T is known to be the adjoint of some operator S in X, then $T^{**} = T$.

It is clear that the above definition and properties of the adjoint operator can be restated "verbatim" for closed operators of a Banach space X into a Banach space Y. We shall tacitly assume this generalization as given, since it will only be used to prove the next corollary.

THEOREM 1.1. *Let T be a closed operator with dense domain from a Banach space X to a Banach space Y. If T^* is one-to-one and has closed range, then the range of T is the entire space Y.*

The proof is divided into several steps.

(A) Consider the weak topology defined on X^* by the elements of $D(T)$, which will be called "the $D(T)$-topology". Since the domain $D(T)$ is dense, and hence total on X^*, the topology is locally convex and a Hausdorff topology. Furthermore, the closed unit sphere S of X^* is compact in the $D(T)$-topology. Indeed, since $D(T)$ is dense in X, the unit sphere S consists of those and only those elements x^* of X^* for which

$$|x^*(x)| \leq |X|, \qquad x \in D(T).$$

Thus the well-known argument using Tychonoff's theorem can be used to construct a proof.

(B) Since the unit sphere S of X^* is compact in the X-topology of X^*

as well, and since the X-topology of X^* is stronger than the $D(T)$-topology, it follows that these two topologies coincide on S. Now consider the bounded X-topology of X^*, namely the strongest topology which coincides with the X-topology on every multiple of the unit sphere S. It is known (cf. Dunford and Schwartz [2], Chapter V) that any linear functional which is continuous in the bounded X-topology is (represented by) an element of X. Using (A) we can define the bounded X-topology by simply restricting the $D(T)$-topology to every multiple of the unit sphere. We shall use this conclusion in step (G).

(C) Consider now on the Banach space $D(T^*)$ (with the T^*-norm) the (weak) $D(T^*)^*$-topology. Every element of Y can be regarded as a linear functional on $D(T^*)$ which is *a fortiori* continuous in the T^*-norm. Thus the $D(T^*)^*$-topology on $D(T^*)$ is stronger than the topology induced on $D(T^*)$ by the Y-topology of Y^*.

(D) We now remark that, since T^* is one-to-one, the range of T is dense in Y. Given an arbitrary element y in Y, we shall construct an element x in $D(T)$ such that $Tx = y$.

Since $D(T^*)$ is total on Y (cf. Lemma 1.4(b)) there is an element y_0^* in $D(T^*)$ for which

$$y_0^*(y) = 1.$$

Now let B be the set of all elements y_1^* of $D(T^*)$ such that $y_1^*(y) = 0$. This set is closed in the $D(T^*)$-topology of $D(T^*)$, since y is a continuous linear functional on $D(T^*)$ (cf. step (C)). Furthermore, every element y^* of $D(T^*)$ can be written uniquely in the form $y^* = y_1^* + \alpha y_0^*$, where y_1^* is in B; we merely have to take $\alpha = y^*(y)$ and $y_1^* = y - \alpha y_0^*$.

(E) We shall now prove that the subspace T^*B is closed in the bounded X-topology of X^*. It suffices to show, by (B), that the intersection of T^*B with the unit sphere S of X^* is closed in the $D(T)$-topology of X^*. Notice first that, since T^* has closed range, its inverse T^{*-1} is continuous as a mapping of X^* (with the usual norm) into $D(T^*)$ with the T^*-norm. Therefore the set $T^{*-1}S$ is contained in some multiple G of the unit sphere of $D(T^*)$, where G is compact in the $D(T^*)^*$-topology of $D(T^*)$.

(F) The operator T^* is continuous from $D(T^*)$ with the $D(T^*)^*$-topology to X^* with the $D(T)$-topology. Indeed, let N be a neighborhood of 0 in the $D(T)$-topology of X^*, which is of the form

$$N = \{x^* : |x^*(x_i)| < \varepsilon,\ x_i \in D(T),\ 1 \leqq i \leqq n\}.$$

Consider the set

$$V = \{y^* : y^* \in D(T^*),\ |y^*(Tx_i)| < \varepsilon,\ 1 \leqq i \leqq n\}.$$

Since the elements Tx_i are continuous linear functionals on $D(T^*)$ with the

T^*-norm, V is a neighborhood of 0 in the $D(T^*)^*$-topology of $D(T^*)$. Furthermore, for y^* in V we have $|(T^*y^*)(x_i)| = |(y^*(Tx_i))| < \varepsilon$, $1 \leqq i \leqq n$. Then TV is contained in N.

(G) It follows from the preceding two steps that the image B under T^* of the compact set G is closed in the $D(T)$-topology of X^*. Thus the set

$$S \cap T^*B = S \cap T^*(B \cap G)$$

is closed in the $D(T)$-topology of X^*, and hence also in the bounded X-topology of X^*, by step (B). Therefore the subspace T^*B is closed in the bounded X-topology of X^*, and is thus X-closed.

(H) The element $T^*y_0^*$ does not lie in the X-closed subspace T^*B. By the Hahn-Banach theorem there exists an element x in X such that

$$(T^*y_0^*)(x) = 1, \qquad (T^*y^*)(x) = 0, \qquad y^* \in B.$$

Let y^* be an arbitrary element of $D(T^*)$. Decompose it into $y_1^* + \alpha y_0^*$, where y_1^* lies in B (cf. step (D)). Then

$$\begin{aligned}(T^*y^*)(x) &= (T^*y_1^*)(x) + \alpha(T^*y_0^*)(x) = \alpha \\ &= y_1^*(y) + \alpha y_0^*(y) = y^*(y).\end{aligned}$$

It follows from Lemma 1.4(c) that x is in $D(T)$ and $Tx = y$.

COROLLARY 1.1. *Let T be a closed operator with dense domain from a Banach space X into itself. Suppose that the range of the adjoint T^* is closed. Then the range of T is closed.*

Proof: Let A be the closure of the range of T, and consider T as a closed operator from X into the Banach space A. By the previous theorem, it suffices to verify that the range of the adjoint $T_1^* : A^* \to X^*$ is closed. Now let $\{u_n\}$ be a sequence in A^* such that $T_1^* u_n$ converges to x^*. Using the Hahn-Banach theorem let u_n be extended to an element y_n^* of X^*. Clearly $\{T^*y_n^*\}$ converges to x^*. Hence there exists an element y^* in $D(T^*)$ such that $T^*y^* = x^*$. Let now u^* be the restiction of y^* to A, then u^* is in $D(T_1^*)$ and $T_1^* u^* = x^*$.

THEOREM 1.2. *Let T be a closed operator with dense domain and finite-dimensional null space from a Banach space X to a Banach space Y. Suppose that the adjoint operator T^* is one-to-one and has closed range. Then the space X^* is the direct sum of $R(T^*)$ and a finite-dimensional subspace. Furthermore, every continuous linear functional on X^* which vanishes on the subspace $R(T^*)$ is (represented by) an element of X.*

Proof: Note that it may be proved, using a method similar to that of Theorem 1.1. that the range of T^* is closed in the X-topology of X^*; it suffices to show that the intersection of $R(T^*)$ with the unit sphere S of X^* is X-closed.

Now let $(x_\alpha)_{\alpha \epsilon A}$ be a Hamel basis for the subspace of those elements of X such that

$$(T^* y^*)(x_\alpha) = 0, \qquad y^* \epsilon D(T^*).$$

Then the set $(x_\alpha)_{\alpha \epsilon A}$ is a subset of $N(T)$ by Lemma 4(b). Hence there exists a finite basis $x_{\alpha_1}, x_{\alpha_2}, \cdots, x_{\alpha_n}$ for the set $(x_\alpha)_{\alpha \epsilon A}$.

Now choose elements $y_1^*, y_2^*, \cdots, y_n^*$ in X^* with the property that

$$y_i^*(x_{\alpha_j}) = \delta_{ij}, \qquad 1 \leqq i \leqq n.$$

Evidently the elements y_i^* are linearly independent modulo $R(T^*)$. Let M be the span of $y_1^*, y_2^*, \cdots, y_n^*$. We say that

$$R(T^*)+M = X^*.$$

Indeed, the space $R(T^*)+M$ is closed in the X-topology of X^*, since it is the sum of an X-closed subspace and a finite-dimensional space. If the last statement were false, there would exist an element x^* not in $R(T^*)+M$, and hence, by the Hahn-Banach theorem, an element x in X such that

$$x^*(x) = 1, \qquad (T^* y^*)(x) = 0, \qquad y^* \epsilon D(T^*).$$

Then x is in $D(T)$ and $Tx = 0$. Thus x is in the span of $x_{\alpha_1}, x_{\alpha_2}, \cdots, x_{\alpha_n}$, a contradiction. It follows from Lemma 1 that X^* is the direct sum of M and $R(T^*)$.

The second assertion is evident.

COROLLARY 1.2. *Let T be a closed operator with dense domain and finite-dimensional null space in a Banach space X. Suppose that the adjoint operator has closed range. Then the conclusion of the previous theorem holds.*

Proof: It suffices to remark that, considering the operator T from X to the Banach space $R(T)$ (cf. Theorem 1.1), the range of the adjoint of this operator coincides with the range of T^*.

2. Differential Operators in Banach Spaces

In the following section the basic definitions and the fundamental theorems about differential operators will be stated and proved. Some of the results in [2] and [3] will be used.

We consider an interval I of the real line, which may be open, half-open or closed. The entire line will be considered open, and a half-line such as $[a, \infty)$ will be considered half-open. Whenever no mention is made of the nature of an interval I, it is understood that the statement is true irrespective of the nature of the interval.

By a *differential operator* on an interval I we shall understand an expression of the form

$$(2.1) \qquad a_n(t)\frac{d^n}{dt^n} + a_{n-1}(t)\frac{d^{n-1}}{dt^{n-1}} + \cdots + a_0(t),$$

where the functions a_i are assumed to be complex-valued and infinitely differentiable in the interval of definition I. Differential operators will be denoted by Greek letters τ, σ, etc. If τ is a differential operator, the symbol τ^* will denote the *Lagrange adjoint* of τ, namely the operator

$$\tau^* = b_n(t)\frac{d^n}{dt^n} + b_{n-1}(t)\frac{d^{n-1}}{dt^{n-1}} + \cdots + b_0(t),$$

where

$$b_i(t) = \sum_{j=i}^{n} (-1)^j \binom{j}{i} \frac{d^{j-i}}{dt^{j-1}} a_j(t).$$

The symbol $A_n(I)$ will denote the vector space of all functions on an interval I which are differentiable $n-1$ times and whose $(n-1)$-st derivative is absolutely continuous. For f in $A_n(I)$ we shall use the notation

$$(\tau f)(t) = a_n(t)f^{(n)}(t) + a_{n-1}(t)f^{(n-1)}(t) + \cdots + a_0(t)f(t),$$

whenever τ is a differential operator of the form given in (2.1). The symbol $K_n(I)$ will denote the vector space of all functions which are differentiable n times and which vanish outside a compact subset of the interior of I that may depend on the function.

For t in I, let

$$F_t^{kj}(\tau) = \sum_{i=j}^{n-k-1} (-1)^i \binom{i}{j} \frac{d^{i-j}}{dt^{i-j}} a_{k+i+1}(t),$$

if $j+k \leqq n-1$, and

$$F_t^{kj}(\tau) = 0$$

otherwise. Let f, g be $n-1$ times differentiable functions in an interval $[a, b]$, and let

$$F_t(f, g) = \sum_{j=0}^{n-1} \sum_{k=0}^{n-1} F_t^{kj}(\tau) f^{(k)}(t) g^{(j)}(t).$$

Then for a, b in I the following identity ("Green's formula") holds:

$$\int_a^b (\tau f)(t) g(t) dt = \int_a^b f(t)(\tau^* g)(t) dt + F_b(f, g) - F_a(f, g).$$

We recall that $(\tau^*)^* = \tau$.

All differential operators will be tacitly assumed to be *regular* on their interval of definition I, that is, it will be assumed that the highest coefficient of τ does not vanish at any point in the interval I. Notice that if τ is a regular differential operator, then for each fixed t in the interval of definition the matrix $(F^{kj}(t))$ is non-singular.

A differential operator τ defined on an interval I with a finite endpoint a will be said to be *singular at a* if the point a does not belong to the interval I. The operator will be called "singular" at any infinite endpoint.

LEMMA 2.1. *Let f be a function in $A_n(I) \cap L_p(I)$, $\infty \geqq p \geqq 1$, and let g be a function in $K_n(I)$. Then the function $(\tau f)(t)g(t)$ is integrable, and*

$$\int_I (\tau f)(t)g(t)dt = \int_I f(t)(\tau^* g)(t)dt.$$

Proof: The statement follows at once from Green's formula. It suffices to notice that, after choosing a compact subinterval $[c, d]$ of I such that

$$g(c) = g'(c) = \cdots = g^{(n-1)}(c) = 0 = g(d) = g'(d) = \cdots = g^{(n-1)}(d)$$

we have

$$\begin{aligned}\int_I (\tau f)(t)g(t)dt &= \int_c^d (\tau f)(t)g(t)dt \\ &= \int_c^d f(t)(\tau^* g)(t)dt + F_d(f, g) - F_c(f, g) \\ &= \int_I f(t)(\tau^* g)(t)dt.\end{aligned}$$

The following lemma is proved in various works in the literature, see for instance the book of Dunford and Schwartz ([2], Chapter XIII, Lemma 2.9).

LEMMA 2.2. *Let f be a function integrable over every compact subinterval of I. Suppose that*

$$\int_I f(t)\tau^* g(t)dt = 0$$

for all functions g in $K_n(I)$. Then f is infinitely differentiable and $\tau f = 0$.

DEFINITION 2.2. Let τ be a regular differential operator of order n on an interval I. We shall denote by $D(T_1(\tau, p))$ the subspace of all functions f in $A_n(I)$ such that f and τf belong to the Banach space $L_p(I)$, $1 \leqq p \leqq \infty$. The linear operator $T_1(\tau, p)$ of $L_p(I)$ into itself is defined as follows: the domain of definition of $T_1(\tau)$ is the subspace $D(T_1(\tau, p))$ of $L_p(I)$, and for f in $D(T_1(\tau, p))$ we have

$$(T_1(\tau, p)f)(t) = (\tau f)(t), \qquad t \in I.$$

When no confusion will arise, the notation $T_1(\tau)$ will be used to designate the operator $T_1(\tau, p)$.

THEOREM 2.1. *Let T be the restriction of the operator $T_1(\tau, p)$, $1 \leqq p \leqq \infty$, of Definition 2.2 to the subspace $K_n(I)$ of $L_p(I)$. Then*

$$T^* = T_1(\tau^*, q),$$

where $p^{-1}+q^{-1} = 1$ if $1 < p < \infty$, $q = \infty$ if $p = 1$ and $q = 1$ if $p = \infty$, and where the adjoint of T in $L_\infty(I)$ is to be taken according to Definition 1.1.

Proof: Notice that the subspace $K_n(I)$ is dense in $L_p(I)$, if $1 \leqq p < \infty$, and L_1-dense in $L_\infty(I)$. Thus an adjoint exists.

Let f lie in $K_n(I)$ and let g lie in $D(T_1(\tau^*, q))$. Then, by Lemma 2.1,

$$\int_I (\tau f)(t) g(t) dt = \int_I f(t)(\tau^* g)(t) dt.$$

Hence $T^*(g) = \tau^* g$, and $D(T_1(\tau^*, q)) \subset D(T^*)$. To show the converse inclusion, let g lie in $D(T^*)$. There exists then a function $h = T^* g$ in $L_q(I)$ such that for all functions f in $K_n(I)$ we have

$$\int_I (\tau f)(t) g(t) dt = \int_I f(t) h(t) dt.$$

Now by the existence theorem for ordinary linear differential equations there exists a function g_0 in $A_n(I)$ such that $\tau^* g_0 = h$. A further application of Lemma 2.1 yields the equality

$$\int_I (\tau f)(t) g(t) dt = \int_I (\tau f)(t) g_0(t) dt$$

and consequently

$$\int_I (\tau f)(t)[g(t) - g_0(t)] dt = 0$$

for all functions f in $K_n(I)$. Lemma 2.2 implies that the function $g - g_0$ is infinitely differentiable. Since the function g_0 lies in $A_n(I)$, it follows that the function g lies in $A_n(I)$. Thus we can apply Lemma 2.1 to obtain

$$\int_I (\tau f)(t) g(t) dt = \int_I f(t)(\tau^* g)(t) dt,$$

and hence

$$\int_I f(t)[h(t) - (\tau^* g)(t)] dt = 0$$

for all functions f in $K_n(I)$. Thus the restriction of the function $h - \tau^* g$ to any compact subinterval I_1 of I is a function in $L_1(I_1)$ such that

$$\int_{I_1} f(t)[h(t) - (\tau^* g)(t)] dt = 0$$

for all functions f in $K_n(I_1)$. Now every essentially bounded function b in I_1 is a limit almost everywhere of functions in $K_n(I_1)$, and hence, by the Lebesgue bounded convergence theorem,

$$\int_{I_1} b(t)[h(t) - (\tau^* g)(t)] dt = 0$$

for all essentially bounded functions b. Therefore $h(t) - (\tau^* g)(t) = 0$ for t in I_1. Since I_1 is arbitrary, $h(t) - (\tau^* g)(t) = 0$ everywhere in I.

COROLLARY 2.1. *The operator T in the space $L_p(I)$, $p > 1$, defined in Theorem* 2.1, *has a closure.*

Proof: Indeed, the domain of the closed operator $T_1(\tau^*, q)$ is $L_p(I)$-

dense. Therefore the operator $T_1(\tau^*, q)$ has an adjoint, and we have

$$\bar{T} = T_1(\tau^*, q)^*.$$

DEFINITION 2.3. We shall denote by $T_0(\tau, p)$ the adjoint of the operator $T_1(\tau^*, q)$. Thus by the preceding theorem and by Lemma 1.4(c) we have

$$T_0(\tau, p)^* = T_1(\tau^*, q), \qquad 1 \leqq p \leqq \infty,$$
$$T_1(\tau, p)^* = T_0(\tau^*, q), \qquad p^{-1} + q^{-1} = 1.$$

The domains of the operators $T_1(\tau, p)$ and $T_0(\tau, p)$ will be denoted by $D_1(\tau, p)$ and $D_0(\tau, p)$, or, when no confusion will arise, simply by $D_1(\tau)$ and $D_0(\tau)$.

A word of caution is necessary here. If $1 \leqq p < \infty$, we have by Lemma 1.4(c) that $T_0(\tau, p)$ is simply the *closure* of the operator T of Theorem 2.1. This is however not the case in $L_\infty(I)$, where the operator $T_0(\tau, \infty)$ is in general a proper extension of the closure of the operator T. This apparent anomaly becomes clear if the method of graphs is used. The graph of the operator $T_0(\tau, \infty)$ is not the closure of the graph of T in the τ-norm, but the closure of the graph of T in the $L_1(I) \oplus L_1(I)$-topology of $L_\infty(I) \oplus L_\infty(I)$. This can be seen as follows: Let S be the graph of T, that is, the set of all pairs $[f, Tf]$ for f in $K_n(I)$. The closure $\bar{S}$ of S in the $L_1(I) \oplus L_1(I)$-topology of $L_\infty(I) \oplus L_\infty(I)$ is simply, in the notation of Section 1, the subspace $(S\perp)\perp$, where $S\perp$ is by definition a subspace of $L_1(I)$. Now $S\perp$ is the set of pairs $[T_1(\tau^*, 1)g, -g]$ for all elements g such that $[g, T_1(\tau^*, 1)g]$ belongs to the graph of g. Therefore the set $S\perp$ consists of all elements of the form $[h, T_1(\tau^*, 1)^*h] = [h, T_0(\tau, \infty)h]$.

An equivalent statement is the following: a function f belongs to $D_0(\tau, \infty)$ if and only if there exists a net $\{f_\alpha\}$ of functions in $K_n(I)$ such that

$$\int_I f_\alpha(t)g(t)dt \to \int_I f(t)g(t)dt, \qquad g \in L_1(I),$$
$$\int_I (\tau f_\alpha)(t)h(t)dt \to \int_I (\tau f)(t)h(t)dt, \qquad h \in L_1(I).$$

The following lemma is a straightforward generalization of a result in [2] in Hilbert space.

LEMMA 2.3. *Let τ be a differential operator of order h on an interval I, let $\{f_n\}$ be a sequence in $D_1(\tau, p)$, $1 \leqq p \leqq \infty$, which converges in the τ-norm, and let J be a compact subinterval of I. Then the sequences $\{f_n\}$, $\{f'_n\}, \cdots, \{f^{(k-1)}_n\}$ converge uniformly on J, and the sequence $\{f^{(k)}_n\}$ converges in $L_p(J)$.*

COROLLARY 2.3. *Suppose that an endpoint a belongs to the interval I. Then every function f in $D_0(\tau, p)$, $1 \leqq p < \infty$, satisfies the condition*

$$f(a) = f'(a) = \cdots = f^{(k-1)}(a) = 0.$$

LEMMA 2.4. *Let τ be a differential operator of order k. Suppose that an endpoint a belongs to the interval I. Then every function f in $D_0(\tau, \infty)$ satisfies the conditions*

$$f(a) = f'(a) = \cdots = f^{(k-1)}(a) = 0.$$

Proof: The discussion following Definition 2.3 shows that there exists a net $\{f_\alpha\}$ of functions in $K_k(I)$ such that for all functions g in $D_1(\tau^*, 1)$ we have

$$\int_I f_\alpha(t)(\tau^* g)(t)dt \to \int_I f(t)(\tau^* g)(t)dt,$$
$$\int_I (\tau f_\alpha)(t)g(t)dt \to \int_I (\tau f)(t)g(t)dt.$$

Now let $I = [a, b)$ and let S be the subspace of $D_1(\tau^*, 1)$ of all functions which vanish in a neighborhood of the endpoint b. By Green's formula we have

$$0 = \int_a^b f_\alpha(t)(\tau^* g)(t)dt - \int_a^b (\tau f_\alpha)(t)g(t)dt$$
$$\to \int_a^b f(t)(\tau^* g)(t)dt - \int_a^b (\tau f)(t)g(t)dt$$
$$= F_a(f, g) = \sum_{i=0}^{k} \sum_{j=0}^{k} F_a^{kj}(\tau) f^{(i)}(a) g^{(j)}(a),$$

$g \in S$. Since the matrix $F_a^{kj}(\tau)$ is non-singular, and since there exist functions in S such that $g(a), g'(a), \cdots, g^{(k-1)}(a)$ take on any arbitrarily assigned set of values, it follows that $f(a) = f'(a) = \cdots = f^{(k-1)}(a) = 0$. The following definition introduces the two notions which will be the object of the present study.

DEFINITION 2.4. (a) A closed operator T in $L_p(I)$, $1 \leqq p \leqq \infty$, is an extension of a differential operator τ on the interval I if

$$T_1(\tau, p) \supset T \supset T_0(\tau, p).$$

An extension of a differential operator τ will sometimes be referred to as a D-operator.

(b) A continuous linear functional B in $D_1(\tau, p)$, $1 \leqq p \leqq \infty$, (with the τ-norm) is a boundary value for the differential operator τ in $L_p(I)$, $1 \leqq p \leqq \infty$, if $B(f) = 0$ for all functions f in $D_0(\tau, p)$.

The following lemma is an immediate consequence of this definition.

LEMMA 2.5. (a) *If T is an extension of a differential operator τ in $L_p(I)$, then there exists a set $(B_\alpha)_{\alpha \in A}$ of boundary values such that*

$$D(T) = \{f : B_\alpha(f) = 0,\ \alpha \in A\}.$$

(b) *If $(B_\alpha)_{\alpha \in A}$ is a set of boundary values of a differential operator τ in $L_p(I)$, then the set*

$$\{f : B_\alpha(f) = 0, \ \alpha \in A\}$$

defines the domain of an extension of τ.

(c) *The adjoint of a D-operator is a D-operator.*

In case (a) we shall say that the extension T is obtained *by the imposition of a set of boundary conditions* $(B_\alpha)_{\alpha \in A}$.

If there exists a Hamel basis $(X_\alpha)_{\alpha \in A}$ of $D(T)$ modulo $D_0(\tau)$, we shall say that the extension T is obtained by extension to the span of the set $(x_\alpha)_{\alpha \in A}$.

3. The Essential Spectrum

We shall introduce some notions connected with the spectrum which, although definable in a more general setting, are useful in the study of some of the spectral properties of differential operators, and which will be the object of this section. Thus the discussion is deliberately restricted to the case of differential operators in the spaces $L_p(I)$, for $1 \leqq p \leqq \infty$.

The principal notion to be defined is the essential spectrum. As the reader will not fail to notice, this concept arises from an attempt to generalize to non-self-adjoint and in particular to differential operators those properties which in the case of self-adjoint operators in Hilbert space are associated with the continuous spectrum. In the extension theory to be developed the essential spectrum will play only a negative role: practically every theorem will have a non-vacuous content only under the assumption that the essential spectrum of a given differential operator shall not cover the entire plane. In the extension theory this assumption is of course necessary irrespective of the particular function space in which the given differential operator is considered. It is however true that many of the results concerning boundary values and adjoints which are contained in this section and the next require this assumption only when a differential operator τ is considered in Banach spaces other than Hilbert space. I do not know thus far whether arguments of a deeper nature could eliminate this hypothesis in the case of an operator acting in one of the spaces $L_p(I)$, $1 < p < \infty$, $p \neq 2$. In $L_1(I)$ and $L_\infty(I)$ the hypothesis seems to be essential. In some instances the cases of differential operators acting on one of these spaces require separate handling, mostly through the theorems developed in Section 1.

DEFINITION 3.1. (a) Let T be a D-operator of a differential operator τ on an interval I in a space $L_p(I)$, $1 \leqq p \leqq \infty$. The essential spectrum $\sigma_e(T)$ will consist of all points λ in the complex plane for which the range of the operator $\lambda I - T$ is not closed. The complement of the essential spectrum shall be called the essential resolvent set $\varrho_e(T)$.

(b) Let τ be a differential operator on an interval I. The essential spectrum $\sigma_e(\tau, p, I)$ of τ in $L_p(I)$ shall be the essential spectrum of the operator $T_0(\tau, p)$. The essential resolvent set $\varrho_e(\tau, p, I)$ of τ shall be the essential resolvent set of the operator $T_0(\tau, p)$.

(c) A complex number λ shall be an *essential eigenvalue* of a differential operator τ in a space $L_p(I)$ it it does not belong to the essential spectrum of τ and is an eigenvalue of the operator $T_0(\tau, p)$.

(d) A complex number λ which belongs neither to the essential spectrum of a differential operator τ nor is an essential eigenvalue of τ shall be called a *regular point* of τ.

The following lemma is an immediate consequence of the definition.

LEMMA 3.1. *If the interval of definition of a differential operator τ contains one of its endpoints, then τ has no essential eigenvalues, and every complex number in the essential resolvent set of τ is a regular point of τ.*

Indeed, let a be such an endpoint. If $\tau f - \lambda f = 0$ for some function f in $D_0(\tau)$, then, by Lemmas 2.3 and 2.4, $f(a) = f'(a) = \cdots = f^{(n-1)}(a) = 0$. Since the operator τ is assumed to be regular, it follows by the uniqueness theorem for solutions of a differential equation that the function f vanishes identically.

The following lemma exhibits the two features described at the beginning of this section: it is valid in Hilbert space irrespective of the nature of the spectrum, and requires a separate proof in the space $L_\infty(I)$.

LEMMA 3.2. *For a given p, $1 \leqq p \leqq \infty$, suppose that the essential resolvent set of a differential operator τ in $L_p(I)$ is non-empty. Then the dimension of the space $D_1(\tau, p)/D_0(\tau, p)$ is finite.*

Proof: Since $D_1(\tau - \lambda, p) = D_1(\tau, p)$ for any complex λ, we can assume without loss of generality that the number 0 belongs to the essential resolvent set of τ.

By Lemma 2.1 the dimension of the quotient space $L_p(I)/R(T_0(\tau, p))$ equals the dimension of the space of all continuous linear functionals on $L_p(I)$ which vanish on $R(T_0(\tau, p))$ if either of these dimensions is finite. The fact that these dimensions are finite follows in the case of $1 \leqq p < \infty$ from Theorem 2.1, and for $p = \infty$ from Corollary 1.2.

Now the operator $T_1(\tau, p)$ induces a bounded one-to-one linear operator of the quotient space $D_1(\tau, p)/[D_0(\tau, p) + N(T_1(\tau, p))]$ into $L_p(I)/R(T_0(\tau, p))$. Thus the dimension of $D_1(\tau, p)/[D_0(\tau, p) + N(T_1(\tau, p))]$ is finite. The assertion follows from the fact that the dimension of the null space $T_1(\tau, p)$ is finite and from Lemma 1.1.

LEMMA 3.3. *For $1 \leqq p \leqq \infty$, suppose that the essential resolvent set of an operator τ in $L_p(I)$ is non-empty. Let T be an arbitrary D-operator obtained* *from τ. Then the essential spectrum of T coincides with the essential spectrum of τ.*

The proof is immediate from the definition and from Lemmas 1,1, 1.2 and 3.2.

The following result is similar to a theorem of Dunford and Schwartz (cf. Chapter XIII, Lemma 7—1).

LEMMA 3.4. *Let T be a closed operator from a Banach space X to a Banach space Y. Suppose that the dimension of the null space $N(T)$ is finite. Then the range of T is a closed subspace of Y if and only if T maps bounded closed sets into closed sets.*

Proof: Suppose that T maps bounded closed sets into closed sets. It will first be shown that the same property holds for the operator T' from the factor space $X' = X/N(T)$ into Y which is induced by T. Let ϕ be the canonical mapping of X onto X', and let B' be a bounded closed set in X'. It suffices to construct a bounded closed set B in H such that $\phi(B) = B'$. Indeed, $T'B' = TB$ and TB is closed by hypothesis. Now suppose for simplicity that

$$\sup \{|x'| : x' \epsilon B'\} = \tfrac{1}{2}.$$

Since the canonical mapping ϕ maps the open unit sphere S of X onto the open unit sphere S' of H', it follows that $\phi(\bar{S}) \supset B'$. Let $B = \phi^{-1}(B') \cap \bar{S}$. Clearly B is the required set.

Thus, to complete the proof of the first part of the theorem, it suffices to show that the range of T' is closed, for, the operator T' on X' has the same range as T.

Suppose now that $T'x_n \to y$, and that y does not coincide with any of the elements x_n. If the sequence $\{x_n\}$ has no convergent subsequence, it forms a closed set in X'. Therefore $T(\{x_n\})$ is a closed set by hypothesis. But this contradicts the fact that $T'x_n \to y$.

To prove the converse a similar approach can be used. It will first be shown that if B is a bounded closed set in X, then $\phi(B)$ is bounded and closed under the canonical mapping $\phi : X \to X' = X/N(T)$. Clearly, the set $\phi(B)$ is closed if the set $B+N(T)$ is closed. Let $\{x_n\}$ be a sequence in $B+N(T)$ converging to x. We can find elements $y_n \epsilon B$ and $z_n \epsilon N(T)$ such that $x_n = y_n + z_n$. Since both $\{x_n\}$ and $\{y_n\}$ are bounded sequences, it follows that $\{z_n\}$ is bounded, and has therefore a convergent subsequence $\{z_{n_i}\}$, $z_{n_i} \to z$. Thus $y_{n_i} \to x - z$ and, since B is closed, $x - z \epsilon B$, $x \epsilon B+N$. Hence ϕ maps bounded closed sets into closed sets, and the proof is reduced to the case of a one-to-one operator T with closed range. But in this case the statement follows at once from the continuity of T^{-1}, which is insured by the closed graph theorem.

COROLLARY 3.4. *Let T be a D-operator. Then the essential resolvent set of T consists of all complex λ such that $\lambda I-T$ maps bounded closed sets into closed sets.*

LEMMA 3.5. *Let τ be a differential operator in $L_p(I)$, $1 \leqq p \leqq \infty$, and let I_1 and I_2 be subintervals whose union is the interval I. Then the essential spectrum $\sigma_e(\tau, p, I)$ is the union of the essential spectra $\sigma_e(\tau, p, I_1)$ and $\sigma_e(\tau, p, I_2)$.*

Proof: The space $L_p(I)$ is the direct sum of the spaces $L_p(I_1)$ and $L_p(I_2)$. Let the complex number λ belong to the essential spectrum $\sigma_e(\tau, p, I_1)$, say. Then there is a bounded subset B of $L_p(I_1) \subset L_p(I)$ such that

$$[\lambda I-T_1(\tau, p, I)]B = [\lambda I-T(\tau, p, I_1)]B$$

is not closed. Thus λ is in $\sigma_e(\tau, p, I)$.

Conversely, suppose that for a bounded closed set B in $L_p(I)$ the image $(\lambda I-T)B$ is not closed. Then there is a sequence $\{f_n\}$ in B which has no convergent subsequence, whereas the sequence $\{[\lambda I-T_1(\tau, p, I)]f_n\}$ converges. Thus one of the sequences $\{f_n|I_1\}$ and $\{f_n|I_2\}$ does not converge, while both sequences $\{[\lambda I-T_1(\tau, p, I_1)]f_n\}$ and $\{[\lambda I-T_1(\tau, p, I_2)]f_n\}$ do. Hence the complex number λ lies in one of the essential spectra $\sigma_e(\tau, p, I_1)$ or $\sigma_e(\tau, p, I_2)$.

LEMMA 3.6. *The essential spectrum $\sigma_e(\tau, p, I)$ is a closed set, $1 \leqq p \leqq \infty$.*

Proof: By the preceding lemma, it suffices to show that the sets $\sigma_e(\tau, p, I_1)$ and $\sigma_e(\tau, p, I_2)$ are closed for a partition of the interval I into subintervals I_1 and I_2. We can furthermore assume without loss of generality that the common endpoint belongs to both I_1 and I_2. Thus, by Lemma 2.1, the operators $\lambda I-T_0(\tau, p, I_1)$ and $\lambda I-T_0(\tau, p, I_2)$ are one-to-one for all complex numbers λ. Suppose that λ does not belong to $\sigma_e(\tau, p, I_1)$, say. Then the closed graph theorem yields a constant $\varkappa$ such that

$$|\lambda I-T_0(\tau)f| \geqq \varkappa|f|, \qquad f \in D(T_0(\tau, I_1)).$$

Thus for $|\varepsilon| \leqq \varkappa$ we have

$$|(\lambda+\varepsilon)I-T_0(\tau)f| \geqq (\varkappa-|\varepsilon|)|f|, \qquad f \in D(T_0(\tau, I_1)),$$

which implies that the range of the operator $(\lambda+\varepsilon)I-T_0(\tau, p, I_1)$ is closed.

THEOREM 3.1. *If a D-operator T in $L_p(I)$, $1 \leqq p < \infty$, has an adjoint, then the essential spectrum of T coincides with the essential spectrum of the adjoint T^*.*

If $1 < p < \infty$ the proof is simple: it suffices to show that the range $R(T^*)$ coincides with the space $N(T)\perp$ of all linear functionals on $L_p(I)$ which vanish on the null space of T. Thus let g be a function in $N(T)\perp$.

For every function f_1 in the range of T, choose a function f in $D(T)$ such that $Tf = f_1$, and define a linear functional F in $R(T)$ as follows:

$$F(f_1) = (f, g), \qquad Tf = f_1, \qquad f \in R(T).$$

The functional F is well-defined. Indeed, if f_2 is a second function in $D(T)$ such that $Tf_2 = Tf = f_1$, then the condition on g implies that $(f_1, g) = (f_2, g)$.

Now since the range of T is closed, there exists a constant $\varkappa$ with the property that for every f in $R(T)$ we can find an h in $D(T)$ such that

$$|h| \leqq \varkappa |f_1|, \qquad Th = f_1.$$

Therefore,

$$\begin{aligned} |F(f_1)| = |(f_1, g)| &\leqq \inf \{|(h, g)| : h \in D(T),\ Th = f_1\} \\ &\leqq |g| \inf \{|h| : h \in D(T),\ Th = f_1\} \leqq \varkappa |g||f_1|, \end{aligned}$$

that is, the linear functional F is bounded on the closed subspace $R(T)$. By the Hahn-Banach theorem, it can be extended to a continuous linear functional on $L_p(I)$, represented by a function g_1 in $L_q(I)$. Consequently we have

$$(Tf, g_1) = (f, g), \qquad f \in D(T),$$

and, by the definition of the adjoint operator, g_1 is in $D(T^*)$ and $T^*g_1 = g$.

If $p = 1$, the same proof yields the result that T^* has closed range if T has closed range. The converse follows from Corollary 1.1.

This theorem is not true without further restrictions in the space $L_\infty(I)$, because of the fact that if T is a closed operator in $L_\infty(I)$ having an adjoint T^* in $L_1(I)$, it is not true in general that $(T^*)^* = T$. It is, however, true in $L_\infty(I)$ if it is assumed that the operator T is the adjoint of some operator $L_1(I)$.

COROLLARY 3.1. *Let τ be a differential operator in $L_p(I)$, $1 \leqq p \leqq \infty$. Then the essential spectrum of τ in $L_p(I)$ coincides with the essential spectrum of τ^* in $L_q(I)$, $p^{-1}+q^{-1} = 1$.*

The computation of the essential spectrum of a given operator τ is in general rather difficult. We give one result which facilitates this calculation, and which seems to be characteristic of differential operators.

THEOREM 3.2. (Spectral mapping) *Let τ be a differential operator on an interval I, and let*

$$P(\lambda) = a_n \lambda^n + a_{n-1} \lambda^{n-1} + \cdots + a_0$$

be a polynomial with constant coefficients. Let $p(\tau)$ be the differential operator

$$P(\tau) = a_n \tau^n + a_{n-1} \tau^{n-1} + \cdots + a_0 .$$

Suppose that the set

$$\{P(\lambda) | \lambda \in \sigma_e(\tau, p, I)\}$$

is not the entire plane. Then

$$\sigma_e(P(\tau), p, I) = \{P(\lambda) | \lambda \in \sigma_e(\tau, p, I)\}.$$

The proof has been divided into several steps. The following convention has been used throughout: whenever N is a finite-dimensional subspace of a Banach space X, we denote by N^0 an arbitrary closed subspace of X with the property that X is the direct sum of N and N^0. It is clear that such a closed subspace N^0 exists; indeed, it suffices to choose a basis $x_1, x_2, \cdots, x_n$ of N, and linear functionals x_i^* such that

$$x_i^*(x_j) = \delta_{ij}, \qquad 1 \leqq i, j \leqq n.$$

Then define

$$N^0 = \{x : x_i^*(x), \ 1 \leqq i \leqq n\} = 0.$$

(A) Let T be an operator in a Banach space X which has a closure $\bar{T}$. Suppose that the null space of T is finite-dimensional and that the range of T is closed. Then T is a closed operator, i.e., $T = \bar{T}$.

Proof: Let T_1 be the restriction of T to $N(T)^0$. If T_1 is closed, then T is closed. Indeed, the graph of T consists of the sum of the graph of T_1 and a finite-dimensional space.

Thus it suffices to show that T_1 is closed. Now the operator T_1^{-1} has a closure, because T_1 does, and has a closed domain. Therefore T_1^{-1} is closed, and hence so is T_1.

(B) Let T be a bounded operator from a Banach space X to a Banach space Y having finite-dimensional null space and closed range. Let M be a closed subspace of X. Then $T(M)$ is a closed subspace of Y.

Proof: Let N be the null space of T, and let T_1 be the one-to-one restriction of T_1 to N^0. By the closed graph theorem, T_1 maps closed sets into closed sets. Now $TM = T_1((M+N) \cap N^0)$, and $(M+N) \cap N^0$ is closed.

(C) Let S and T be D-operators with finite-dimensional null spaces and closed ranges. Then ST is a closed D-operator with closed range.

Proof: We shall prove the fact that ST has closed range first. Once it is shown that $D(S) \cap R(T)$ is closed in $D(S)$ (with the S-norm), the first assertion will follow from (B). Let $\{x_n\}$ be a sequence in $D(S) \cap R(T)$ which is a Cauchy sequence in the S-norm of $D(S)$. Since Sx_n converges to Sx, and x_n converges to x, it follows that x is in $D(S)$ and in $R(T)$.

Since S and T are D-operators, it follows that the operator ST has finite-dimensional null space. Hence by (A), ST is a closed operator. To show that ST is a D-operator, let S be defined by means of the differential operator σ, and let T be defined by means of the operator τ. Suppose that σ is of order m and τ of order n. From the fact that

$$D(S) \supset K_m(I), \qquad D(T) \supset K_n(I),$$

we deduce that $D(ST) \supset K_{m+n}(I)$. Since ST is closed, it follows that

$$ST \supset T_0(\sigma\tau).$$

Similarly, from the fact that

$$S \subset T_1(\sigma), \qquad T \subset T_1(\tau),$$

we have

$$ST \subset T_1(\sigma)T_1(\tau).$$

Furthermore, it is clear from the definition that $T_1(\sigma)T_1(\tau) \subset T_1(\sigma\tau)$. Hence

$$T_0(\sigma\tau) \subset ST \subset T_1(\sigma\tau).$$

(D) Let $T_1, T_2, \cdots, T_n$ be D-operators which commute in the following sense: if $i_1, i_2, \cdots, i_n$ is an arbitrary permutation of the indices $1, 2, \cdots, n$, then $D(T_1T_2 \cdots T_n) = D(T_{i_1}T_{i_2} \cdots T_{i_n})$, and

$$T_1T_2 \cdots T_n x = T_{i_1}T_{i_2} \cdots T_{i_n} x$$

for every element x in the common domain. Then:

(a) if the range of the operator $T_1T_2 \cdots T_n$ is closed, then $T_1T_2 \cdots T_n$ is a closed D-operator.

(b) The range of the operator $T_1T_2 \cdots T_n$ is closed if and only if the range of T_i is closed, $1 \leqq i \leqq n$.

Proof: The first assertion follows as in the case of two operators, proved in step (C). If the ranges of the T_i, $1 \leqq i \leqq n$, are closed, then an iteration of the result of (C) yields the fact that the range of $T_1T_2 \cdots T_n$ is closed.

Suppose now that $R(T_1T_2 \cdots T_n)$ is closed. Let $\tau_1, \tau_2, \cdots, \tau_n$ be the differential operators from which $T_1, T_2, \cdots, T_n$ are obtained. Then clearly $T_1T_2 \cdots T_n \subset T_1(\tau_1\tau_2 \cdots \tau_n)$ and, by (a), $T_1T_2 \cdots T_n$ is a closed operator. Hence $T_1T_2 \cdots T_n \supset T_0(\tau_1\tau_2 \cdots \tau_n)$, as in step (C), and $(T_1T_2 \cdots T_n)^* \subset T_1(\tau_n^*\tau_{n-1}^* \cdots \tau_0^*)$. Thus the complement of the range of $T_1T_2 \cdots T_n$ is finite-dimensional. Therefore, the range of T_1 consists of the sum of the range of $T_1T_2 \cdots T_n$ and a finite-dimensional space. It follows that the range of T_1 is closed. By commuting the operators and repeating the same argument, we conclude that the range of T_i is closed for every i.

(E) The proof of the theorem can be completed as follows:

It is assumed that there is a point λ_0 in the plane for which none of the roots of the equation $p(z) = \lambda_0$ lies in the essential spectrum of τ. Let these roots be $\lambda_1^0, \lambda_2^0, \cdots, \lambda_n^0$. Then the range of the operator $\lambda_i^0 I - T_0(\tau)$, $1 \leqq i \leqq n$, is closed. It follows by step (D) (a) that the operator

$$\lambda_0 I - P(T_0(\tau)) = [\lambda_1^0 I - T_0(\tau)][\lambda_2^0 I - T_0(\tau)] \cdots [\lambda_n^0 I - T_0(\tau)]$$

has a closed range and is an extension of the differential operator $\lambda - P(\tau)$.

Therefore, the operator $P(T_0(\tau))$ is closed and is an extension of the operator $P(\tau)$, that is,

$$P(T_0(\tau)) \supset T_0(P(\tau)).$$

We conclude that for any complex λ, either both the ranges of $\lambda I - P(T_0(\tau))$ and of $\lambda I - T_0(P(\tau))$ are closed, or else neither range is closed. It will thus henceforth suffice to deal with the operator $P(T_0(\tau))$.

Suppose first that the range of the operator $\lambda - P(T_0(\tau))$ is closed. Let $\lambda_1, \lambda_2, \cdots, \lambda_n$ be the roots of the equation $p(z) = \lambda$, and decompose the operator into the product of the commuting operators

$$\lambda - P(T_0(\tau)) = [\lambda_1 I - T_0(\tau)][\lambda_2 I - T_0(\tau)] \cdots [\lambda_n I - T_0(\tau)].$$

By step (D) we conclude that each of the operators $\lambda_i I - T_0(\tau)$, $1 \leqq i \leqq n$, has closed range. Conversely, if each of the operators $\lambda_i I - T_0(\tau)$ has closed range, it again follows by step (D) that the operator $\lambda - P(T_0(\tau))$ has closed range. The proof is therefore complete.

4. Structure of Extensions

We are now ready to weave together the various threads and develop the results which constitute the main object of the present study.

The problem to be studied is the following: given a differential operator τ on an interval I, find an extension T of τ in $L_p(I)$, $1 \leqq p \leqq \infty$, for which the spectrum is minimal. The restriction we shall impose on the differential operator τ is that the essential resolvent set shall not be empty in $L_p(I)$. This seemingly natural condition in effect considerably restricts the class of differential operators. For instance, the operator $e^{it} d/dt$ on the interval $[0, \infty)$ has an empty essential resolvent set in $L_p(0, \infty)$, $1 < p < \infty$. The "qualitative" results concerning the spectrum of extensions and the distribution of essential eigenvalues are developed in this section. Later we shall study some of the questions concerning the analytic nature of boundary values and the relationship with the spectrum of the adjoint operator. We collect in the following lemma the properties of boundary values which have already been implicitly proved, and which will be the only ones to be used in this section.

LEMMA 4.1. *Suppose that the differential operator τ in $L_p(I)$, $1 \leqq p \leqq \infty$, has a non-empty essential resolvent set. Then the dimension of the space of boundary values in $L_p(I)$ is finite and equals the dimension of the quotient space $D_1(\tau, p)/D_0(\tau, p)$. Every extension T of τ is obtained by the imposition of a finite set of linearly independent boundary values, or equivalently, by extending the domain $D_0(\tau, p)$ to the span of a finite set of linearly independent functions in $D_1(\tau, p)$.*

The notion of an extension which is "best possible" for our problem is given by the following

DEFINITION 4.1. An extension T of a differential operator τ in $L_p(I)$, $1 \leqq p \leqq \infty$, is a (local) maximal extension of τ at a point λ in the complex plane if λ lies in the resolvent set of the operator T (i.e., if the operator $\lambda I - T$ has a bounded everywhere defined inverse).

It is clear that there exists no maximal extension of a differential operator τ at any point in the essential spectrum of τ. If the operator τ is singular at both endpoints of the interval, then clearly there can be no maximal extension at any essential eigenvalue.

The following terms will be used: Suppose that the range of a D-operator T in $L_p(I)$, $1 \leqq p \leqq \infty$, is closed, and let $g_1, g_2, \cdots, g_n$ be a basis for the elements of $L_q(I)$ which vanish on $R(T)$, that is, a basis for the null space $N(T^*)$. We shall call the functions $f_1, f_2, \cdots, f_n$ of $L_p(I)$ a dual basis of the set $\{g_i\}$ if

$$(g_i, f_j) = \delta_{ij}.$$

If such is the case, then the space $L_p(I)$ is the direct sum of $R(T)$ and the span of the elements $f_1, f_2, \cdots, f_n$. Notice that this notion is well defined in $L_\infty(I)$, in virtue of Corollary 1.2.

The following simple lemma determines the structure of local maximal extensions.

LEMMA 4.2. (a) *Let τ be a differential operator in $L_p(I)$, $1 \leqq p \leqq \infty$, and let λ be an arbitrary complex number which is a regular point of τ in $L_p(I)$ and of τ^* in $L_q(I)$, $p^{-1}+q^{-1} = 1$. Then there exists a maximal extension T_0 of τ at λ.*

(b) *Let the functions $f_1, f_2, \cdots, f_n$ be chosen in $D_1(\tau, p)$ so that the functions*

$$\lambda f_i - \tau f_i, \qquad 1 \leqq i \leqq m,$$

form a dual basis for $N(\lambda I - T_1(\tau^*, q))$. Let $g_1, g_2, \cdots, g_m$ be arbitrary functions in $D_1(\tau, p)$ (not necessarily distinct and possibly null) such that

$$\lambda g_i - \tau g_i = 0 \qquad 1 \leqq i \leqq m.$$

Then the functions $f_1+g_1, f_2+g_2, \cdots, f_m+g_m$ are linearly independent modulo $D_0(\tau)$, and the extension of τ to the span of $f_1+g_1, f_2+g_2, \cdots, f_m+g_m$ is a maximal extension at λ.

Conversely, every maximal extension T of τ at λ is obtained by such an extension of the domain.

(c) *Let T be a maximal extension of τ at λ obtained by the imposition of the linearly independent boundary conditions*

$$B_1(f) = B_2(f) = \cdots = B_{n_1}(f) = 0.$$

Let $f_1, f_2, \cdots, f_n$ *be a basis for* $N(\lambda I - T_1(\tau))$. *Then* $n = n_1$, *and the determinant of the matrix*

$$(4.1) \qquad (B_i(f_j))$$

is different from zero.

Conversely, given n linearly independent boundary values $B_1, B_2, \cdots, B_n$, *the extension T of* τ *obtained by the imposition of the boundary conditions*

$$B_1(f) = B_2(f) = \cdots = B_n(f) = 0$$

is a maximal extension at λ *if the determinant of the matrix* (4.1) *does not vanish.*

(d) *Let* $f_1, f_2, \cdots, f_m$ *be a basis for* $D_1(\tau)$ *modulo* $D_0(\tau) + N(\lambda I - T_1(\tau))$. *Then the extension of* τ *to the span of* $f_1, f_2, \cdots, f_m$ *is a maximal extension at* λ.

Conversely, let T be a maximal extension of τ *at* λ, *and let* $\{f_1, f_2, \cdots, f_m\}$ *be a basis for* $D(T)$ *modulo* $D_0(\tau)$. *Then* $\{f_1, f_2, \cdots, f_m\}$ *is a basis for* $D_1(\tau)$ *modulo* $D_0(\tau) + N(\lambda I - T_1(\tau))$.

Proof: The assumption that λ is a regular point of τ means that

$$N(\lambda I - T_1(\tau)) \cap D_0(\tau) = \{0\}$$

and the assumption that λ is a regular point of τ^* means that the range of the operator $\lambda I - T_1(\tau, p)$ is the entire space $L_p(I)$. Therefore, we can decompose the space $D_1(\tau, p)$ into a direct sum

$$(4.2) \qquad D_1(\tau, p) = D_0(\tau, p) \oplus N(\lambda I - T_1(\tau, p)) \oplus S,$$

where S is a finite-dimensional space with basis $f_1, f_2, \cdots, f_m$. It is now clear that the set

$$\{\lambda f_i - f_i\}, \qquad 1 \leqq i \leqq m,$$

is a dual basis of $N(\lambda I - T_1(\tau^*, q))$. Hence the extension T_0 of τ to the span of S is a maximal extension at λ.

(b) *Let* $g_1, g_2, \cdots, g_m$ be functions such that

$$\lambda g_i - \tau g_i = 0, \qquad 1 \leqq i \leqq m.$$

Then the functions $f_1 + g_1, f_2 + g_2, \cdots, f_m + g_m$ are linearly independent modulo $D_0(\tau, p)$, as follows from the direct sum decomposition (4.2). Let T be the extension of τ to the span of the functions $f_1 + g_1, f_2 + g_2, \cdots, f_m + g_m$. The range of $\lambda I - T$ coincides with the range of T_0 (cf. (A)); thus $\lambda I - T$ maps onto $L_p(I)$.

In order to show that $\lambda I - T$ is one-to-one, we suppose that

$$(\lambda I - T)f = 0,$$

for some f in $D(T)$. Then f can be written in the form $h_1 + h_2$, where h_1 lies in S and h_2 lies in $N(\lambda I - T_1(\tau, p))$. Hence

$$(\lambda I - T)h_1 = [\lambda I - T_1(\tau, p)]h_1 = 0$$

and $h_1 = 0$, since $\lambda I - T_1(\tau, p)$ is one-to-one on S.

To see that every maximal extension T of τ at λ is of the form described, notice that there exist functions $h_1, h_2, \cdots, h_m$ in $D(T)$ such that

$$(\lambda I-T)h_i = \lambda f_i-\tau f_i, \qquad 1 \leqq i \leqq m;$$

consequently there exist functions $g_1, g_2, \cdots, g_m$ such that $\tau g_i-\lambda g_i = 0$, $1 \leqq i \leqq m$, and

$$h_i = f_i+g_i.$$

Suppose that there exists a function f in $D(T)$ not lying in the subspace spanned by $D(T_0(\tau))$ and f_i+g_i, $1 \leqq i \leqq m$. Then

$$(T-\lambda I)(f+g) = 0$$

for some g in this subspace, since $\lambda I-T$ is an "onto" map. The operator $T-\lambda I$ is one-to-one; it follows that $x = -g$ and hence

$$D(T) = \text{span}\,\{D(T_0(\tau)), x_i+y_i,\ 1 \leqq i \leqq m\}.$$

(c) Let us first verify that $n_1 = n = \dim N(T_1(\tau)-\lambda I)$. Let $m = \dim N(T_1(\tau^*)-\lambda I)$. Clearly we have the equality

$$n_1 = \dim \frac{D_1(\tau)}{D(T)}.$$

Furthermore, by (b),

$$\dim \frac{D(T)}{D_0(\tau)} = m$$

and

$$\begin{aligned} n+m &= \dim \frac{D_1(\tau)}{D(T_0(\tau))} \\ &= \dim \frac{D_1(\tau)}{D(T)} + \dim \frac{D(T)}{D_0(\tau)} \\ &= n_1+m. \end{aligned}$$

Suppose now that the determinant of $(B_i(f_j))$ vanishes. Then there exists a non-zero linear combination f of $f_1, f_2, \cdots, f_n$ such that

$$B_1(f) = B_2(f) = \cdots = B_n(f) = 0.$$

Thus f lies in $D(T)$ and $Tf-\lambda f = 0$. Therefore $f = 0$.

To verify the converse we notice first that, since the boundary values $B_1, B_2, \cdots, B_n$ are linearly independent, we have

$$\dim \frac{D(T_1(\tau))}{D(T)} = n,$$

and thus, by (B),

$$\dim \frac{D(T)}{D_0(\tau)} = n+m - \dim \frac{D_1(\tau)}{D(T)} = m.$$

Secondly, we notice that the operator $T-\lambda I$ is one-to-one. Otherwise there would exist a non-zero function $f = \sum_{i=1}^{n} \alpha_i f_i$ such that

$$B_1(f) = B_2(f) = \cdots = B_n(f) = 0,$$
$$Tf - \lambda f = 0.$$

Consequently, the rows of the matrix $(B_i(f_j))$ would be linearly dependent.

Now let $u_1, u_2, \cdots, u_m$ in $D(T)$ be functions which are linearly independent modulo $D(T_0(\tau))$. Then the functions $Tu_1 - \lambda u_1$, $Tu_2 - \lambda u_2$, $\cdots$, $Tu_m - \lambda u_m$ are linearly independent modulo $R(T_0(\tau) - \lambda I)$. Indeed, if

$$\sum_{i=1}^{m} \beta_i (Tu_i - \lambda u_i) = T_0(\tau) f - \lambda f,$$

then

$$(T - \lambda I)\left(\sum_{i=1}^{m} \beta_i u_i - f\right) = 0$$

and hence $\sum_{i=1}^{m} \beta_i u_i = f$. Since f lies in $D_0(\tau)$, it follows that $\sum_{i=1}^{m} \beta_i u_i = 0$, and thus $\beta_1 = \beta_2 = \cdots = \beta_m = 0$. Finally, the fact that the dimension of the complement of $R(T_0(\tau) - \lambda I)$ is m implies that the range of the operator $\lambda I - T$ is the entire space $L_p(I)$.

(d) It is clear from Lemma 4.1 that

$$\text{span}\,\{f_1, f_2, \cdots, f_m, D_0(\tau)\}$$

is the intersection of n linearly independent hyperplanes $B_1(f) = 0$, $B_2(f) = 0, \cdots, B_n(f) = 0$, where $B_1, B_2, \cdots, B_n$ are boundary values. The assumption implies that the condition of (c) is satisfied. Thus T is a maximal extension at λ.

The converse is also immediate from (c).

Let the operator T be a maximal extension of the differential operator τ at λ. Since the resolvent $(\lambda I - T)^{-1}$ is analytic in λ, there exists a neighborhood $N(\lambda)$ of λ such that the operator $\mu I - T$ has a bounded everywhere defined inverse for μ in $N(\lambda)$. Thus T is a maximal extension of τ at μ. The problem, therefore, arises of finding the relationship between the resolvent set of the maximal local extension T and the essential resolvent set of τ. The following theorem opens the way to an answer to this problem.

We recall that the set of those points in the complex plane which are regular points of both τ in $L_p(I)$ and τ^* in $L_q(I)$ is an open set (as follows at once from the closed graph theorem). We shall denote this set as R_p. Clearly $R_p = R_q$, $p^{-1} + q^{-1} = 1$.

THEOREM 4.1. *Let A be a connected component of the set R_p. Then for λ in A, the dimension of the space of functions f in $D_1(\tau, p)$ such that*

$$\lambda f - \tau f = 0$$

is a constant.

Proof: It suffices to show that, given a point λ_0 in A, the integer

$$\dim N(\lambda I - T_1(\tau, p))$$

is a constant for λ in a neighborhood of λ_0. For, if such is the case, then the integer-valued function

$$\lambda \to \dim N(\lambda I - T_1(\tau, p))$$

is continuous, and takes therefore a constant value on each connected component of R_p.

Let T be a local extension of τ^* at λ_0 in $L_q(I)$ (cf. Lemma 4.2 (a)). Since the resolvent set of T is open, it follows that T is also a local extension of τ^* at λ for λ in a neighborhood of λ_0. Let $f_1, f_2, \cdots, f_m$ be a basis for $D(T)$ modulo $D_0(\tau, q)$. By Lemma 4.2 (b),

$$m = \dim N(\lambda I - T_0(\tau, p)).$$

If T is also an extension at λ, then the functions

$$\lambda f_i - T f_i, \qquad 1 \leqq i \leqq m,$$

are linearly independent modulo $R(\lambda I - T_0(\tau^*, q))$. Furthermore,

$$R(\lambda I - T_0(\tau^*, q)) + \text{span}\,\{\lambda f_i - T f_i,\ 1 \leqq i \leqq m\} = L_q(I).$$

Thus the functions $\lambda f_i - T f_i$, $1 \leqq i \leqq m$, span a complement of the closed subspace $R(\lambda I - T_0(\tau^*, q))$, for λ in a neighborhood of λ_0. As has been remarked at the beginning of this section, the dimension of a complement to the closed subspace $R(\lambda I - T_0(\tau^*, q))$ equals the dimension of the nullspace of the adjoint operator $[\lambda I - T_0(\tau^*, q)]^* = \lambda I - T_1(\tau, p)$. Therefore, for λ in a neighborhood of λ_0, we have

$$\dim N(\lambda I - T_1(\tau, p)) = m.$$

The result stated in the preceding theorem, which will be extended in Theorem 4.3, motivates

DEFINITION 4.2. Let λ be a point in the set R_p of a differential operator τ on the interval I and let A be a connected component of the set R_p of τ.

(a) The non-negative integer $\dim N(\lambda I - T_1(\tau, p))$ will be referred to as the *first deficiency index of τ at λ.*

(b) The non-negative integer $\dim N(\lambda I - T_1(\tau^*, q))$ will be referred to as the *second deficiency index of τ at λ.*

(c) The non-negative integer $\dim N(\lambda I - T_1(\tau, p))$ for any λ in A will be referred to as the *first deficiency index of τ in the component A.*

(d) The non-negative integer $\dim N(\lambda I - T_1(\tau^*, q))$ for any λ in A will be referred to as the *second deficiency index of τ in the component A.*

Now the question arises: what are the requirements of an extension T which is "best possible" in the sense that the spectrum of T is as "near" the essential spectrum of τ as possible? A consideration of the well-known case of a differential operator on a compact interval, where the essential spectrum is empty in the finite plane, shows that in general we cannot expect to reduce the spectrum of T to the essential spectrum of τ and at the same time prevent the appearance of eigenvalues. In the case of a compact interval of definition, where the essential spectrum reduces to the "point at infinity", the eigenvalues appearing in an extension converge to the point at infinity in the extended plane. We may therefore expect that in the more general case a "best possible" extension will display a sequence of eigenvalues all of whose limit points lie in the essential spectrum. Theorem 4.2 indicates that this is indeed the case, and in fact provides a simple method for obtaining an extension which is "best possible" in a given component of the set R_p of regular points.

LEMMA 4.3. *Let T be a maximal extension of τ at λ, and let A be the connected component of the set R_p of regular points to which the point λ belongs. Then for every point μ in A, either the operator T is a maximal extension at μ, or μ is an eigenvalue of T.*

Proof: We exclude the case when μ belongs to the residual spectrum of T. Suppose that the range of the operator $\mu I - T$ is not the entire space $L_p(I)$. Let $h_1, h_2, \cdots, h_m$ be a basis of $D(T)$ modulo $D_0(\tau)$. It follows from Lemma 4.2 (b) that the functions $\mu h_i - \tau h_i$ do not form a basis of $L_p(I)$ modulo $R(T_0(\tau))$. Theretore, there exists a function f in $D_0(\tau)$ and a linear combination h of the functions h_i such that

$$\mu f - \tau f = \mu h - \tau h,$$

or

$$(\mu I - T)(f - h) = 0,$$

that is, μ is an eigenvalue of T.

THEOREM 4.2. *Let T be a maximal extension in $L_p(I)$ of the differential operator τ at a point λ in the set R_p. Let A be the connected component of the set R_p to which the point λ belongs. Then the spectrum of T in the component A consists of (at most) a sequence of eigenvalues all of whose limit points lie in the boundary of A.*

Proof: Let C be the boundary of the resolvent set of T. It suffices to show that $C \cap A$ is an isolated set.

Suppose λ_0 lies in $C \cap A$. Let T_0 be a maximal extension of τ at λ_0, and let $f_1, f_2, \cdots, f_m$ be a basis of $D(T)$ modulo $D_0(\tau)$. Clearly, the second deficiency index of τ in A is equal to m. Thus by Theorem 4.1 and by Lemma 4.2 (b) there exists a basis of m elements $g_1, g_2, \cdots, g_m$ of $D(T_0)$ modulo $D_0(\tau)$. Let B_1 be the space spanned by this basis. By Lemmas 1.1 (a) and 4.1 the space $D_1(\tau)$ can be decomposed into a direct sum:

$$D_1(\tau) = D_0(\tau) \oplus B_1 \oplus B_2 = D_0(\tau) \oplus B,$$

where $B = B_1 \oplus B_2$. By Lemma 1.1 (c) there exists a projection E of $D_1(\tau)$ onto B which is bounded in the τ-norm and such that

$$Ef = 0, \qquad f \in D_0(\tau).$$

Now there exists a neighborhood $N(\lambda_0)$ of λ_0 such that for μ in $N(\lambda_0)$, the operator-valued function of μ

$$(\mu I - T_0)^{-1} \tag{4.3}$$

is analytic when the operator (4.3) is considered as a bounded operator of $L_p(I)$ onto $D(T_0)$. Therefore, the m vector-valued functions

$$F_i(\mu) = E(\mu I - T_0)^{-1}(\mu f_i - T f_i), \qquad 1 \leqq i \leqq m,$$

with values in $D_1(\tau)$ are analytic for μ in $N(\lambda_0)$, since they are "products" of analytic functions. Notice further that the functions $F_i(\mu)$ take their values in the finite-dimensional subspace B_1. Thus

$$F_i(\mu) = \sum_{j=1}^{m} F_{ij}(\mu) g_j, \qquad 1 \leqq i \leqq m,$$

and the functions $F_{ij}(\mu)$ are analytic for μ in $N(\lambda_0)$. It follows that the determinant

$$D(\mu) = \|F_{ij}(\mu)\|, \qquad \mu \in N(\lambda_0),$$

is an analytic function of μ.

It will now be shown that for μ in $N(\lambda_0)$ the value of the analytic function $D(\mu)$ is zero if and only if the point μ is in the spectrum of T. This will complete the proof. Indeed, if such is the case, then $D(\lambda_0) = 0$. Now the analytic function D does not vanish identically, because λ_0 lies on the boundary of the resolvent set of T. Therefore λ_0 is an isolated zero and, since λ_0 was arbitrarily chosen, every point of A which is in the spectrum of T is isolated, as was to be proved.

Thus, suppose first that a point μ in $N(\lambda_0)$ is in the spectrum of T. By Lemma 4.3, the point μ is an eigenvalue. There are, therefore, two non-null functions, f in the span of $f_1, f_2, \cdots, f_m$ and g in $D_0(\tau)$, such that

$$\mu(f+g) - T(f+g) = \mu(f+g) - \tau(f+g) = 0.$$

Suppose that

$$f = \sum_{i=1}^{m} \alpha_i f_i$$

where not all the coefficients α_i vanish. Since μ is in the resolvent set of T_0, there exist uniquely defined $h_1, h_2, \cdots, h_m$ in $D(T_0)$ such that

$$(\mu I - T_0) h_i = (\mu I - T) f_i, \qquad 1 \leqq i \leqq m.$$

We claim that the vectors $Eh_i = F_i(\mu)$, $1 \leqq i \leqq m$, are linearly dependent. Indeed,

$$\begin{aligned}\sum_{i=1}^{m} \alpha_i E h_i &= \sum_{i=1}^{m} \alpha_i E(\mu I - T_0)^{-1}(\mu I - T) f_i \\ &= \sum_{i=1}^{m} \alpha_i E(\mu I - T_0)^{-1}(\mu I - T) f_i + E g \\ &= \sum_{i=1}^{m} \alpha_i E(\mu I - T_0)^{-1}(\mu I - T) f_i + E(\mu I - T_0(\tau))^{-1}(\mu I - T_0(\tau)) g \\ &= \sum_{i=1}^{m} \alpha_i E(\mu I - T_0)^{-1}(\mu I - T) f_i + E(\mu I - T_0)^{-1}(\mu I - T) g \\ &= E(\mu I - T_0)^{-1}\left[\sum_{i=1}^{m} \alpha_i(\mu I - T) f_i + (\mu I - T) g\right] \\ &= E(\mu I - T_0)^{-1}(\mu I - T)(f+g) = 0.\end{aligned}$$

Therefore the determinant $||F_{ij}(\mu)||$ of the vectors $F_i(\mu)$, $1 \leqq i \leqq m$, vanishes.

Conversely, suppose that the point μ is in the resolvent set of T. Then by Lemma 4.2 (b) the functions

$$(\mu I - T) f_i, \qquad 1 \leqq i \leqq m,$$

are linearly independent modulo $R(\mu I - T_0(\tau))$. Therefore the functions

$$(\mu I - T_0)^{-1}(\mu I - T) f_i, \qquad 1 \leqq i \leqq m,$$

are linearly dependent modulo $D_0(\tau)$. Since the projection E vanishes on $D_0(\tau)$, it follows that the vectors

$$F_i(\mu) = E(\mu I - T_0)^{-1}(\mu I - T) f_i, \qquad 1 \leqq i \leqq m,$$

are linearly independent in the space B_1. Hence the determinant $D(\mu)$ does not vanish.

In the case of a differential operator τ defined on a half-closed interval, the set R_p coincides with the essential resolvent set (cf. Lemma 2.1). We therefore have

COROLLARY 4.2. *Let τ be a differential operator on a half-closed interval I. Let λ be an arbitrary point in the essential resolvent set of τ in $L_p(I)$, and let A be the connected component of the essential resolvent set to which the point λ belongs. Let T be a maximal extension of τ at λ in $L_p(I)$. Then the spectrum of T in A consists (at most) of a sequence of eigenvalues with limits in the essential spectrum of τ.*

In the case of an operator which is singular at both endpoints of its interval of definition, the appearance of essential eigenvalues renders the structure of the extension more complex. We begin to investigate the distribution of essential eigenvalues with a theorem which introduces a generalization of the notion of "deficiency indices" which will be the basis of several results in the last section.

THEOREM 4.3. *Let τ be a differential operator in $L_p(I)$, and let A be a connected component of the essential resolvent set of τ. Then for every extension T of τ in $L_p(I)$ the integer*

$$K(\lambda, T) = \dim N(\lambda I - T) - \dim N(\lambda I - T^*)$$

is constant for λ in A.

Proof: For simplicity, assume $\lambda = 0$. It suffices, as in Theorem 4.1. to show that $K(\lambda, T)$ remains constant for λ in a neighborhood of zero. Clearly there exists a direct decomposition

$$D(T) = N(T) \oplus S \tag{4.4}$$

since the null space $N(T)$ is finite-dimensional. Let T' be the restriction of $T_1(\tau, p)$ to the subspace S. It is clear that T' is a closed one-to-one operator whose range coincides with the range of T.

(A) The first assertion we shall prove is the following: there exists a neighborhood $V(0)$ of 0 such that for λ in $V(0)$ the dimension of the quotient space

$$\frac{L_p(I)}{R(\lambda I - T')}$$

is constant.

Set

$$m = \dim \frac{L_p(I)}{R(T')} .$$

Choose m functions $z_1, z_2, \cdots, z_m$ in $L_p(I)$ which are linearly independent modulo $D_1(\tau, p)$, e.g. characteristic functions of disjoint compact subintervals of I, and let S_1 be their span. Now choose a complementary basis $f_1, f_2, \cdots, f_m$ of $R(T')$ in L_p (cf. Lemma 1.1). Define the operator T'' on the domain $S+S_1$ as follows:

$$T''f = T'f, \qquad f \in S,$$
$$T''z_i = f_i, \qquad 1 \leqq i \leqq m.$$

The operator T'' is well defined and is a one-to-one onto closed operator, since its graph is the sum of the graph of T' and of a finite-dimensional space. There exists therefore a neighborhood $V(0)$ of 0 such that for λ in $V(0)$ the operator $\lambda I - T''$ has a bounded everywhere defined inverse. Thus for λ in $V(0)$, a complement in $L_p(I)$ of the closed subspace

$$(\lambda I - T'')S = R(\lambda I - T')$$

is the m-dimensional subspace spanned by the elements $(\lambda I - T'')z_i$, $1 \leqq i \leqq m$. The assertion now follows from Lemma 1.1 (b).

(B) An immediate consequence of (A) is

$$\dim N(T^*) = \dim N(T'i^*) = \dim \frac{L_p(I)}{R(\lambda I - T')}$$

for λ in $V(0)$.

It therefore suffices to find a neighborhood $N(0)$ of 0 contained in $V(0)$ such that for λ in $N(0)$,

$$\dim \frac{L_p(I)}{R(\lambda I - T')} = \dim \frac{L_p(I)}{R(\lambda I - T)} - \dim N(\lambda I - T) + \dim N(T). \tag{4.5}$$

(C) There exists a neighborhood $N(0)$ of 0 contained in $V(0)$ such that for λ in $N(0)$ none of the solutions of $\lambda I - T$ lies in the subspace S. This follows at once from the closed graph theorem applied to the operator $\lambda I - T'$. Indeed, since the operator T' is one-to-one and has closed range, there is a constant K such that

$$|T'f| \geqq K|f|, \qquad f \in S.$$

Let $N(0) = \{\lambda : \lambda \in V(0),\ |\lambda| < K\}$. Then, for λ in $N(0)$,

$$|(\lambda I - T')f| \geqq (K - |\lambda|)|f|$$

and hence $\lambda I - T'$ is one-to-one. Now if any root of $\lambda I - T$ were to lie in S, it would be a root of $\lambda I - T'$.

(D) For λ in $N(0)$, let $f_1(\lambda), f_2(\lambda), \cdots, f_n(\lambda)$ be a basis for the null space of $\lambda I - T$. From (4.4) we obtain a uniquely defined decomposition

$$f_i(\lambda) = g_i(\lambda) + h_i(\lambda), \qquad 1 \leqq i \leqq m$$
$$g_i(\lambda) \in S, \qquad h_i(\lambda) \in N(T).$$

We claim that the functions $h_i(\lambda)$, $1 \leqq i \leqq n$, are linearly independent. Otherwise we should have

$$\sum_{i=1}^{n} \alpha_i h_i(\lambda) = 0, \qquad \text{some } \alpha_i \neq 0.$$

Then

$$0 \neq \sum_{i=1}^{n} \alpha_i f_i(\lambda) = \sum_{i=1}^{n} \alpha_i [g_i(\lambda) + h_i(\lambda)] = \sum_{i=1}^{n} \alpha_i g_i(\lambda),$$

contradicting (C).

(E) Since the $h_i(\lambda)$, $1 \leqq i \leqq n$, are linearly independent, they can be completed to a basis $h_1(\lambda), h_2(\lambda), \cdots, h_r(\lambda)$ of $N(T)$. Now since

$$D(\lambda I - T') = D(T') \subset D(T) = D(\lambda I - T)$$

we have

$$R(\lambda I - T') \subset R(\lambda I - T), \qquad \lambda \in N(0).$$

Both ranges are closed subspaces and thus we have a direct decomposition

$$R(\lambda I - T) = M(\lambda) \oplus R(\lambda I - T').$$

Let $k(\lambda)$ be the (finite) dimension of $M(\lambda)$. Then

$$\dim \frac{L_p(I)}{R(\lambda I - T')} = \dim \frac{L_p(I)}{R(\lambda I - T)} + k(\lambda)$$

and therefore, by (4.5), the assertion of the theorem reduces to the identity

$$k(\lambda) = \dim N(T) - \dim N(\lambda I - T).$$

(F) In $N(T)$, let F_1 be the subspace spanned by $h_1(\lambda), h_2(\lambda), \cdots, h_n(\lambda)$, and let F_2 be the subspace spanned by $h_{n+1}(\lambda), h_{n+2}(\lambda), \cdots, h_r(\lambda)$. We have

$$\dim N(T) = r$$

and, by (D),

$$\dim N(\lambda I - T) = n = \dim F_1 .$$

Thus, to complete the proof of the theorem it suffices to show that

$$k(\lambda) = \dim M(\lambda) = \dim F_2 .$$

Now

$$R(\lambda I - T) = R(\lambda I - T') + (\lambda I - T)F_1 + (\lambda I - T)F_2$$

and, by (D),

$$(\lambda I - T)F_1 \subset R(\lambda I - T').$$

Furthermore, the operator $\lambda I - T$ is one-to-one on the space F_2. Suppose that for some function h of F_2 and for some function g in $D(T')$ we have

$$(\lambda I - T')g = (\lambda I - T)h, \qquad h = \sum_{i=n+1}^{r} \beta_i h_i(\lambda).$$

Then the function $g-h$ lies in the nullspace of $\lambda I - T$, and thus

$$g - h = \sum_{i=1}^{n} \alpha_i f_i(\lambda) = \sum_{i=1}^{n} \alpha_i [g_i(\lambda) + h_i(\lambda)].$$

It follows that

$$g - \sum_{i=1}^{n} \alpha_i g_i(\lambda) = \sum_{i=1}^{n} \alpha_i h_i(\lambda) + \sum_{i=n+1}^{r} \beta_i h_i(\lambda),$$

hence

$$T'(g - \sum_{i=1}^{n} \alpha_i g_i(\lambda)) = \tau(g - \sum_{i=1}^{n} \alpha_i g_i(\lambda))$$
$$= (\tau)(\sum_{i=1}^{n} \alpha_i h_i(\lambda) + \sum_{i=n+1}^{n} \beta_i h_i(\lambda)) = 0$$

and, since T' is one-to-one, $g - \sum_{i=1}^{n} \alpha_i g_i(\lambda) = 0$. Therefore

$$\sum_{i=1}^{n} \alpha_i h_i(\lambda) + \sum_{i=n+1}^{r} \beta_i h_i(\lambda) = 0,$$

and hence $\alpha_1 = \alpha_2 = \cdots = \alpha_n = \beta_{n+1} = \cdots = \beta_r = 0$. We conclude that

$$R(\lambda I - T') \cap (\lambda I - T)F_2 = \{0\},$$

and finally that

$$\dim M(\lambda) = \dim (\lambda I - T)F_2 = \dim F_2 .$$

The following theorem gives a "qualitative" description of the distribution of essential eigenvalues for a differential operator which is singular at both endpoints of its interval of definition.

THEOREM 4.4. *Let τ be a differential operator in $L_p(I)$, and let A be a connected component of the essential resolvent set of τ. Suppose that there exists one point λ in A which is a regular point of both τ in $L_p(I)$ and τ^* in $L_q(I)$. Then*:

(a) *The points in A which are essential eigenvalues of either τ in $L_p(I)$ or of τ^* in $L_q(I)$ form a countable set all of whose limit points lie in the essential spectrum $\sigma_e(\tau, p, I)$.*

(b) *For every extension T of τ in $L_p(I)$ which is maximal at some regular point of A, $\sigma(T) \cap A$ consists of a countable set (including all essential eigenvalues) all of whose limit points lie in the essential spectrum $\sigma_e(\tau, p, I)$.*

Proof: Let λ be a point in A which is a regular point of both τ in $L_p(I)$ and τ^* in $L_q(I)$. By Lemma 4.2 (a) there exists a maximal extension T of τ at λ.

The method of proof will be similar to that used in Theorem 4.2. We consider a point λ_0 on the boundary of the spectrum of T, and prove that this point is an isolated eigenvalue in the spectrum of T. If λ_0 is a regular point of τ and of τ^*, then the proof of Theorem 4.2 will yield the result.

Now if λ_0 is an essential eigenvalue of τ^*, then it is an eigenvalue of the adjoint operator T^*. By Theorem 4.3 we have

$$\begin{aligned} 0 &= \dim N(\lambda I - T) - \dim N(\lambda I - T^*) \\ &= \dim N(\lambda_0 I - T) - \dim N(\lambda_0 I - T^*), \end{aligned}$$

and hence λ_0 is an eigenvalue of T.

Thus it suffices to consider the case in which λ_0 is an essential eigenvalue of either τ or τ^*; as we have seen, in both these cases λ_0 is an eigenvalue of T. Furthermore, assertion (a) follows from (b).

A second consequence of Theorem 4.3 and of the assumption that the component A contains a regular point is the following: Let λ_0 be an essential eigenvalue of τ in A and let λ be a regular point in A. Then, applying Theorem 4.3 to the operator $T_0(\tau, p)$, and recalling that $T_0(\tau, p)^* = T_1(\tau^*, q)$ (cf. Theorem 2.1) we have

$$\begin{aligned} &\dim N(\lambda_0 I - T_1(\tau^*, q)) - \dim N(\lambda_0 I - T_0(\tau, p)) \\ (4.6) \qquad &= \dim N(\lambda I - T_1(\tau^*, q)) - \dim N(\lambda I - T_0(\tau, p)) \\ &= \dim N(\lambda I - T_1(\tau^*, q)). \end{aligned}$$

We can decompose the space $D(T)$ into a direct sum

$$D(T) = D_0(\tau, p) \oplus M_1,$$

where, by Lemma 4.2 (B),

$$\dim M_1 = \dim N(\lambda I - T_1(\tau^*, q)).$$

Now the space $D_0(\tau, p)$ can be decomposed into a direct sum

$$D_0(\tau, p) = D_{00} \oplus M_2;$$

here M_2 is the null space of the operator $\lambda_0 I - T_0(\tau, p)$. Thus

$$D(T) = D_{00} \oplus M,$$

where $M = M_1 \oplus M_2$ and, by (4.6),

$$\begin{aligned} m = \dim M &= \dim N(\lambda I - T_1(\tau^*, q)) + \dim N(\lambda_0 I - T_0(\tau, p)) \\ &= \dim N(\lambda_0 I - T_1(\tau^*, q)). \end{aligned}$$

Let T' be the closed operator obtained by restricting the domain of $T_1(\tau, p)$ to the subspace D_{00}. Since T' has closed range (by Lemma 1.2) it follows from the closed graph theorem that there exists a neighborhood $V(\lambda_0)$ of λ_0 such that for μ in $N(\lambda_0)$ the operator $\mu I - T'$ is one-to-one and has closed range. Now choose a dual basis $f_1, f_2, \cdots, f_m$ for the null space of the operator $\lambda_0 I - T_1(\tau^*, q)$. As pointed out in the remarks at the beginning of this section, the span of this dual basis provides a complement to the closed subspace $R(\lambda_0 I - T') = R(\lambda_0 I - T_0(\tau, p))$. Choose, as in the proof of Theorem 4.2, m functions $g_1, g_2, \cdots, g_m$ in $L_p(I)$ which are linearly independent modulo $D_0(\tau, p)$, and define an operator T_0 on the space spanned by D_{00} and $g_1, g_2, \cdots, g_m$ by the conditions

$$\begin{aligned} (\lambda_0 I - T_0) g_i &= f_i, \qquad & 1 \leqq i \leqq m, \\ T_0 f &= T' f, & f \in D_{00}. \end{aligned}$$

Then T_0 is a well-defined closed operator, and $\mu I - T_0$ has a bounded inverse for μ in a neighborhood $N(\lambda_0)$ of λ_0 contained in $V(\lambda_0)$. Furthermore, the Banach space $D(T)$ can be decomposed into a direct sum

$$D(T) = D_{00} \oplus S,$$

where S is the span of the functions $g_1, g_2, \cdots, g_m$.

By Lemma 1.1 (c), we can construct a projection E of $D(T)$ onto S which is bounded in the T-norm and which vanishes on D_{00}. Now choose a basis $h_1, h_2, \cdots, h_m$ for the space M and consider the functions

$$F_i(\mu) = \sum_{j=1}^{m} F_{ij}(\mu)g_j = E(\mu I - T_0)^{-1}(\mu h_i - Th_i), \qquad 1 \leq i \leq m.$$

It follows, as in Theorem 4.2, that the determinant

$$D(\mu) = ||F_{ij}(\mu)||$$

is an analytic function of μ for μ in $N(\lambda_0)$. The proof can then be concluded as in Theorem 4.2.

The last question to arise is the following: when can an extension be obtained of a differential operator τ which is maximal at two points lying in different components of the essential resolvent set? The following lemma provides the simple answer.

LEMMA 4.4. *Let $\{\lambda_n\}_1^\infty$ be a countable set of points which are regular points of both τ in $L_p(I)$ and τ^* in $L_q(I)$ and whose first deficiency indices are equal. Then there exists an extension T of τ which is maximal at λ_n, $n = 1, 2, \cdots$.*

Proof: It follows from Theorem 4.3 that equality of the first deficiency indices implies equality of the second deficiency indices. Choose a direct decomposition

$$D_1(\tau, p) = D_0(\tau, p) \oplus M$$

and let E be a bounded projection of $D_1(\tau, p)$ onto M which vanishes on $D_0(\tau, p)$. Furthermore, let

$$M_n = EN(\lambda_n I - T_1(\tau, p))$$

and let

$$m = \dim M_n = \dim N(\lambda_n I - T_1(\tau, p)), \qquad n = 1, 2, \cdots.$$

We shall inductively define linearly independent functions $f_1, f_2, \cdots, f_k$ in M spanning a subspace F of dimension

$$\dim F = k = \dim N(\lambda_n I - T_1(\tau^*, q)), \qquad n = 1, 2, \cdots,$$

with the further property that

$$F \cap M_n = \{0\}, \qquad n = 1, 2, \cdots.$$

It then follows from Lemma 4.2 (b) that the extension of τ to the space of $f_1, f_2, \cdots, f_k$ is maximal at λ_n, $n = 1, 2, \cdots$.

Suppose that the functions $f_1, f_2, \cdots, f_i$, $i \geq 0$, in M have thus been defined and are linearly independent modulo M_n, $n = 1, 2, \cdots$. Let $M_n^{(i)}$ be the span of M_n and $f_1, f_2, \cdots, f_i$. Then

$$\dim M_n^{(i)} = m+i, \qquad n = 1, 2, \cdots.$$

If $m+i < m+k$, then $M_n^{(i)}$ is a proper closed subspace of M, and consequently a closed nowhere dense subset in the relative topology of M. By the Baire category theorem, there exists a (non-zero) function f_{i+1} not lying in

$$\bigcup_{n=1}^{\infty} M_n^{(i)} .$$

Therefore the elements $f_1, f_2, \cdots, f_{t+1}$ are linearly independent modulo M_n, $n = 1, 2, \cdots$.

All the threads being in our hands, we shall collect them in

THEOREM 4.5. *Let τ be a differential operator in $L_p(I)$, $1 \leqq p \leqq \infty$, and let $\{A_n\}$ be a collection of connected components of the essential resolvent set of τ having equal deficiency indices, each one of which contains a point which is regular for both τ in $L_p(I)$ and τ^* in $L_q(I)$, $p^{-1}+q^{-1} = 1$. Then there exists an extension T of τ in $L_p(I)$ obtained by the imposition of a finite set of linearly independent boundary conditions such that the spectrum of T in A_n consists of a denumerable set of eigenvalues all of whose limit points lie in the essential spectrum of τ.*

5. Adjoints and Boundary Values

Among the several questions which remain to be answered the following two arise naturally to complete a description of the "qualitative" aspects of extension theory of differential operators. The first concerns the structure of the space of boundary values. Thus far the notion of boundary value, given by Definition 2.3, which appeared to be essential in the theorems of the preceding section, has hardly been explored and exploited to the length which is necessary for a workable extension theory. This will be the first task. Analytic "expressions" for boundary values, vaguely reminiscent of the classical notion connected with the term, will be derived, and several corollaries of "practical" interest will be stated.

The second question to be studied will be the effect of an extension upon the adjoint of a differential operator. Here the study is limited to workable criteria for the determination of the adjoint of a given extension.

Finally, several results of varying interest connecting a D-operator with its adjoint operator will be stated; among these the so-called "Fredholm alternative" will be found to be valid for all regular differential operators.

The following lemma is a simple consequence of the results of Section 2.

LEMMA 5.1. *Let τ be a differential operator of order n in $L_p([a, b])$, $1 \leqq p \leqq \infty$, $-\infty < a < b \leqq \infty$. Then the linear functionals*

$$B(f) = f^{(i)}(a), \qquad 0 \leqq i \leqq n-1,$$

are boundary values for τ in $L_p([a, b])$.

The proof is immediate from Lemmas 2.4 and 2.5.

LEMMA 5.2. *Suppose that the essential resolvent set of a differential* *operator τ in $L_p(I)$ is not empty. Then the dimension of the space of boundary values of τ is finite.*

Proof: Assume that the vector 0 lies in the essential resolvent set. By Lemma 1.1 the dimension of the space of boundary values of τ equals the dimension of the quotient space $D_1(\tau, p)/D_0(\tau, p)$, which is finite by Lemma 2.5.

A stronger result is given by

THEOREM 5.1. *Suppose that the essential resolvent set of a differential operator τ in $L_p(I)$ is non-empty. Then:*

(a) *The dimension of the space of boundary values of τ in $L_p(I)$, $1 \leqq p \leqq \infty$, equals the dimension of the space of boundary values of τ^* in $L_q(I)$, $p^{-1}+q^{-1} = 1$, p or q possibly infinite.*

(b) *For any point λ in the essential resolvent set of τ in $L_p(I)$ the dimension of the space V of boundary values for τ is given by the expression*

$$\dim V = \dim N(\lambda I - T_1(\tau, p)) - \dim N(\lambda I - T_0(\tau, p)) + \dim N(\lambda I - T_1(\tau^*, q)) - \dim N(\lambda I - T_0(\tau^*, q)).$$

Proof: Notice that the expression given in (b) for the dimension of the space of boundary values is symmetric in τ and τ^*. Therefore, the first statement follows from the second one, which we now proceed to prove.

Let N be the subspace consisting of all functions f lying in $D_1(\tau, p)$ but not in $D_0(\tau, p)$ such that $\lambda f - T_1(\tau, p) f = 0$. Then we have a direct sum decomposition

$$D_1(\tau, p) = D_0(\tau, p) \oplus N \oplus C.$$

Choose a dual basis to $N(\lambda I - T_1(\tau^*, q))$, say $f_1, f_2, \cdots, f_n$, in $L_p(I)$, of which the functions $f_1, f_2, \cdots, f_k$ form a dual basis for the space $N(\lambda I - T_0(\tau^*, q))$. Let M_1 be the span of $f_1, f_2, \cdots, f_k$, and let M_2 be the span of $f_{k+1}, f_{k+2}, \cdots, f_n$. Then we have a direct decomposition

$$L_p(I) = R(\lambda I - T_0(\tau, p)) \oplus M_1 \oplus M_2,$$

where

$$R(\lambda I - T_1(\tau, p)) = R(\lambda I - T_0(\tau, p)) \oplus M_2.$$

Hence

$$\dim M_2 = \dim C = \dim N(\lambda I - T_1(\tau^*, q)) - \dim N(\lambda I - T_0(\tau^*, q)),$$

and consequently

$$\begin{aligned} \dim V &= \dim N + \dim C \\ &= \dim N(\lambda I - T_1(\tau, p)) - \dim N(\lambda I - T_0(\tau, p)) \\ &\quad + \dim N(\lambda I - T_1(\tau^*, q)) - \dim N(\lambda I - T_0(\tau^*, q)). \end{aligned}$$

The following corollary is useful for the determination of the essential spectrum.

COROLLARY 5.1. *Let τ be a differential operator on a half-closed interval. If the sum of the first and second deficiency indices of τ in $L_p(I)$, $1 \leqq p \leqq \infty$, at λ is not equal to the dimension of the space of boundary values, then λ belongs to the essential spectrum of τ in $L_p(I)$.*

DEFINITION 5.1. (a) Let τ be a differential operator in $L_p(I)$, $1 \leqq p \leqq \infty$. For f in $D_1(\tau, p)$ and g in $D_1(\tau^*, q)$ let

$$\{f, g, \tau, p\} = (g, T_1(\tau, p)f) - (T_1(\tau^*, q)g, f).$$

We shall refer to this bilinear form as "the boundary form" and abbreviate it as $\{f, g\}$ when no confusion will arise.

(b) Given a differential operator τ in $L_p(a, b)$, $1 \leqq p \leqq \infty$, we say that a boundary value B of τ is a boundary value at a (at b) if B vanishes on all functions in $D_1(\tau, p)$ which vanish in a neighborhood of a (of b).

LEMMA 5.3. *Suppose that the essential resolvent set of a differential operator τ of order k in $L_p(a, b)$ is non-empty. Then:*

(a) *The space of boundary values of τ in $L_p(a, b)$, $1 \leqq p \leqq \infty$, is a direct sum of the subspaces of boundary values at a and at b.*

(b) *For $1 \leqq p \leqq \infty$ there exists a basis of $D_1(\tau, p)$ modulo $D_0(\tau, p)$ each of whose elements is a function vanishing either in a neighborhood of a or in a neighborhood of b.*

(c) *Let $g_1, g_2, \cdots, g_n$ be a basis of $D_1(\tau^*, g)$ modulo $D_0(\tau^*, q)$, $1 \leqq p \leqq \infty$, $p^{-1}+q^{-1} = 1$, p or q possibly infinite. Suppose that the point λ belongs to the essential resolvent set of τ. Then the linear functionals*

$$B_i(f) = \{g_i, f, \lambda - \tau, p\}, \qquad 1 \leqq i \leqq n,$$

form a basis for the space of boundary values of τ in $L_p(I)$, $1 \leqq p \leqq \infty$.

(d) *Under the assumptions of (c), suppose that the function g in $D_1(\tau^*, q)$ vanishes in a neighborhood of b (of a). Then the linear functional*

$$B(f) = \{f, g, \lambda - \tau, p\}$$

is a boundary value for τ at a (at b). Conversely, every boundary value B of τ in $L_p(a, b)$, $1 \leqq p \leqq \infty$, at a is given by

$$B(f) = \{g, f\}, \qquad f \in D_1(\tau, p),$$

where g is a function in $D_1(\tau, q)$ which vanishes in a neighborhood of b.

Proof: Choose two functions f_1 and f_2 in $C^\infty(a, b)$ with the following properties:

(1) f_1 vanishes in a neighborhood of a.

(2) f_2 vanishes in a neighborhood of b.

(3) $f_1(t)+f_2(t) = 1, \quad a < t < b$.

Define the (closed, and hence continuous) linear operators

$$(S_i f)(t) = f_i(t) f(t), \qquad i = 1, 2,$$

of $D_1(\tau, p)$ into itself, for $1 \leqq p < \infty$.

(a) For $1 \leqq p < \infty$ it suffices to remark that the mapping

$$B \to BS_1$$

is a continuous linear operator of the space of boundary values into the space of boundary values at b, and similarly the mapping

$$B \to BS_2$$

is a continuous linear operator of the space of boundary values into the space of boundary values at a.

Furthermore

$$BS_1 + BS_2 = B.$$

The conclusion is now evident.

(b) Let $g_1, g_2, \cdots, g_n$ be a basis of $D_1(\tau, p)$ modulo $D_0(\tau, p)$, $1 \leqq p < \infty$. We claim that a subset of the set

$$\{S_j g_i\}, \qquad j = 1, 2, \qquad 1 \leqq i \leqq n,$$

each one of whose elements vanishes either in the neighborhood of a or in the neighborhood of b, forms a basis for $D_1(\tau, p)$ modulo $D_0(\tau, p)$. Indeed, if this were not the case, then a boundary value B could be found such that

$$B(S_j(g_i)) = 0, \qquad j = 1, 2, \qquad 1 \leqq i \leqq n.$$

The preceding discussion shows that the functionals BS_1 and BS_2 are boundary values and, moreover, that $BS_1 = 0$ and $BS_2 = 0$. Hence $B = BS_1 + BS_2 = 0$.

(c) Assume for the sake of brevity that $\lambda = 0$; then the functional

$$B_i(f) = \{g_i, f\}, \qquad 1 \leqq i \leqq n,$$

is continuous in the τ-norm of $D_1(\tau, p)$, $1 \leqq p \leqq \infty$, as follows immediately from Definition 5.1. Furthermore, for f in $D_0(\tau, p)$ we have

$$\begin{aligned}(g_i, f) &= (g_i, T_0(\tau, p)f) - (T_1(\tau^*, q)g_i, f) \\ &= (T_1(\tau^*, q)g_i, f) - (T_1(\tau^*, q)g_i, f) = 0.\end{aligned}$$

For $i \neq j$ the functionals B_i and B_j are linearly independent, for, if

$$\{g, f\} = 0, \qquad f \in D_1(\tau, p),$$

for some g in $D_1(\tau^*, q)$, then, by Theorem 2.1, g lies in $D_0(\tau^*, q)$ contrary to our hypothesis. From Lemma 5.2 we infer that $B_1, B_2, \cdots, B_n$ form a basis for the space of boundary values.

(d) Let g be a function in $D_1(\tau^*, q)$, $p^{-1} - q^{-1} = 1$, p or q possibly infinite, $1 \leqq p \leqq \infty$, which vanishes for $c \leqq t \leqq d$. Let f in $D_1(\tau, p)$ vanish for $a \leqq t \leqq c$. Then, by Green's formula,

$$\{g, f\} = \int_a^b (\tau f)(t) g(t) dt - \int_a^b f(t)(\tau^* g)(t) dt = \int_e^d - \int_e^d = 0$$

The converse is immediate for $1 < p \leqq \infty$ by (c) and (b).

We still have to prove (a) and (b) for $p = \infty$, and the converse to (d) for $p = 1$. Now (a) for $p = \infty$ is an immediate corollary of (b) for $p = 1$ and of the first part of (d). Furthermore, (b) for $p = \infty$ is an immediate consequence of the converse to (d) for $p = 1$. Thus we only have to prove (d).

Let B be a boundary value for τ at a in $L_1(a, b)$. By (c), there exists a function g in $D_1(\tau^*, \infty)$ such that $B(f) = \{g, f\}$, for f in $D_1(\tau, 1)$. Now choose a function h in $C^\infty(a, b)$ which is identically equal to one in a neighborhood of a and which vanishes identically in a neighborhood of b. Then the function gh lies in $D_1(\tau, \infty)$. Let

$$B'(f) = \{gh, f\}, \qquad f \in D_1(\tau, 1),$$

then B' is a boundary value at a. Now choose a function h_1 in $C^\infty(a, b)$ which is identically equal to one in a neighborhood of a and such that $h_1(t) = 0$ for all t such that $h(t) \neq 1$. For every function f in $D_1(\tau, 1)$ the function fh_1 is in $D_1(\tau, 1)$, and since B and B' are boundary values at a we have

$$B'(f) = B'(fh_1) = \{gh, fh_1\} = \{g, fh_1\} = B(fh_1) = B(f).$$

The following lemma gives an "analytic expression" for a boundary value of a differential operator τ in $L_p(I)$, $1 \leqq p \leqq \infty$. The result is less satisfactory in $L_\infty(I)$, owing to the closure properties of the subspace $D_0(\tau, \infty)$.

LEMMA 5.4. *Let τ be a differential operator of order k in (a, b).*

(a) *Suppose that the functions $h_0, h_1, \cdots, h_{k-1}$ have the property that the limit*

$$\lim_{t \to a} \sum_{i=0}^{k-1} h_i(t) f^{(i)}(t) = B(f)$$

exists for all f in $D_1(\tau, p)$, $1 \leqq p < \infty$. Then B is a boundary value for τ at a in $L_p(a, b)$. Conversely, every boundary value for τ at a in $L_p(a, b)$, $1 \leqq p < \infty$, is of this form. Furthermore, the functions $h_0, h_1, \cdots, h_{k-1}$ can be chosen to vanish in an arbitrary neighborhood of b not containing a.

(b) *Let B be a boundary value for τ in a in $L_\infty(a, b)$. Then there exist functions $h_0, h_1, \cdots, h_k$ such that*

$$B(f) = \lim_{t \to a} \sum_{i=0}^{k-1} h_i(t) f^{(i)}(t), \qquad f \in D_1(\tau, \infty).$$

The functions $h_0, h_1, \cdots, h_{k-1}$ can be chosen to vanish in an arbitrary neighborhood of b not containing a.

Proof: (a) The first assertion is immediate from the closed graph theorem and from the fact that the space $D_0(\tau, p)$ is the closure in the τ-norm of the subspace $K_k(a, b)$. To prove the converse and (b), let B be a boundary value for τ at a in $L_p(a, b)$, $1 \leqq p \leqq \infty$. By Lemma 5.3 (d) there exists a function g in $D_1(\tau^*, q)$ such that

$$B(f) = \{g, f, \tau\} \qquad f \in D_1(\tau, p).$$

It is clear from the proof of Lemma 5.3 that the function g can be chosen to vanish in an arbitrarily assigned neighborhood of b which does not contain the endpoint a. Suppose that $g(t) = 0$ for $a < c \leqq t < b$. For any point s such that $a < s < c$ we have, by Green's formula,

$$\int_s^b (\tau f)(t)g(t)dt - \int_s^b f(t)(\tau^* g)(t)dt = \int_s^c - \int_s^c$$

$$F_s(f, g) = \sum_{i=0}^{k-1}\sum_{j=0}^{k-1} F_s^{ij}(\tau)g^{(i)}(s)f^{(j)}(s),$$

and

$$\{g, f\} = \int_a^b (\tau f)(t)g(t)dt - \int_a^b f(t)(\tau^* g)(t)dt = \int_a^c - \int_a^c$$

$$= \lim_{s\to a}\left(\int_s^c - \int_s^c\right) \qquad \lim_{s\to a}\sum_{i=0}^{k-1}\sum_{j=0}^{k-1} F_s^{ij}(\tau)g^{(i)}(s)f^{(j)}(s).$$

Thus we can take

$$h_j(t) = \sum_{i=0}^{k-1} F_t^{ij}(\tau)g^{(i)}(t), \qquad 0 \leq j \leq k-1.$$

The following lemma gives one of the several criteria which can be stated in terms of the boundary form for the determination of the adjoint of a given D-operator.

LEMMA 5.5. *Suppose that the essential resolvent set of a differential operator τ in $L_p(a, b)$, $1 \leqq p \leqq \infty$, is non-empty and let $B_1, B_2, \cdots, B_n$ be a basis for the space of boundary values of τ in $L_p(a, b)$.*

(a) *Let $B_1^*, B_2^*, \cdots, B_n^*$ be a basis for the space of boundary values of τ^* in the conjugate space $L_q(a, b)$. Then for every point λ in the complex plane the boundary form can be expressed as*

$$\{g, f, \tau-\lambda, p\} = \sum_{i=1}^{n}\sum_{j=1}^{n} a_{ij}B_i(f)B^*(g), \qquad f \in D_1(\tau, p), \qquad g \in D_1(\tau^*, q),$$

where the matrix (a_{ij}) is uniquely determined by the basis $B_1^*, B_2^*, \cdots, B_n^*$ and is independent of λ.

(b) *Let T be an extension of τ in $L_p(a, b)$ obtained by extending the domain of τ to the space of the functions $f_1, f_2, \cdots, f_m$. Then the adjoint T^* of τ^* in the conjugate space $L_q(a, b)$ is obtained by the imposition of the boundary values*

$$B_i^*(g) = \{g, f_i\}, \qquad g \in D_1(\tau^*, q), \qquad 1 \leqq i \leqq m.$$

(c) *Let T be an extension of τ obtained by the imposition of boundary values*

$$B_i(f) = \{g_i, f\}, \quad f \in D_1(\tau, p), \quad g \in D_1(\tau^*, q), \quad 1 \leqq i \leqq m.$$

Then the adjoint T^* is the extension of τ^* to the space of the functions $g_1, g_2, \cdots, g_m$.

Proof: (a) Notice that

$$\{g, f, \tau-\lambda, p\} = \int_a^b ((\tau-\lambda)f)(t)g(t)dt - \int_a^b f(t)((\tau-\lambda)g)(t)dt$$
$$= \{g, f, \tau, p\}.$$

We shall assume that $\lambda = 0$. Let $f_1, f_2, \cdots, f_n$ be a basis for $D_1(\tau, p)$ modulo $D_0(\tau, p)$ and let $g_1, g_2, \cdots, g_n$ be a basis for $D_1(\tau^*, q)$ modulo $D_0(\tau^*, q)$, chosen (cf. Lemma 5.3) so that

$$\begin{aligned} B_j(f) &= \{g_j, f\}, \quad f \in D_1(\tau, p), \\ B_i(g) &= \{g, f_i\}, \quad g \in D_1(\tau^*, q), \end{aligned} \qquad 1 \leqq i, \quad j \leqq n.$$

Assume that

$$g = \sum_{i=1}^{n} \beta_i g_i .$$

The linear functional $F_i(g) = \beta_i$ is continuous on $D_1(\tau^*, q)$ and vanishes on $D_0(\tau^*, q)$. Therefore

$$F_i = \sum_{j=1}^{n} \alpha_{ij} B_j^* ,$$

and hence

$$\{g, f\} = \sum_{i=1}^{n} \beta_i g_i , \quad f = \sum_{i=1}^{n} \beta_i B_i(f) = \sum_{i=1}^{n} \sum_{j=1}^{n} \alpha_{ij} B_j^*(g) B_i(f).$$

(b) and (c). The domain of the adjoint T^* of T can be characterized as the set of all g in $D_1(\tau^*, q)$ such that

$$\{g, f\} = 0, \qquad f \in D(T).$$

From this (b) and (c) follow immediately.

The following property of the adjoint—the "Fredholm alternative"—is a consequence of Theorem 4.3.

THEOREM 5.2. *Let T be an extension of a differential operator τ in $L_p(I)$ which is maximal at some point λ in the essential resolvent set of τ, and let A be the connected component of the essential resolvent set of τ to which the point λ belongs. Then every eigenvalue of T in A is an eigenvalue of T^* in A, and the multiplicities are equal.*

Proof: Since T is maximal at λ in A we have

$$\dim N(\lambda I - T) = 0 = \dim N(\lambda I - T^*).$$

Therefore the constant $K(T)$ of Theorem 4.3 is zero. Thus at any point μ in A we have

$$\dim N(\mu I - T) - \dim N(\mu I - T^*) = 0.$$

Acknowledgement

The author wishes to express his thanks to Professor J. Schwartz and Professor N. Dunford for their encouragement and advice.

Bibliography

[1] Bade, W. G., *Unbounded spectral operators*, Pacific J. Math., Vol. 4, 1954, pp. 373-392.
[2] Dunford, N., and Schwartz, J., *Linear Operators. Part I, General Theory*, Interscience Publishers, Inc., New York, 1957.
[3] Dunford, N., and Schwartz, J., *Linear Ooperators. Part II, Spectral Theory*, to appear.
[4] Dunford, N., *Spectral theory II. Resolution of the identity*, Pacific J. Math., Vol. 2, 1952, pp. 559-614.
[5] Dunford, N., *Spectral operators*, Pacific J. Math., Vol. 4, 1954, pp. 321-354.
[6] Hille, E., *The abstract Cauchy problem and Cauchy's problem for parabolic differential operators*, J. Analyse Math., Vol. 3, 1954, pp. 81-196.
[7] Schwartz, J., *Perturbations of spectral operators and applications I. Bounded perturbations*, Pacific J. Math., Vol. 4, 1954, pp. 415-458.
[8] Weyl, H., *Über gewöhnliche Differentialgleichungen mit Singularitäten und die zugehörigen Entwicklungen willkürlicher Funktionen*, Math. Ann., Vol. 68, 1910, pp. 220-259.
[9] Achieser, N. I., and Glasmann, I. M., *Theorie der linearen Operatoren im Hilbert-Raum*, Akademie-Verlag, Berlin, 1954.
[10] Calkin, J. W., *Symmetric Transformations in Hilbert Space*, Duke J., Vol. 7, 1940, p. 504.

Received May, 1957.

American Mathematical Monthly, **67** (1960), 835–841

ON THE COMPLETENESS OF STURM-LIOUVILLE EXPANSIONS

GARRETT BIRKHOFF, Harvard University, AND GIAN-CARLO ROTA,* Massachusetts Institute of Technology

1. Historical remarks. A very old difficulty in the theory of ordinary differential equations is the proof of the fundamental assertion about the eigenfunctions of a Sturm-Liouville system: that such a sequence of functions is complete. In the special case of $y''+\lambda y=0$, this difficulty can be resolved by elementary methods taught in advanced calculus (See, for example, [**4**], Ch. IX or [**10**], Ch. XIII). However, it should be remembered that, even in this case, the difficulty perplexed such mathematical giants as Euler and Lagrange, being finally resolved rigorously by Dirichlet and Riemann† only after a century of controversy.

The general case is correspondingly more difficult, though the difficulty is commonly glossed over by physicists. In his celebrated *Theory of Sound* (Art. 187), Rayleigh discussed the number of zeros of the remainder of an "arbitrary function" after the first n terms of the Sturm-Liouville series have been subtracted from it. He then asserted that "the application of this theorem to demonstrate the possibility of expanding an arbitrary function in an infinite series of normal functions would proceed exactly as in Art. 142," that is, as in the case of Fourier series. Sommerfeld (*Partial Differential Equations*, p. 179) refers to this result as the *Ohm-Rayleigh Principle.*

Apparently Bôcher queried Lord Rayleigh about his statement. His question received the following reply, in 1899; a photostat of it is reproduced in Figure 1 because of its historical interest.

> Dear Prof. Bôcher
>
> I doubt if I had any special method in my mind when I wrote §187. At any rate if I had then, I have not now.
>
> I think I took this case as covered by the general theory as in Ch. IV. I have often wished that mathematicians would devote themselves to improving the rigour of this and similar demonstrations, rather than to inventing special methods for the particular cases.
>
> Prof. Klein, who was here a few months ago, spoke of some recent work of Poincaré in this direction. Probably you know of this. I think Klein specially mentioned something published at Palermo.

Actually, Rayleigh could have justified his statement quite easily, had he combined already available results with some simple observations about Hilbert space. It is the purpose of this note to supply the needed argument; it follows lines originally suggested by George D. Birkhoff [1].

2. Role of Hilbert space. Let $\mathfrak{C}$ be the Euclidean vector space ([**1**], p. 189) of all real functions $y=f(x)$ continuous on $a\leq x\leq b$. Given $f\in\mathfrak{C}$ and a sequence

* Professor Rota's work was partially done under Contract 7667, Office of Naval Research.

† See [**6**]. Fourier's arguments were hardly rigorous; he referred loosely to "all possible functions" (Analytical Theory of Heat, end of Art. 224).

$\{\phi_k\}$ of normalized orthogonal functions, one can define the formal expansion

$$f \sim c_1\phi_1 + c_2\phi_2 + \cdots, \qquad c_k = (f, \phi_k) = \int_a^b f\phi_k dx. \tag{1}$$

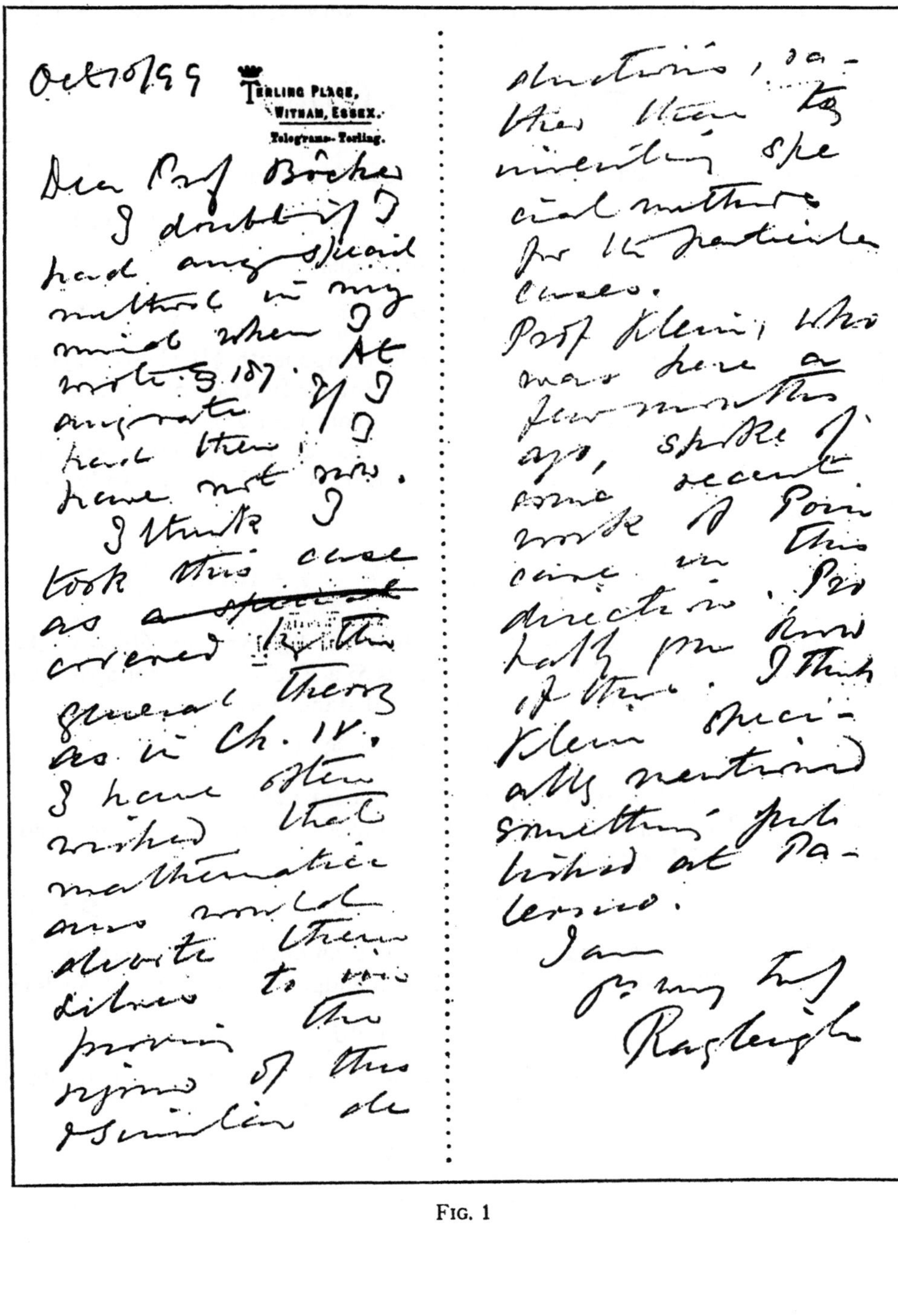

Oct 10/99

TERLING PLACE,
WITHAM, ESSEX.
Telegrams- Terling.

Dear Prof Bôcher

I doubt if I had any special method in my mind when I wrote § 187. At any rate if I had then I have not now.

I think I took this case as ~~a special~~ covered by the general theory as in Ch. IV.

I have often wished that mathematicians would devote themselves to the proving of this and similar demonstrations, rather than to inventing special methods for the particular cases.

Prof Klein, who was here a few months ago, spoke of some recent work of Poincaré in this direction. Probably you know of this. I think Klein specially mentioned something published at Palermo.

I am
yours very truly
Rayleigh

FIG. 1

The coefficients c_k always satisfy the Bessel inequality

$$\sum_1^\infty c_k^2 \leqq \int_a^b f^2(x)dx < +\infty. \tag{2}$$

The sequence $\{\phi_k\}$ is called *complete* when, for every $f \in \mathfrak{C}$, the Bessel inequality (2) is an equality (the Parseval equality):

$$\sum_1^\infty c_k^2 = \sum_1^\infty (f, \phi_k)^2 = \int_a^b f^2(x)dx. \tag{2*}$$

It is proved in advanced calculus that the cosine functions

$$\phi_k = \sqrt{(2/\pi)} \cos kx \qquad (k = 0, 1, \cdots)$$

are complete on $0 \leqq x \leqq \pi$, as are the sine functions

$$\bar{\phi}_k = \sqrt{(2/\pi)} \sin kx \qquad (k = 1, 2, \cdots);$$

these are correlated with the even and odd components $\frac{1}{2}[f(x)+f(-x)]$ and $\frac{1}{2}[f(x)-f(-x)]$, respectively, of a general continuous function on $-\pi \leqq x \leqq \pi$.

The expansion (1) establishes a linear isomorphism $\theta: f \rightarrow (c_1, c_2, \cdots)$ between $\mathfrak{C}$ and a subspace $\theta(\mathfrak{C}) = \mathfrak{G}$ of the space $\mathfrak{H}$ of all infinite sequences $(c_1, c_2, \cdots)$ satisfying $\sum_1^\infty c_k^2 < +\infty$. Moreover, if (real) inner products are defined in $\mathfrak{H}$ by

$$(\gamma, \bar{\gamma}) = \sum_1^\infty c_k \bar{c}_k, \qquad \gamma = (c_1, c_2, \cdots), \qquad \bar{\gamma} = (\bar{c}_1, \bar{c}_2, \cdots), \tag{3}$$

then the mapping θ preserves inner products: if $\theta(f) = \gamma$ and $\theta(\bar{f}) = \bar{\gamma}$, then $\int_a^b f(x)\bar{f}(x)dx = (\gamma, \bar{\gamma})$. The definition (3) makes $\mathfrak{H}$ a Euclidean vector space, and in fact a *Hilbert space* l_2; relative to the usual distance formula

$$\|\gamma - \bar{\gamma}\| = (\gamma - \bar{\gamma}, \gamma - \bar{\gamma})^{1/2} \tag{3'}$$

defined in $\mathfrak{H}$ by (3), θ is distance-preserving (an *isometry*).

Now consider a general regular Sturm-Liouville system. It suffices to consider the differential equation

$$y'' + (\lambda - q(x))y = 0, \qquad 0 \leqq x \leqq \pi, \tag{4}$$

since other cases can be reduced to this by Liouville's transformation ([9], p. 238; [5], pp. 291–295.) We assume *separated* endpoint conditions, of the form

$$\alpha f(0) + \alpha_1 f'(0) = \beta f(\pi) + \beta_1 f'(\pi) = 0. \tag{4'}$$

As essentially shown by Liouville, one can then derive the following asymptotic estimate ([9], p. 244; [3], Part B) for the eigenfunctions of (4)–(4′):

$$\psi_k(x) = \sqrt{(2/\pi)} \cos kx + O(1/k), \tag{5}$$

provided $\alpha_1 \neq 0$ and $\beta_1 \neq 0$. (If $\alpha_1 = 0$ or $\beta_1 = 0$, use sine in place of cosine.) As a corollary, we have the relation $\|\psi_k-\phi_k\|^2=O(1/k^2)$, where ϕ_k is an orthonormal basis* of cosine or sine functions. Hence, in particular,

$$\sum_{k=1}^{\infty}\|\psi_k-\phi_k\|^2<+\infty. \tag{5'}$$

3. Main theorem. The completeness of the ψ_k now follows from a general theorem about bases in Hilbert space. This can be made to follow from more general theorems on biorthogonal systems ([**8**], p. 206), but we have not seen a simple direct proof in the literature.

THEOREM 1. *Let ϕ_k $(k=0, 1, \cdots)$ be an orthonormal basis in Hilbert space $\mathfrak{H}$ and let ψ_k $(k=0, 1, \cdots)$ be an orthonormal sequence in $\mathfrak{H}$ satisfying*

$$\sum_{k=1}^{\infty}\|\psi_k-\phi_k\|^2<+\infty. \tag{6}$$

Then the sequence $\{\psi_k\}$ is also an orthonormal basis.

LEMMA. *Let $\phi_0, \phi_1, \cdots$ be an orthonormal basis in the Hilbert space $\mathfrak{H}$. Let $\psi_0, \psi_1, \cdots$ be an orthonormal sequence in $\mathfrak{H}$ satisfying the conditions*

$$\sum_{k=1}^{\infty}\|\psi_k-\phi_k\|^2<1. \tag{7}$$

Then the sequence $\{\psi_k\}$ is also an orthonormal basis in $\mathfrak{H}$.

Proof. If the sequence $\{\psi_k\}$ were not a basis, then we could find a nonzero function g orthogonal to every ψ_k. For this function, $(g, \phi_k)=(g, \psi_k)+(g, \phi_k-\psi_k)=(g, \phi_k-\psi_k)$. Squaring and using the Schwarz inequality,

$$(g, \phi_k)^2=(g, \phi_k-\psi_k)^2\leqq\|g\|^2\|\phi_k-\psi_k\|^2. \tag{7'}$$

Summing with respect to k, we get

$$\sum_{k=0}^{\infty}(g, \phi_k)^2\leqq\|g\|^2\sum_{k=0}^{\infty}\|\phi_k-\psi_k\|^2<\|g\|^2,$$

the $\{\phi_k\}$ being a basis, in evident violation of Parseval's equality.

COROLLARY 1. *Replace condition* (7) *above by the weaker condition*

$$\sum_{k=N+1}^{\infty}\|\psi_k-\phi_k\|^2<1$$

for some integer N. Then every element of $\mathfrak{H}$ orthogonal to $\phi_0, \phi_1, \cdots, \phi_N$ and to $\psi_{N+1}, \psi_{N+2}, \cdots$ must vanish.

* By a *basis* for the Hilbert space $\mathfrak{H}$, we mean a subset, whose finite linear combinations are dense in $\mathfrak{H}$.

Proof. Any such element g would satisfy the inequality (7') for all $k>N$. Hence, summing over all k, since $(g, \phi_k)=0$ for $k=1, \cdots, N$, we have

$$\sum_{k=1}^{\infty} (g, \phi_k)^2 = \sum_{k=N+1}^{\infty} (g, \phi_k)^2 \leqq \|g\|^2 \sum_{k=N+1}^{\infty} \|\phi_k - \psi_k\|^2 < 1,$$

again contradicting Parseval's equality (2*).

COROLLARY 2. *Under the assumptions of Corollary* 1, *every element of* $\mathfrak{H}$ *orthogonal to* $\psi_{N+1}, \psi_{N+2}, \cdots$ *and to the elements*

$$\eta_n = \phi_n - \sum_{k=N+1}^{\infty} (\phi_n, \psi_k)\psi_k, \qquad n = 0, 1, \cdots, N, \tag{8}$$

must vanish.

Proof. For any such element g we have

$$(g, \phi_n) = (g, \eta_n) + \sum_{k=N+1}^{\infty} (\phi_n, \psi_k)(g, \psi_k) = 0; \quad n = 0, 1, \cdots, N.$$

Thus g also satisfies the conditions of Corollary 1, and therefore vanishes.

The proof of Theorem 1 can now be completed as follows.

Choose an integer N so that

$$\sum_{k=N+1}^{\infty} \|\psi_k - \phi_k\|^2 < 1.$$

By Corollary 2, any element g of $\mathfrak{H}$ which is orthogonal to the elements ψ_{N+1}, $\psi_{N+2}, \cdots$ and to the elements η_n $(n=0, 1, \cdots, N)$ defined by (8) must vanish. Denote by S the set of all elements of $\mathfrak{H}$ orthogonal to $\psi_{N+1}, \psi_{N+2}, \cdots$. Evidently S is a vector space containing $\eta_0, \eta_1, \cdots, \eta_n$. By virtue of the above remark, the vector space S contains only the linear combinations of these elements. In other words, S is a finite-dimensional vector space whose dimension is at most N.

But the elements $\psi_1, \cdots, \psi_N$ also belong to the vector space S, and they are linearly independent (they are an orthonormal sequence!). Therefore the elements $\psi_1, \cdots, \psi_N$ are a *basis* for the vector space S ([2], pp. 168–9). Hence the elements $\eta_0, \eta_1, \cdots, \eta_N$ are linear combinations of $\psi_1, \cdots, \psi_N$. It follows that any element of $\mathfrak{H}$ which is orthogonal to all the ψ_k must vanish, because such an element is also orthogonal to $\eta_0, \eta_1, \cdots, \eta_N$ and to $\psi_{N+1}, \psi_{N+2}, \cdots$. We conclude that the sequence $\{\psi_k\}$ is complete.

4. Further discussion. As Rayleigh, like most of his contemporaries, probably interpreted "completeness" as referring to pointwise or uniform convergence, it is relevant to discuss the relation between the "mean-square convergence" defined by the Parseval equality, and pointwise or uniform convergence. Some facts can be proved easily, without an appeal to the Lebesgue integral.

Thus, from the Weierstrass approximation theorem (surely known to Rayleigh!), it follows immediately that the set $\mathfrak{P}$ of *polynomial* functions is dense in $\mathfrak{H}$. Therefore, the functions of class $\mathfrak{C}'$, $\mathfrak{C}''$, etc. (the number of primes referring to continuous differentiability) are also dense in $\mathfrak{H}$. By a slight modification near the endpoints, one can however approximate any such function by one satisfying the separated endpoint conditions (4′), arbitrarily closely in $\mathfrak{H}$. We now observe

THEOREM 2. *If $f \in \mathfrak{C}''$ satisfies the endpointcon ditions (4′), then the kth term of the expansion of f in the eigenfunctions ψ_k of the Sturm-Liouville system (4)–(4′) is $O(1/k^2)$.*

Proof. Equation (4) implies

$$c_k = \int_0^\pi f(x)\psi_k(x)dx = (1/\lambda_k)\int_0^\pi [q(x)\psi_k(x) - \psi_k''(x)]f(x)dx. \tag{9}$$

On the other hand, by the Lagrange identity and (4′)

$$\int_0^\pi (\psi_k f'' - f\psi_k'')dx = [\psi_k f' - f\psi_k']_{x=0}^{x=\pi} = 0,$$

since f and the ψ_k both satisfy (4′). Hence (9) implies

$$c_k = (1/\lambda_k)\int_0^\pi [q(x)\psi_k(x) - \psi_k(x)f''(x)]dx = O(1/k^2), \tag{9*}$$

where the last equality follows since $\lambda_k = O(k^2)$, while the ψ_k, q, and f'' are uniformly bounded, all by (5). Uniform convergence follows from (9*) by the Weierstrass M-test.

So far, Lebesgue integration has not been invoked. This is needed to identify $\mathfrak{H}$ as constructed above with the class $L_2(0, \pi)$ of functions whose squares are Lebesgue integrals. Specifically, given any formal sum $\sum_1^\infty c_k\phi_k(x)$ with $\sum_1^\infty c_k^2 < +\infty$, one can extract from the partial sums $\sigma_n(x) = \sum_1^n c_k\phi_k(x)$ a subsequence $\sigma_{n(k)}(x)$ converging almost everywhere to a (Lebesgue integrable) limit $\sigma_\infty(x)$; this is the essence of the proof of the Riesz-Fischer Theorem ([7], p. 185); any two such limits are equal a.e., and determine the same σ_∞ in $\mathfrak{H}$. Using these facts, one can prove

THEOREM 4. *If the nth partial sums $\sigma_n(x) = \sum_1^n c_k\phi_k(x)$ in (1), converge pointwise to $f(x)$ then they converge in the mean square.*

For, by hypothesis, the $\sigma_{n(k)}(x)$ mentioned above converge pointwise to $f(x)$; hence $f(x)$ coincides with $\sigma_\infty(x)$ almost everywhere, implying the Parseval equality (2*). We omit the details.

COROLLARY. *If the nth partial sums $\sigma_n(x)$ in (1) converge pointwise to $f(x)$, for all f in any dense subset $\mathfrak{D}$ of $\mathfrak{H}$, then $\{\phi_k\}$ is complete.*

It is not clear that the above results can be proved without using the Lebesgue integral. They provide a final link between classical pointwise convergence and the modern theory of mean square convergence.

References

1. George D. Birkhoff, A theorem on theories of orthogonal functions with an application to Sturm-Liouville theories, Proc. Nat. Acad. Sci. U.S.A., vol. 3, 1917, pp. 656–659; Collected Mathematical Papers, vol. 1, New York, pp. 90–93.
2. G. Birkhoff and S. MacLane, Survey of Modern Algebra (rev. ed.), New York, 1952.
3. G. Birkhoff and G.-C. Rota, Lecture Notes on Ordinary Differential Equations, Boston, 1960–61.
4. R. Courant, Differential and Integral Calculus, vol. 1, New York, 1937.
5. R. Courant and D. Hilbert, Methoden der Mathematischen Physik, (2nd ed.) Berlin, 1931.
6. R. E. Langer, Fourier's series, this MONTHLY, vol. 54, 1947, Part II, pp. 1–86.
7. E. J. McShane, Integration, Princeton University Press, 1944.
8. F. Riesz and B. Sz-Nagy, Lecons d'Analyse Fonctionelle (2nd ed.), Budapest, 1953.
9. G. Sansone, Equazioni Differenziali nel Campo Reale, vol. 1, Bologna, 1948.
10. D. V. Widder, Advanced Calculus, New York, 1947.

OPERATORS AND THE r-WAY CROSSED CLASSIFICATION

J. D. BANKIER, McMaster University

1. Introduction. The development of the analysis of variance for the r-way crossed classification usually involves some tedious algebra. It is our purpose to clear much of this algebra away by means of an operational approach to the problem. The resulting presentation has been used successfully with undergraduates.

2. The 2-way crossed classification. It will be easier to understand what follows if we first review some of the results for the case where $r=2$. Here we make observations, X_{ijk}, and are concerned about the effect upon these observations of two factors, temperature and pressure, say. We make a selection of p temperatures and q pressures and make n observations at each of the pq possible combinations of temperature and pressure. We assume that the expected value of X_{ijk} is given by the equation

$$(1)\quad E(X_{ijk}) = \mu + \alpha_i + \beta_i + (\alpha\beta)_{ij} \quad (i = 1, \cdots, p; j = 1, \cdots, q; k = 1, \cdots, n),$$

where μ is a constant, α_i and β_j are the *main effects* of the ith temperature and jth pressure while $(\alpha\beta)_{ij}$ is called the *interaction* between the ith temperature and the jth pressure. As we shall see in the general case, we may assume without loss of generality that

On the Spectra of Singular Boundary Value Problems

GIAN-CARLO ROTA*

Communicated by L. CESARI

1. The first difficulty in trying to establish a non-selfadjoint eigenfunction expansion (cf. [4], [6], [9]) is the *location of the spectrum.* This difficulty arises to a much lesser degree in the selfadjoint case, since there one at least knows that the spectrum is real. There are two steps to the problem: the location of eigenvalues, which depend upon the particular boundary conditions, and the location of that subset of the spectrum which corresponds to the physicists' "band spectrum" and is independent of the boundary conditions, namely the *essential spectrum.* Definitions and results used in the present paper can be found in [11]. There are strong reasons for discarding the older notion of continuous spectrum in favor of the essential spectrum, *cf.* [12].

The purpose of this note is to point out an exceedingly simple criterion which can be applied to locate the essential spectra of a large class of non-selfadjoint ordinary differential operators, including the cases of most frequent occurrence, namely differential operators with rational coefficients. The criterion is then applied to the study of the variation of the essential spectrum *with the norm* for the RIEMANN differential operator, to point out a phenomenon which has no counterpart in selfadjoint spectral theory.

2. Consider a formal differential operator

$$L = a_n(t)\left(\frac{d}{dt}\right)^n + a_{n-1}(t)\left(\frac{d}{dt}\right)^{n-1} + \cdots + a_0(t), \qquad t \in I$$

on an interval I, open or closed, possibly infinite, assuming regularity, namely $a_n(t) \neq 0$ for t in I. The function $a_k(t)$ $(0 \leqq k \leqq n)$ is complex-valued, and for simplicity assume it is k times continuously differentiable (this is a convenience: the differentiability of the coefficient functions is largely irrelevant to the

* The work of the author was partially supported by the Office of Naval Research and partially by Contract No. Da 19020 ORD-3778 with the Office of Ordnance Research.

problem). The case of interest is when the coefficient functions a_k are singular at one or both endpoints of the interval I. Since the influence of each endpoint of the interval upon the essential spectrum is independent of the other endpoint (*cf.* [11], Lemma 3.5), we can and shall assume that the interval I is half-closed ($[a, b)$). Consider the *closed* operator $T_1(L)$ acting on a LEBESGUE space $L_p(I)$, for *fixed* p ($1 \leqq p \leqq \infty$), having as domain the graph-closure in $L_p(I)$ of the subspace of all functions f of class $L_p(I)$ such that Lf is defined and of class $L_p(I)$, and such that $T_1(L)f(t) = (Lf)(t)$ ([11], p. 32).

The *essential spectrum* of L in $L_p(I)$ is the set of all complex numbers λ for which the operator $\lambda I - T_1(L)$ *does not have a closed range in* $L_p(I)$ ([11], pp. 36–43).

It is known that the essential spectrum of every extension of the operator L, obtained by imposition of (possibly singular) boundary conditions, coincides with the essential spectrum of L in each L_p. Any criterion for the location of the essential spectrum which is at all workable must be based upon information about the operator which can be effectively obtained from the coefficients. The one given in [11] (Lemma 3.4) is of difficult verification. Other criteria in the literature deal with special cases, mostly with the second-order operator

$$Lf(t) = f''(t) + q(t)f(t).$$

The selfadjoint case, for example, is treated in the series of papers by HARTMAN and WINTNER on the American Journal of Mathematics, 1946–54. We shall be content with a *sufficient* criterion. It is probably also necessary. We can however at present only prove the necessity only for special cases, such as selfadjoint differential operators, rational coefficients, *etc.* To this end the following notion proves to be useful.

A basis of solutions $f_1(\lambda, t), f_2(\lambda, t), \cdots, f_n(\lambda, t)$ of the differential equation $(\lambda - L)f = 0$ is said to be a *norm-analytic basis* at a given point λ_0 in the complex plane when there exists a neighborhood N of λ_0 such that: (a) the functions f are analytic in λ for λ in N; (b) they form a basis of solutions for the differential equation $(\lambda - L)f = 0$ for each λ in N; (c) there is an integer k such that for λ in N the subspace of all those solutions of the differential equation $(\lambda - L)f = 0$ which are of class $L_p(I)$ is of dimension k and is spanned by $f_1(\lambda, t), f_2(\lambda, t), \cdots, f_k(\lambda, t)$.

Intuitively, this means that at λ_0 no solution "ceases to be" or "becomes" of class $L_p(I)$ unless it was already such for λ in the neighborhood of λ_0.

The desired criterion is the following: *if at* λ *either the differential operator* L *in* $L_p(I)$ *or its Lagrange adjoint* L^* *in* $L_q(I)$ (*where* $p^{-1} + q^{-1} = 1, 1 \leqq p \leqq \infty$) *does not have a norm-analytic basis, then* λ *belongs to the essential spectrum of* L.

The proof is easily derived from general results (*cf.* [11]) on extension of differential operators. If λ_0 does not belong to the essential spectrum of L in $L_p(I)$, then take an extension T of L which is maximal in $L_p(I)$ at the point λ_0. Since we are dealing with a half-closed interval, it follows that the resolvent $(\lambda I - T)^{-1}$ exists as a bounded operator in $L_p(I)$ and is holomorphic for λ in a neighborhood of λ_0. Now let $f_1(\lambda_0, t), \cdots, f_k(\lambda_0, t)$ be a basis of solutions of class $L_p(I)$ at λ_0. We continue these analytically in a neighborhood of λ by setting

$$f(\lambda, t) = f(\lambda_0, t) - (\lambda - \lambda_0)((\lambda - T)^{-1}f(\lambda_0, t).$$

Clearly $f(\lambda, t)$ is of class $L_p(I)$ and is analytic in λ whenever the resolvent exists. The verification that such a function is indeed a solution of the differential equation $(\lambda - L)f = 0$ is easy. We have

$$\begin{aligned}(\lambda - L)f(\lambda, t) &= [(\lambda - \lambda_0) + (\lambda_0 - L)]f(\lambda, t)\\ &= (\lambda - \lambda_0)f(\lambda_0, t) - (\lambda - \lambda_0)(\lambda - L)(\lambda - T)^{-1}f(\lambda_0, t)\\ &= (\lambda - \lambda_0)f(\lambda_0, t) - (\lambda - \lambda_0)f(\lambda_0, t) = 0.\end{aligned}$$

It is now trivial to complete these functions to an analytic basis—just add any $n - k$ solutions holomorphic in λ, say $f_{k+1}(\lambda, t), \cdots, f_n(\lambda, t)$, which together with $f_1(\lambda, t), \cdots, f_k(\lambda, t)$ form a linearly independent set at $\lambda = \lambda_0$. The statement about the LAGRANGE adjoint L^* follows from the fact that the essential spectra of L in $L_p(I)$ and L^* in $L_q(I)$ coincide ([11], Theorem 3.1). This completes the proof.

3. The preceding criterion permits to effectively compute the essential spectra of differential operators with rational coefficients. We wish presently to apply it to the study of the spectra of the boundary value problems obtained from the *Riemann differential equation.* This will reveal an unusual property of this equation, which seems to have so far escaped attention. The selfadjoint eigenfunction expansion associated with this differential equation are worked out in [3], Ch. 4 and [4], Ch. XIII. The analogous problems in LEBESGUE spaces seem not to have been considered before.

There are exactly three forms of the RIEMANN differential equation which lead to operators which can be studied without any deeper results than those pertaining to hypergeometric functions. These operators L are determined by the requirement that for any λ the differential equation $Lf - \lambda f = 0$ be a RIEMANN differential equation. An easy computation, which we forego, shows that the following are the only three possible operators:

$$L_1(f)(t) = t(t - 1)f''(t) + (at + b)f'(t) + \frac{et^2 + dt + e}{t(t - 1)} f(t), \tag{1}$$

$$L_2(f)(t) = t^2(t - 1)f''(t) + t(at + b)f'(t) + \frac{ct^2 + dt + e}{t - 1} f(t), \tag{2}$$

$$L_3(f)(t) = t(t - 1)^2f''(t) + (t - 1)(at + b)f'(t) + \frac{ct^2 + dt + e}{t} f(t). \tag{3}$$

One sees by inspection that L_2 and L_3 are obtained from each other by the change of variable $u = 1 - t$. A slightly more involved computation shows that L_1 is obtained from L_2 by interchanging the singularities at 0 and ∞. Thus it suffices to study the operator L_1. We omit a study of this operator on the interval (0, 1). The spectrum of the operator in this interval is *discrete* in all $L_p(I)$, $1 \leqq p \leqq \infty$, and is the same in all these spaces. The expansion theory is much like that of the classical orthogonal polynomials. The L_p-theory has been studied in the well-known works of H. Pollard. A more interesting interval is $(-\infty, 0)$, or the interval $(1, \infty)$, which gives the same behaviour. The preceding criterion will be used to determine the essential spectrum of the operator L_1 in the interval $(-\infty, 0)$ in all Lebesgue spaces $L_p(-\infty, 0)$. In the course of the discussion the (generalized) *deficiency indices* (*cf.* [11], p. 48) will also be determined. Let $\alpha_1, \alpha_2, \beta_1, \beta_2, \nu_1, \nu_2$ be the exponents at the regular singular points 0, 1, ∞ respectively of the operator $\lambda - L_1$, and $e_1, e_2, m_1, m_2, n_1, n_2$ those of its Lagrange adjoint $\lambda - L_1^*$. It is easily checked that the adjoint differential equation $(\lambda - L_1^*) f = 0$ is a Riemann differential equation for all λ. These exponents are of course dependent upon the parameter λ and satisfy the Riemann identity. The operator $\lambda - L_1$ takes the form

$$(L_1 - \lambda)f(t) = \frac{d}{dt}\, t(t-1)\frac{df(t)}{dt} + [(\nu_1 + \nu_2 - 1)t + (\alpha_1 + \alpha_2)]f'(t)$$
$$+ \left[-\frac{\alpha_1\alpha_2}{t} + \frac{\beta_1\beta_2}{t-1} + \nu_1\nu_2\right]f(t), \qquad \lambda = \nu_1\nu_2,$$

whereas its adjoint is the operator

$$(L_1^* - \lambda)f(t) = \frac{d}{dt}\, t(t-1)\frac{df(t)}{dt} + [(n_1 + n_2 - 1)t + (e_1 + e_2)]f'(t)$$
$$+ \left[-\frac{e_1e_2}{t} + \frac{m_1m_2}{t-1} + n_1n_2\right]f(t),$$

where $e_i = -\alpha_i$, $m_i = -\beta_i$ and $n_i = 1 - \nu_i$. Clearly only the exponents ν_i and n_i depend upon the parameter λ; all other exponents remain constant. It follows that at any point λ there is a norm-analytic basis of solutions in $L_p(-1, 0)$, that is, in a neighborhood of the singular point 0. The general criterion given above need therefore be applied only by studying norm-analytic bases for the operator L in the half-closed interval $(-\infty, -1)$. Writing $A = 1 - \alpha_1 - \alpha_2 - \beta_1 - \beta_2$, we have the condition $\nu_1 + \nu_2 = A$, $\nu_1\nu_2 = \lambda$, giving two linearly independent solutions $f_1(\lambda, t)$ and $f_2(\lambda, t)$ depending analytically upon λ, which at infinity are asymptotic to $t^{-\nu_1}$ and $t^{-\nu_2}$ respectively. For the results on the hypergeometric differential equation used here see [14], Ch. II. Note that the logarithmic case $\nu_1 = \nu_2$ causes no difficulty here.

These solutions are of class L_p at infinity if and only if $\operatorname{Re} \nu_i > 1/p$. Therefore these solutions form a norm-analytic basis at all points λ except those for which $\operatorname{Re} \nu_1 = 1/p$ or $\operatorname{Re} \nu_2 = 1/p$. To determine the set of these λ's, set $\nu_1 = 1/p + ir$, where r is a real parameter. Then the above conditions give the parametric equation of the curve

$$\lambda = (1/p + ir)(A - 1/p - ir), \qquad -\infty < r < \infty \tag{1}$$

depending upon the real parameter r.

This curve therefore belongs to the essential spectrum of the operator L_1 in $L_p(-\infty, 0)$. A similar argument with the adjoint operator L_1^* leads to the same curve as the set where there is no norm-analytic basis for the solutions of the differential equation $(\lambda - L_1^*) f = 0$. This follows from the fact that $\text{Re}\, n_i = 1/q$ whenever $\text{Re}\, \nu_i = 1/p$.

4. The preceding argument can now be strengthened to show that *the curve* (1) *is actually all of the essential spectrum of the differential operator L_1*. This is done by establishing that at any point not on the curve (1) a maximal extension can be found.

It is easily verified that for all values of the constant A the curve (1) is a parabola with its axis parallel to the axis $\text{Re}\, \lambda = 0$ and tending to infinity along positive values of $\text{Re}\, \lambda$. The focus lies on the abscissa if and only if $\text{Im}(A_2) = 0$. The parabola degenerates to a half-line if and only if $\text{Im}\, A = 0$ and $\text{Re}\, A = 2/p$. For λ in the interior of this parabola, the differential equation $(\lambda - L_1) f = 0$ has the exponents $\nu_i = +A/2 \pm (\sqrt{A^2 - 4\lambda})/2$, and these give two solutions, both or neither of which are of class L_p. By going over to the adjoint if necessary, we can assume that *both* are of class L_p. Denoting these functions by $f_1(t, \lambda)$ and $f_2(t, \lambda)$ and by $g_i(t, \lambda)$ the solutions of the adjoint differential equation with exponents n_i, we obtain ([2], p. 87) the following expression for the *Green's kernel* in the interval $(-\infty, -1)$:

$$\int_{-\infty}^{-1} G(t, s, \lambda)h(s)\, ds = f_1(t, \lambda) \int_t^{-1} g_1(s, \lambda)h(s)\, ds + f_2(t, \lambda) \int_t^{-1} g_2(s, \lambda)h(s)\, ds. \tag{2}$$

The range of the operator $\lambda I - T_1(L)$ is closed if the operator (2) is bounded. To show this, by the closed graph theorem it suffices to prove that this operator maps functions of class L_p into functions of the same class. We have to estimate the L_p-norms of each of the two terms on the right side for $h \in L_p(-\infty, -1)$. This is done by using a slightly modified form of a well-known inequality of G. H. Hardy [15]. Using the estimates

$$f_i(t, \lambda) = O(|t|^{-\text{Re}\, \nu_i})$$

and

$$g_i(t, \lambda) = O(|t|^{\text{Re}\, \nu_i - 1})$$

and setting $\bar{h}(s) = h(-s)$, $\alpha = \operatorname{Re} \nu_i$ we have:

$$\left[\int_{-\infty}^{-1} \left| f_i(t, \lambda) \int_t^{-1} g_i(s, \lambda) h(s)\, ds \right|^p dt \right]^{1/p}$$

$$= \left[\int_1^{\infty} \left| O(1) t^{-\alpha} \int_1^t s^{\alpha-1} O(1) \bar{h}(s)\, ds \right|^p dt \right]^{1/p}$$

$$= \left[\int_1^{\infty} \left| O(1) \int_{1/t}^1 u^{\alpha-1} \bar{h}(ut)\, du \right|^p dt \right]^{1/p}$$

$$= O\left(\int_0^1 u^{\alpha-1} \left(\int_1^{\infty} |\bar{h}(ut)|^p\, dt\right)^{1/p} du\right) = O\left(\int_0^1 u^{\alpha-1-1/p}\, du\right) = O(1),$$

since $\alpha = \operatorname{Re} \nu_i > 1/p$ by hypothesis. Here, as well as in the next estimate, it is assumed that $1 \leqq p < \infty$, since the case $p = \infty$ is trivial. This proves that (2) is bounded and that the region interior to the parabola (1) is free of essential spectrum.

A similar estimate can be used to show that the region exterior to the parabola (1) is free of essential spectrum. For λ in this region the differential equation $(\lambda - L) f = 0$ $((\lambda - L^*) g = 0)$ has exactly *one* solution $f_1(t, \lambda)$ $(g_1(t, \lambda))$ of class $L_p(L_q)$ at infinity. The GREEN's kernel is therefore of the form

$$\int_{-\infty}^{-1} G(t, s) h(s)\, ds = f_1(t, \lambda) \int_t^{-1} g_2(s, \lambda) h(s)\, ds + f_2(t, \lambda) \int_{-\infty}^t g_1(s, \lambda) h(s)\, ds,$$

where f_2 and g_2 are other linearly independent solutions not of class L_p or L_q. It is sufficient to consider the second term. By the same inequality of HARDY we have, setting $\beta = \operatorname{Re} n_2$:

$$\left[\int_{-\infty}^{-1} \left| f_2(t, \lambda) \int_{-\infty}^t g_2(s, \lambda) h(s)\, ds \right|^p dt \right]^{1/p}$$

$$= \left[\int_1^{\infty} \left| O(1) t^{\beta-1} \int_t^{\infty} s^{-\beta} \bar{h}(s)\, ds \right|^p dt \right]^{1/p}$$

$$= \left[\int_1^{\infty} \left|\left| O(1) \int_{-\infty}^1 u^{-\beta} \bar{h}(ut)\, du \right|^p dt \right|\right]^{1/p} = O\left(\int_1^{\infty} u^{-\beta-1/p}\, du\right) = O(1)$$

since $\beta > 1/q$.

It has thus been shown that *the essential spectrum of the Riemann operator* L_1 *is in all cases the curve* (1).

The most striking conclusion which can be drawn from this is that the *essential spectrum of the Riemann differential operator varies with the Leqesgue space* L_p. This is at odds with "traditional" ideas of harmonic analysis: the classical theory of the FOURIER transform depends upon the fact that for the operator id/dt on $(-\infty, \infty)$ the essential spectrum remains invariant with p. LEON EHRENPREIS has recently suggested that the above phenomenon for the RIEMANN differential equation is closely related to a similar occurrence for the FOURIER transform on the unimodular group.

Note that since the deficiency indices in the two regions into which the essential spectrum divides the λ-plane are unequal, it is impossible to obtain a maximal extension without residual spectrum except when the parabola reduces to a straight line. This does not only happen in the selfadjoint case. In fact when $A = 2/p$ one has such an extension in L_p *but not in Hilbert space*!

5. As an application of the preceding theory we shall settle in the negative a conjecture raised a few years ago by A. E. TAYLOR. The conjecture is whether or not a real selfadjoint integral operator has a purely real spectrum in L_p for $p \neq 2$. A selfadjoint integral operator whose spectrum in L_p goes "off the real axis" can be constructed as follows. Consider the *Legendre operator*

$$\frac{d}{dt}\left(t(1 - t)\frac{d}{dt}\right) = \Lambda$$

on the interval $(-\infty, 0)$. This is a particular case of the RIEMANN operator L_1 with $\nu_1 + \nu_2 = 1$, $\alpha_1 = \alpha_2 = \beta_1 = \beta_2 = 0$. Here $A = 1$ and the essential spectrum in $L_p(-\infty, 0)$, given by (1), is the curve

$$\lambda = (p^{-1} + ir)(1 - p^{-1} - ir).$$

It is *real* only for $p = 2$ (the selfadjoint case). In all other cases it is a parabola. There are two boundary conditions at the endpoint 0 for all $L_p(I)$. By imposing one real boundary condition at $\lambda = 0$ we can get a maximal extension T_2 in all $L_p(I)$. This extension in general has a discrete sequence of eigenvalues on the negative real axis. If λ is a negative real number not equal to any of these eigenvalues, then the resolvent $(\lambda I - T_2)^{-1}$ can be represented as an integral operator (the GREEN kernel) which is bounded in all L_p, has purely real spectrum in $L_2(-\infty, 0)$ and whose spectrum in $L_p(-\infty, 0)$ is an ellipse with foci on the real axis for all $p \neq 2$.

By varying this example one can easily obtain examples of bounded real integral operators (GEEEN's kernels) whose spectrum is real in L_{p_0} for some p_0 and becomes an ellipse with foci on the real axis for other LEBESGUE norms.

BIBLIOGRAPHY

[1] G. D. BIRKHOFF, Boundary value and expansion problems of ordinary differential equations, *Transactions of American Mathematical Society*, 9 (1908) 373–395.

[2] N. LEVINSON & E. A. CODDINGTON, *Theory of ordinary differential equations*, McGraw-Hill, New York, (1955).

[3] E. C. TITCHMARSH, *Eigenfunction expansions associated with second-order differential equations*, Oxford, University press, Vol. I (1946) and II (1958).

[4] N. DUNFORD & J. T. SCHWARTZ, *Linear operators*, Vol. II (spectral theory), Interscience, New York (forthcoming).

[5] F. BRAUER, Spectral theory for the differential equation $Lu = \lambda Mu$, *Canadian Journal of Mathematics*, 10 (1958) 431–446.

[6] R. R. D. KEMP, A singular boundary-value problem for a non-selfadjoint differential operator, *Canadian Journal of Mathematics*, 10 (1958) 447–462.
[7] H. P. KRAMER, Perturbation of differential operators, Technical Report No. 5, Contract DA-04-200-ORD-171 Task Order 5 for Office of Ordnance Research, University of California, Berkeley, California, 1954.
[8] D. MCGARVEY, Differential operators with periodic coefficients, thesis, Yale University, 1957.
[9] J. MOSER, Störungstheorie des kontinuirlichen Spektrums für gewöhnliche Differentialgleichungen zweiter Ordnung, *Math. Annalen*, 125 (1953) 366–393.
[10] M. ROSENBLUM, Perturbation of continuous spectrum and unitary equivalence, *Pacific Journal of Mathematics*, 7 (1957) 997–1010.
[11] G. C. ROTA, Extension theory of differential operators, *Communications in Pure and Applied Mathematics*, XI (1958) 23–65.
[12] N. ARONSZAJN, On a problem of Weyl in the theory of singular Sturm-Liouville equations, *American Journal of Mathematics*, LXXIX, (1957) 597–610.
[13] HALBERG & TAYLOR, On the spectra of linked operators, *Pacific Journal of Mathematics*, 6 (1956) 283–290.
[14] ERDELYI, et. al., *Higher Transcendental Functions*, McGraw-Hill, New York, 1953, Vol. I.
[15] HARDY, LITTLEWOOD & POLYA, *Inequalities*, Cambridge, 1952, p. 229 ff.

Massachusetts Institute of Technology

Rota's Work on Differential Operators

Bernd Schultze

The paper "Extension theory of differential operators I" was Rota's first publication based on his Ph. D. thesis. In this paper, Rota develops an extension theory for ordinary differential operators by analogy to the classical one leading to John von Neumann's extension theory for symmetric linear operators in Hilbert spaces. He considers, on arbitrary intervals I, general formally non-self-adjoint differential expressions of the form

$$\tau = \sum_{j=0}^{n} a_j(t) \frac{d^j}{dt^j}$$

and their realizations as differential operators (D-operators) in general $L_p(I)$ spaces with $1 \leq p \leq \infty$. Generalizing in a direct way methods and results of Chapter XIII of Dunford and Schwartz [3], to which he also contributed essential sections, this paper represents a new approach for non-self-adjoint problems, offering an alternative to the theory of spectral operators that was developed contemporaneously (see Dunford and Schwartz, Part III [3]) and that uses the self-adjoint results via perturbation theory.

Rota's definitions of the maximal operator $T_1(\tau, p)$ and the minimal operator $T_0(\tau, p)$ generated by τ follow the pattern of the Hilbert space case, but use the "preconjugate" operator for $p = \infty$. These definitions were adopted in a slightly generalized way in Section VI of Goldberg [4]. With these definitions, the so-called D-operators are defined as closed operators T with $T_0(\tau, p) \subset T \subset T_1(\tau, p)$. Their domains are generated by boundary values as in the self-adjoint case.

The *essential spectrum*, defined by

$$\sigma_e(\tau) := \{\lambda \in \mathbb{C} : \text{range}(\lambda I - T_0(\tau, p)) \text{ is not closed}\},$$

is the principal notion for the ensuing results in which it is mostly assumed that $\sigma_e(\tau) \neq \mathbb{C}$. In this case, the closed set $\sigma_e(\tau)$ remains invariant if in the above definition, the minimal operator $T_0(\tau, p)$ is replaced by any other D-operator generated by τ and coincides with the essential spectrum for the adjoint expression τ^*. Dissecting the interval I as in Lemma 3.5 of Rota's paper, the decomposition principle holds. Moreover, a most remarkable "spectral-mapping" theorem holds: If P is a polynomial with constant coefficients, then $\sigma_e(P(\tau)) = P(\sigma_e(\tau))$. See also Balslev and Gamelin [2].

Ignoring here the technical complications which create the possible existence of essential eigenvalues (eigenvalues of $T_0(\tau, p)$) – which do not appear if I contains one of its endpoints – the main results can be summarized as follows:

> In the complement $\mathbb{C} \backslash \sigma_e(\tau)$ in each one of the (at most countably many) connected components the nullities of τ and τ^*, i. e. $\dim N(\lambda I - T_1(\tau, p))$

and $\dim N(\lambda I - T_1(\tau^*, q))$ resp., are constant and their sum is equal to the dimension of the space of boundary values ($p^{-1} + q^{-1} = 1$). This property, as is known from the self-adjoint case, allows the assignment to each such connected component these integers as the two deficiency indices.

Furthermore a locally defined *maximal extension* T in some $\lambda \in \mathbb{C} \setminus \sigma_e(\tau)$, which is a D-operator having λ in its resolvent set, turns out to be globally maximal on the connected component containing λ with the exception of an at most countable number of eigenvalues there, having limit points in $\sigma_e(\tau)$.

With respect to several such connected components having equal deficiency indices then again global maximal extensions for the entirety of these connected components exist that are realized by the imposition of a finite set of boundary conditions. The spectrum of these restrictions of $T_1(\tau, p)$ consists of $\sigma_e(\tau)$ taken together with an at most countable number of eigenvalues all of whose limit points lie in $\sigma_e(\tau)$. This result is a really striking generalization of the self-adjoint case.

In the last section the space of boundary values is investigated more extensively and a "Fredholm-alternative-like" result is shown: for a maximal D-operator T and its adjoint T^* the eigenvalues and their multiplicities are equal.

This well-rounded theory is only lacking in interesting applications and examples. One reason for this is that in general $\sigma_e(\tau)$ and the deficiency indices are very difficult to compute. For the singular Riemann differential expression

$$\tau f = t(t-1)f'' + (at+b)f' + \frac{ct^2 + dt + e}{t(t-1)} f$$

for $t \in (-\infty, 0)$, this task is carried through in Rota's second and last paper in this area "On the spectra of singular boundary value problems." Rota proves here at first the following general sufficient criterion for $\lambda \in \mathbb{C}$ to be a point of the essential spectrum: If the solution space of either $(\lambda - \tau)f = 0$ or $(\lambda - \tau^*)f = 0$ has no norm-analytic basis at $\lambda \in \mathbb{C}$, then $\lambda \in \sigma_e(\tau)$. For $(\lambda - \tau)f = 0$ to have a norm-analytic basis at $\lambda_0 \in \mathbb{C}$ means that for every λ in a whole neighborhood of λ_0 there exists a basis of the solution space of $(\lambda - \tau)f = 0$ depending analytically on λ and for which a fixed number of elements (independent of λ) is in $L_p(I)$.

Using also the fact that the singularities of the Riemann expression are regular singular points, their deficiency indices and essential spectrum are obtained. The representation of the latter allowed the surprising conclusion that the essential spectrum can be different for different $p \in (1, \infty)$. In Agapito [1] it is shown that in the weighted Hilbert spaces $L_2{}^w(I)$ it can even depend on the weight function.

Generalizations of these results were given in Balslev and Gamelin [2] for $1 < p < \infty$ and in Goldberg [4] including $p = 1$ and $p = \infty$. They determined for $I = [1, \infty)$ the essential spectrum and the deficiency indices for the whole class of expressions with constant coefficients and for the Euler-expressions

$$\tau = \sum_{j=0}^{n} b_j t^j \frac{d^j}{dt^j}.$$

For $1 < p < \infty$, they characterized for these two classes also the relatively compact perturbations, thus significantly broadening the applicability of their results.

For not necessarily self-adjoint expressions in the Hilbert space $L_2(I)$, these results are generalized for a substantially larger class in Schultze [5], admitting polynomial coefficients together with lower degree perturbations. The essential spectrum and the deficiency indices are determined in this class by an interesting pattern generated by the degrees of these polynomials. In Schultze [6], a quite general sufficient condition for the existence of the essential spectrum is given and it is shown, by an example, how sensitive to small perturbations this subset of $\mathbb{C}$ can be.

References

[1] J. C. Agapito, *On Relatively Compact Perturbations of Special Expressions in* $L_2^w(I)$. Ph. D. Thesis, Dept. of Mathematics, University of the Philippines, Diliman, Manila, 1993.

[2] E. Balslev and T. W. Gamelin, The essential spectrum of a class of ordinary differential operators. *Pacific J. Math.* **14** (1964), 755–776.

[3] N. Dunford and J. T. Schwartz, *Linear Operators (Part I: General Theory, Part II: Spectral Theory, Part III: Spectral Operators).* Wiley Interscience, New York, 1957, 1963, 1971.

[4] S. Goldberg, *Unbounded Linear Operators.* McGraw-Hill, New York, 1966.

[5] B. Schultze, Spectral properties of not necessarily self-adjoint linear differential operators. *Adv. Math.* **83** (1990), 75–95.

[6] B. Schultze, Regions of spectral stability and instability for singular ordinary differential operators. In T. Sunada and P. W. Sy, eds., *Functional Analysis and Global Analysis (Quezon City, 1996),* Springer-Verlag, Singapore, 1997, pp. 250–257.

Sturm-Liouville Expansions

Norton Starr

This paper verified an assertion of Lord Rayleigh about the completeness of the eigenfunctions of a Sturm-Liouville system. As a key step in justifying Rayleigh's claim, Birkhoff and Rota showed in the paper "On the Completeness of Sturm-Liouville Expansions" that an orthonormal set in Hilbert space is complete provided it is suitably close to a complete orthonormal set. This latter theorem is of independent interest, and the direct proof due to Birkhoff and Rota has been further simplified in separate approaches by Urata [8] and by Tsao [7]. (See also Hochstadt [4].) Moreover, Urata noted that Nina Bari proved the completeness result [1] in 1941, a priority recognized in the second and third editions of Birkhoff and Rota's differential equations text [2,3]. Similarly, Tsao pointed out that Paley and Wiener's A. M. S. Colloquium monograph [6, p. 100] included a specialized version of Birkhoff and Rota's completeness theorem.

Bari's completeness results are also referred to by Iséki [5], who cites work by others in the Soviet Union, Sweden and the United States. See also p. 122 of Bari's obituary in *Russian Mathematical Surveys* [9], to which Urata refers.

In the *Zentralblatt* review [10] of this paper, M. S. P. Eastham remarked that "on page 840 of the paper, the relation $\lambda_k \sim k^2$, rather than $\lambda_k = O(k)^2$, is required."

References

[1] N. Bary, Sur la stabilité de la propriété d'être un système complet de fonctions. *C. R. (Doklady) Acad. Sci. USSR (N. S.),* **37** (1942), 83–87.

[2] G. Birkhoff and G.-C. Rota, *Ordinary Differential Equations*, 2nd edition, Blaisdell, Waltham, MA, 1969.

[3] G. Birkhoff and G.-C. Rota, *Ordinary Differential Equations*, 3rd edition, Wiley, New York, 1978.

[4] H. Hochstadt, On perturbations of complete sets. *SIAM Review,* **12** (1970), 103–106.

[5] K. Iséki, On complete orthonormal sets in Hilbert space. *Proc. Japan Acad.,* **33** (1957), 450–452.

[6] R. E. A. C. Paley and N. Wiener, *Fourier Transforms in the Complex Domain,* Amer. Math. Soc., New York, 1934.

[7] N. Tsao, Approximate bases in a Hilbert space. *Amer. Math. Monthly,* **75** (1968), 750.

[8] Y. Urata, A theorem of Bari on the completeness of orthonormal systems. *Proc. Japan Acad.,* **39** (1963), 160–161.

[9] *Russian Mathematical Surveys,* **17** (1962), 119–131.

[10] *Zentralblatt für Mathematik und Ihre Grenzgebiete,* **132** (1967), 55.

CHAPTER 2

Theory of Linear Operators

Rendiconti del Circolo Matematico di Palermo, Serie II
8 (1959), 182–184

NOTE ON THE INVARIANT SUBSPACES OF LINEAR OPERATORS (1)

by **G. C. Rota** (Cambridge, Massachusetts, U. S. A.)

A problem of long standing in linear analysis is whether or not every bounded linear operator in separable Hilbert space — or, more generally, in separable Banach space — admits a proper closed invariant subspace (2). The simple result proved below sheds some light on this problem. For simplicity the discussion in confined to Hilbert space; the result is easily extended to more general linear spaces.

We recall that a *closed* subspace M of separable Hilbert space H is said to be an *invariant subspace* of a bounded linear operator T in M when Tx is in M for every element x in M. If S is a second bounded linear operator in H, we say that *S is a model* for the linear operator T in H when there exist (a) an invariant subspace M of S and (b) a bounded one-to-one linear operator I_T in H whose range is the closed subspace M, such that T coincides with the operator $I_T^{-1} S I_T$. Several special cases of the notion of «model» have recently received attention; for example, necessary and sufficient conditions are known that T shall have a normal model (3). Intuitively, an operator S is a model for T when the restriction of S to some one of its invariant subspaces gives an operator *similar* to T.

---

(1) This work was partially sponsored by the Office of Ordinance Research, Contract No. Da 19020 ORD-3778. I wish to thank Professor J. Wermer for a very enlightening conversation on the present topic.

(2) In an inseparable Banach space the problem is trivial.

(3) For a thorough treatment, see the appendix to the second edition of (3).

Among the obvious properties of the relation « S is a model for T » we shall use the following below: if S is a model for T and if the bounded linear operator R is similar to T, then S is a model for R.

Our result is the following:

Theorem. There exists a « universal model » for all bounded linear operators in separable Hilbert space H. That is, there exists a bounded linear operator in H which is a model for a multiple of *every* bounded linear operator in H.

Proof. Consider the « tensor product » space H' consisting of all sequences $x = (x_1, x_2, \ldots)$ of elements x_k of H which are square-summable. Clearly H' is a Hilbert space with the inner product

$$(x, y) = \sum_{k=1}^{\infty} (x_k, y_k).$$

If H is separable, so is H', so that H' is isomorphic to H. We thus identify H' with H; in particular, every bounded linear operator in H' can be considered in a natural way as an operator in H.

We define the « shift » operator U in H' by the equation $Ux = y$, where $x = (x_1, x_2, x_3, \ldots)$ and $y = (x_2, x_3, \ldots)$. Clearly the operator U is bounded, in fact of norm one.

We say that U is the desired universal operator. Indeed, let T be an arbitrary bounded linear operator. Since any such operator has the same invariant subspaces as a linear operator of norm less than one (simply multiply the operator by the inverse of the norm, if need be), we may assume that $\|T\| < 1$. Consider the set $M = M(T)$ of all sequences of the form

$$(x) = (x, Tx, T^2x, \ldots), \qquad x \in H.$$

The inequality

$$\|x\|^2 = \sum_{k=0}^{\infty} \|T^k x\|^2 \leqslant \|x\| \sum_{k=0}^{\infty} \|T^k\|^2 \leqslant \|x\|/(1 - \|T\|^2) \tag{1}$$

shows that M is a subset of H', indeed a *subspace* of H'. M is in fact a *closed* subspace of H'. To see this, let (x_n) be a Cauchy sequence in M. Then the « first coordinates » x_n form a Cauchy sequence in H, wich converges to x in H, say, and the inequality

$$\|x_n - x\|^2 = \sum_{k=0}^{\infty} \|T^k x_n - T^k x\|^2 \leqslant \frac{\|x_n - x\|^2}{1 - \|T\|^2}$$

shows that (x_n) converges to (x) in H'.

Now, define the operator I_T by the equation $I_T x = (x)$. From (1) it follows that I_T is bounded, one-to-one and onto M. Finally, we have the identity

$$I_T^{-1} U I_T x = I_T^{-1} U(x) = I_T^{-1}(Tx, T^2 x, \ldots) = Tx,$$

and the proof is complete.

It is hardly surprising that the operator U is not easily amenable to «spectral» study. If the Hilbert space H is represented as a space l_2 of sequences $(a_1, a_2, \ldots)$ of complex numbers a_k, and if S is the shift operator in l_2 analogous to U, then $U = S \otimes I$, where I is the identity operator in H. The operator S has been completely studied by Beurling; essential new difficulties arise from tensoring it with the identity operator, even when I acts only in a finite-dimensional space.

The preceding result has an interesting interpretation in lattice-theoretic terms:

Corollary. There exists a bounded linear operator S in separable Hilbert space such that the lattice of closed invariant subspaces of *every* bounded linear operator is an *ideal* (in the lattice sense) in the lattice of closed invariant subspaces of S.

Cambridge, (Massachusetts, U. S. A.), June 1959.

REFERENCES

(1) Dunford and Schwartz, *Linear Operators,* vol. I, New York, 1958.

(2) Riesz and Nagy, *Leçons d'Analyse Fonctionelle,* Budapest, 1952.

(3) A. Beurling, *On two problems concerning linear transformations in Hilbert space.* Acta Mathematica, vol. 81, pp. 239-255.

COMMUNICATIONS ON PURE AND APPLIED MATHEMATICS, VOL. XIII, 469–472 (1960)

On Models for Linear Operators*

GIAN-CARLO ROTA
Massachusetts Institute of Technology

§ 1. The notion of a *model* for a linear operator in Hilbert space has been introduced in the author's note [3]. In this note we shall give some remarkable applications which can be obtained from this notion, combined with some exceedingly simple imbedding techniques. Contrary to [3], Hilbert space is not assumed to be separable here.

We recall that a bounded linear operator T in[1] Hilbert space H is said to be *similar* to a second linear operator T_1 if one can find a bounded linear operator V with a bounded inverse in H such that $T = V^{-1}T_1V$. A bounded linear operator T is said to be a *contraction* operator when $\|T\| \leqq 1$. A *closed* subspace M of Hilbert space H is said to be an *invariant subspace* of a bounded linear operator T in M if Tx is in M for every element x in M. If S is a second bounded linear operator in H, we say that *S is a model for the linear operator T in H* when there exist (a) an invariant subspace M of S and (b) a bounded one-to-one linear operator I_T in H whose range is the closed subspace M, such that T coincides with the operator $I_T^{-1}SI_T$. Several special cases of the notion of "model" have recently received attention.[2] Intuitively, an operator S is a model for T when the restriction of S to some one of its invariant subspaces gives an operator *similar* to T.

Among the obvious properties of the relation "S is a model for T" we shall use the following: if S is a model for T and if the bounded linear operator R is similar to T, then S is a model for R.

§ 2. Our main result is the following[3]:

THEOREM 1. *There exists a universal model for all bounded linear operators in H whose spectra lie in the interior of the unit disc. More precisely, there exists*

*The present work was begun while the author was a Temporary Member at the Institute of Mathematical Sciences. It was later sponsored by Contract No. 7667, Office of Naval Research, at the Massachusetts Institute of Technology. Reproduction in whole and in part permitted for any purpose of the United States Government.

I wish to thank Dr. Adam Koranyi of Harvard University for some stimulating conversations on the present topic.

[1]That is, mapping H into itself.

[2]See the reference in [3]. For example, necessary and sufficient conditions are known that T shall have a *normal* model.

[3]This result greatly generalizes the one obtained in the author's note quoted under [3].

a single bounded linear operator U in H which is a contraction operator and which is a model for all bounded linear operators whose spectra lie in the interior of the unit disc.

Proof: Consider the "tensor product" space H' consisting of all sequences $\mathbf{x} = (x_1, x_2, \cdots)$ of elements x_k of H which are square-summable in the norm. Clearly H' is a Hilbert space with the inner product

$$(\mathbf{x}, \mathbf{y}) = \sum_{k=1}^{\infty} (x_k, y_k).$$

If the dimension of H is the cardinal number $\alpha \geqq \aleph_0$, then the dimension of H' is the cardinal number $\alpha \cdot \aleph_0 = \alpha$. Since H and H' have the same dimension, we can identify H' with H; in particular, every bounded linear operator in H' can be considered in a natural way as an operator in H.

We define the "shift" operator U in H' by the equation $U\mathbf{x} = \mathbf{y}$, where $\mathbf{x} = (x_1, x_2, x_3, \cdots)$ and $\mathbf{y} = (x_2, x_3, \cdots)$. Clearly the operator U is bounded, in fact it is a contraction operator. We say that U is the desired universal operator. Indeed, let T be an arbitrary bounded linear operator whose spectrum lies inside the unit disc. It is well-known[4] that

$$\lim_{n\to\infty} ||T^n||^{1/n} = r < 1.$$

Letting

$$f(z) = \sum_{n=0}^{\infty} ||T^n|| z^n,$$

we see that the function f is holomorphic for $|z| < r^{-1}$, that is, in a disc properly including the unit disc. It follows that the integral

$$\int_0^{2\pi} f(e^{i\theta})\overline{f(e^{i\theta})}d\theta$$

is finite. But such an integral is equal to a multiple of $\sum_{n=0}^{\infty} ||T^n||^2$; we conclude that this sum is also finite. Let

$$K(T) = \sum_{n=0}^{\infty} ||T^n||^2.$$

Consider the set $M = M(T)$ of all sequences of the form

$$\mathbf{x} = (x, Tx, T^2x, \cdots), \qquad x \in H.$$

The inequality

$$(*) \qquad ||\mathbf{x}||^2 = \sum_{k=0}^{\infty} ||T^k x||^2 \leqq ||x||^2 \sum_{k=0}^{\infty} ||T^k||^2 = K(T)||x||^2$$

shows that M is a subset of H', indeed a subspace of H'. In fact M is a closed

[4] Cf. [1], Ch. VII, § 3.5.

subspace of H'. To see this, let $(\mathbf{x}_n)$ be a Cauchy sequence in M. Then the "first coordinates" (x_n) form a Cauchy sequence in M, which converges to x in M, say, and the inequality

$$||\mathbf{x}_n-\mathbf{x}||^2 = \sum_{k=0}^{\infty} ||T^k x_n - T^k x||^2$$
$$\leqq ||x_n-x||^2 \sum_{k=0}^{\infty} ||T^k||^2 = K(T)||x_n-x||^2$$

shows that $(\mathbf{x}_n)$ converges to $\mathbf{x}$ in M. Now, define the operator I_T by the equation $I_T x = \mathbf{x}$. From (*) it follows that I_T is bounded, one-to-one and onto M. By the closed-graph theorem the inverse I_T^{-1} exists and is bounded. Finally, we have the identity

$$I_T^{-1} U I_T x = I_T^{-1}(Tx, T^2x, \cdots) = Tx,$$

which shows that U is a model for T.

What is remarkable in this theorem is that the construction of a universal model can be carried out within a Hilbert space *of the same infinite dimension*. It would be easy to construct models by "adding up" sufficiently many replicas of Hilbert spaces, one for each operator. Thus the theorem can be interpreted as a restriction on the number of equivalence classes of linear operators under similarity.

§ 3. As a first application of Theorem 1 we prove the following

THEOREM 2. *Every bounded linear operator in Hilbert space whose spectrum lies in the interior of the unit disc is similar to an operator of norm less than one (that is, to a proper contraction operator).*

Proof: Evidently it is sufficient to prove that such an operator T is similar to an operator T_1 of norm less than *or equal to* one. The general result follows from this by applying the weaker result to the operator $(1+\varepsilon)T$ for sufficiently small ε. To prove the weaker statement, it suffices to remark that the restriction of the universal operator S in Theorem 1 to any one of its invariant subspaces is a contraction operator.

As a second application we prove a result established by altogether different techniques by Nagy (cf. [2]). The present proof has the merit of brevity.

THEOREM 3. *Every compact operator in Hilbert space whose spectrum lies in the closed unit disc and the norms of whose iterates are uniformly bounded* ($||T^n|| \leqq M$, $n = 1, 2, 3, \cdots$) *is similar to a contraction operator.*

Proof: The idea is that such an operator T has at most a finite number of eigenvalues on the unit circumference. In view of the assumption, it is easy to see (by a "finite-dimensional" argument!) that these are simple.

By taking the projection E on the span of the eigenvectors belonging to such eigenvalues, we are reduced to considering the operator $(I-E)T(I-E)$ on the subspace $(I-E)H$. But this operator satisfies the hypothesis of Theorem 2.

The following is a rather unexpected consequence of Theorem 2.

COROLLARY. *Every quasi-nilpotent bounded linear operator is similar to a bounded linear operator of arbitrarily small norm.*

Proof: Let N be such an operator ($\lim_{n\to\infty} ||N^n||^{1/n} = 0$). Then for any $\varepsilon > 0$ the operator εT satisfies the hypothesis of Theorem 2 and is therefore similar to a contraction S. But then N is similar to the operator εS which is of norm ε.

Bibliography

[1] Dunford, N., and Schwartz, J. T., *Linear Operators*, Vol. 1, Interscience Publishers, New York, 1959.

[2] Sz.-Nagy, B., *Completely continuous operators with uniformly bounded iterates*, Publ. Math. Inst. Hungarian Acad. Sci., Vol. 4, 1959, pp. 89–93.

[3] Rota, G. C., *Note on the invariant subspaces of linear operators*, Rend. Circ. Mat. Palermo, Ser. II, Vol. 7, 1959, pp. 182–184.

Received December, 1959.

KONINKL. NEDERL. AKADEMIE VAN WETENSCHAPPEN – AMSTERDAM
Reprinted from Proceedings, Series A, **63**, No. 4 and Indag. Math., **22**, No. 4, 1960

MATHEMATICS

A NOTE ON THE JOINT SPECTRAL RADIUS

BY

GIAN-CARLO ROTA AND W. GILBERT STRANG [1])

(Communicated by Prof. H. FREUDENTHAL at the meeting of April 30, 1960)

The notion of joint spectral radius of a set of elements of a normed algebra, introduced below, was obtained in the course of some work in matrix theory. It was later noticed that the same considerations are valid in any normed algebra, irrespective of dimension. The notion seems to be useful enough in certain contexts to warrant the following elementary discussion.

Let $\mathbf{B}$ be any bounded subset of the normed algebra $\mathfrak{A}$ with identity e. Let P_n be the set of all elements of $\mathfrak{A}$ which are the products of n elements of $\mathbf{B}$. The *joint spectral radius* of the set $\mathbf{B}$ is defined to be the non-negative number

$$r(\mathbf{B}) = \lim_{n\to\infty} \sup_{T\in P_n} \|T\|^{1/n}.$$

That this number is well-defined follows just as in the by now classical case of the spectral radius of a single element, to which this notion reduces when the set $\mathbf{B}$ consists of a single element. Indeed, notice that $\log \sup_{T\in P_n} \|T\|$ is a subadditive function of n, and apply a well known result of Polya and Szego (cf. HILLE [1], p. 124, Lemma 4.7.1). Intuitively, the joint spectral radius is related to the commutativity of the elements of $\mathbf{B}$.

We prove below a result connecting the joint spectral radius with the *equivalent norms* the algebra may be provided with. Let $\mathfrak{N}$ be the family of norms on $\mathfrak{A}$ under which $\mathfrak{A}$ is a normed algebra, equivalent to the given norm on $\mathfrak{A}$ in the sense that $k\|T\| \leqslant N(T) \leqslant k^{-1}\|T\|$ for some constant $k \neq 0$ and for all T in $\mathfrak{A}$. Note that all the norms in $\mathfrak{N}$ are assumed to be multiplicative.

Proposition 1. $$r(\mathbf{B}) = \inf_{N\in\mathfrak{N}} \sup_{T\in\mathbf{B}} N(T).$$

In other words, the joint spectral radius of a bounded set $\mathbf{B}$ is equal to the infimum of the "norm" of the set, ranging over all equivalent norms. This result will be a consequence of the following

[1]) Work by G.-C. ROTA done by Contract with the Office of Naval Research and by W. G. STRANG under Air Force Contract AF-49(638)-42, monitored by the Air Force Office of Scientific Research of the Air Research and Development Command, both at the Massachusetts Institute of Technology, Cambridge, Mass., U.S.A.

Lemma. Let $\mathbf{S}$ be any subset of $\mathfrak{A}$. A necessary and sufficient condition that there exist an equivalent norm N such that $N(T) \leqslant 1$ for all $T \in \mathbf{S}$ is that there exist a constant K such that $\|T_1 T_2 \ldots T_n\| \leqslant K$ for any $T_1, T_2, \ldots, T_n$ in $\mathbf{S}$ and for all n.

Proof. The necessity of the condition is trivial. To prove that it is sufficient, define the norms N_1 and N_2 as

$$N_1(T) = \sup_{i,n \geqslant 0} \|T_1 T_2 \ldots T_i T T_{i+1} \ldots T_n\|,$$

where the supremum is taken over all products of elements T_j of $\mathbf{S}$, including the empty product, which gives $\|T\|$ on the right side; and let

$$N_2(T) = \sup_{\substack{T_1 \in \mathfrak{A} \\ T_1 \neq 0}} [N_1(TT_1)/N_1(T_1)].$$

Obviously both N_1 and N_2 are vector norms. We must verify that $\mathfrak{A}$ becomes a normed algebra under either of them. To see this, we have first

$$\|T_1 T_2 \ldots T_i T T_0 T_{i+1} \ldots T_n\| \leqslant \|T_1 T_2 \ldots T_i T\| \, \|T_0 T_{i+1} \ldots T_n\| \leqslant$$
$$\leqslant N_1(T) N_1(T_0),$$

hence taking the sup on the left, $N_1(TT_0) \leqslant N_1(T) N_1(T_0)$ follows. Next, $N_1(TT_0T_1)/N_1(T_1) = [N_1(TT_0T_1)/N_1(T_0T_1)][N_1(T_0T_1)/N_1(T_1)] \leqslant N_2(T)N_2(T_0)$ for $T \neq 0$. Hence taking the sup on the left it again follows that $N_2(TT_0) \leqslant N_2(T)N_2(T_0)$.

Secondly, N_2 is an equivalent norm to the given norm on the algebra $\mathfrak{A}$. Indeed,

$$N_2(T) \geqslant \frac{N_1(T)}{N_1(e)} \geqslant \frac{\|T\|}{N_1(e)} \quad \text{for all } T,$$

Where e is the identity element, and $N_1(TT_1)/N_1(T_1) \leqslant N_1(T) \leqslant \|T\| (1+K^2)$, hence taking the sup on the left, $N_2(T) \leqslant \|T\| (1+K^2)$. Lastly, since it is clear that $N_1(TT_1) \leqslant N_1(T_1)$ for T in $\mathbf{S}$, we have $N_2(T) \leqslant 1$.

Now to the proof of Proposition 1. Let N be an equivalent norm ($N \in \mathfrak{N}$). For T in P_n, so that $T = T_1 T_2 \ldots T_n$, we have

$$N(T)^{1/n} \leqslant (N(T_1) \ldots N(N_n))^{1/n} \leqslant \sup_{T \in \mathbf{B}} N(T);$$

since $N(T)^{1/n} \geqslant (k\|T\|)^{1/n}$ for some k, it follows taking limits that

$$r(\mathbf{B}) \leqslant \sup_{T \in \mathbf{B}} N(T);$$

hence

$$r(\mathbf{B}) \leqslant \inf_{N \in \mathfrak{N}} \sup_{T \in \mathbf{B}} N(T).$$

On the other hand, for any $\varepsilon > 0$ consider the set $\mathbf{S} = \{T/(r(\mathbf{B}) + \varepsilon) | T \in \mathbf{B}\}$. Then $r(\mathbf{S}) < 1$, and $\mathbf{S}$ satisfies the condition of the preceding Lemma. We can therefore find a norm N in $\mathfrak{N}$ such that $N(T_0) \leqslant 1$ for T_0 in $\mathbf{S}$, that is, $N(T) \leqslant r(\mathbf{B}) + \varepsilon$ for T in $\mathbf{B}$. This completes the proof.

Corollary. A necessary and sufficient condition that there exist an equivalent norm N such that $\sup_{T \in B} N(T) = r(\mathbf{B})$ is that

$$\sup_{T \in P_n} \|T\| = 0(r(\mathbf{B})^n) \quad \text{as} \quad n \to \infty .$$

Proof. The condition is obviously necessary. Conversely, if it is satisfied, then the set $\mathbf{S} = \{T/r(\mathbf{B}) | T \in \mathbf{B}\}$ satisfies the hypothesis of the Lemma. Therefore any norm N provided by the Lemma will do.

In the special case when $\mathfrak{A}$ is a subalgebra of the algebra of bounded operators on a Banach space X, there is an alternative construction for the norms. Let

$$M(x) = \sup_{n \geqslant 0} \|T_1 T_2 \dots T_n x\|, \quad T_j \in \mathbf{S}, \; x \in X$$

and let

$$N_2'(T) = \sup_{x \neq 0} M(Tx)/M(x).$$

For a single operator on Hilbert space a much stronger result can be established, namely

$$r(\mathbf{B}) = \inf \|S^{-1}TS\|$$

where the inf is taken over all invertible bounded operators S. The proof of this result is an immediate Corollary of the main Theorem in [3].

A second application of the above Lemma, of independent interest, is the following: given a norm N' on a subalgebra $\mathfrak{A}'$ of a normed alge a $\mathfrak{A}$, is it possible to extend it to the entire algebra? In particular, is it possible to find an equivalent multiplicative norm N defined on all of $\mathfrak{A}$ such that $N(T) \leqslant N'(T)$ for T in $\mathfrak{A}'$? The following result provides an answer.

Proposition 2. Let N' be a norm on the subalgebra $\mathfrak{A}'$ of $\mathfrak{A}$ which on $\mathfrak{A}'$ is equivalent to the restriction of the given norm on $\mathfrak{A}$. Then there is an equivalent norm $N \in \mathfrak{N}$ on $\mathfrak{A}$ such that $N(T) \leqslant N'(T)$ for T in $\mathfrak{A}'$.

The proof is immediate from the Lemma. Consider the set

$$\mathbf{S} = \{T | T \in \mathfrak{A}', \; N'(T) = 1\}$$

and apply the conclusion of the Lemma.

Massachusetts Institute of Technology
Cambridge, Mass. U.S.A.

BIBLIOGRAPHY

1. Dunford, N. and J. T. Schwartz, Linear Operators, Part I, Interscience, New York, 1959.
2. Hille, F. and R. S. Phillips, Functional Analysis and Semigroups, AMS, 1958.
3. Rota, G.-C., On Models for Linear Operators, Communications in Pure and Applied Mathematics, (to appear) (1960).

Notices of the American Mathematical Society 8(1961), 272–273.

Abstract 61T-125. G.-C. Rota: Multiplicative extensions of positive linear operators.

Let (S, Σ, μ) be a probability space, and let T be a linear operator defined on bounded real-valued measurable functions $f(s)$, $s \in S$ such that $Tf \geq 0$, and $T1 = 1$, where 1 is the function identically equal to one. *Theorem.* There exist: (a) a σ-field Σ' containing Σ and a measure μ defined on Σ' and extending μ; (b) a measurable transformation $\phi : S \to S$ such that $\phi^{-1}(E) \in \Sigma'$ for $E \in \Sigma'$; such that for all integers $n \geq 0$ and for all $f \in L_\infty(S, \Sigma, \mu)$ we have $T^n f = EV^n f$, where V is defined by $Vg(S) = g(\phi(s))$ and E is the conditional expectation operator (projection) defined on $L_\infty(S, \Sigma', \mu)$ and projecting onto $L_\infty(S, \Sigma, \mu)$. Applications are made to probability theory and ergodic theory. In particular, a proof of the ergodic theorem for general Markov processes of E. Hopf is obtained as a simple consequence of the classical ergodic theorem of G. D. Birkhoff.

(Received March 21, 1961.)

Dilations of Positive Operators

GIAN-CARLO ROTA

Let (S, Σ, μ) be a measure space, and let P be a linear order-preserving operator defined on $L_1(S, \Sigma, \mu)$ and $L_\infty(S, \Sigma, \mu)$, with the following properties:

(a) $\int |Pf| d\mu \leq \int |f| d\mu$,

(b) $P1 = 1$,

(c) P^* is also defined in $L_1(S, \Sigma, \mu)$.

We are concerned with the spectral theory of P on the unit circle. To this end, a useful technique is the study of the *dilations* of P into a measure-preserving transformation, by a method adapted from the theory of stochastic processes (cf. Doob, *Stochastic Processes*, and Nelson, *The Adjoint Markov Process*), which generalizes the Halmos-Nagy theory of dilations of contractive operators in Hilbert space. Specifically, one can embed (S, Σ, μ) into a measure space $(S, \bar{\Sigma}, \mu)$ (where $\bar{\Sigma} \supset \Sigma$), and obtain a measure-preserving transformation Q on $(S, \bar{\Sigma}, \mu)$ such that

(a) $P^n = EQ^n$ and $P^{*n} = EQ^{-n}$ for $n \geq 0$, where E is the conditional expectation operator of $L_p(S, \bar{\Sigma}, \mu)$ onto $L_p(S, \Sigma, \mu)$.

(b) $E(f_1 Q f_2 Q^2 f_3 \ldots Q^n f_{n+1}) = f_1 P[f_2 P[f_3 P \ldots P f_{n+1}] \ldots]$.

(c) If $QF = \lambda F$ for some $F \in L_p(S, \bar{\Sigma}, \mu)$, and for $|\lambda| = 1, p < \infty$, then $F \in L_p(S, \Sigma, \mu)$ and $PF = \lambda F$.

(d) If $Pf = \lambda f$ for $|\lambda| = 1$ and $f \in L_p(S, \Sigma, \mu)$, $p < \infty$, then $Qf = \lambda f$.

Using this dilation, one can establish the following properties of P:

(1) (Generalized Perron-Frobenius theorem) If $f \in L_p(S, \Sigma, \mu)$, $f = |f|g$, with $|g| = 1$, and $Pf = \lambda f$, with $|\lambda| = 1$, then

$$P(|f|g^n) = \lambda^n |f| g^n.$$

Unpublished manuscript, Massachusetts Institute of Technology, no date, circulated around 1961.

(2) If (S, Σ, μ) is a normal probability space, then P can be written as a direct integral of ergodic operators with the same properties as P (as defined above).

(P is *ergodic* if $Pf = f$ implies $f = 1$.)

(3) Call an ergodic P of the type being considered a *diffusion* operator, when for no non-trivial set $e \in \Sigma$ is it true that

$$\|P^n \chi_e\|_1 = \|P^{*n} \chi_e\|_1 = \mu(e).$$

If (S, Σ, μ) is a probability space, then a diffusion operator can be shown to be *mixing*, that is,

$$(\chi_g, P^n \chi_e) \to \mu(g)\mu(e) \quad \text{for } e, g \in \Sigma.$$

(4) Every P can be represented in the form $P = RP'$, where R is a measure-preserving transformation and P' is a direct integral of diffusion operators, in other words as a semidirect product of a measure-preserving transformation and diffusion operators.

Problems

(I) Given a sequence of operators P_n, under what necessary and sufficient conditions can we assert that $P_n = EQ^n$ for some measure-preserving transformation Q?

(II) Suppose that P is as above, $P1 = 1$, 1 is a simple eigenvalue (P is ergodic) and the rest of the spectrum of P lies in the *interior* of the unit disk. Is the dilation of Q conjugate to a shift operator? (This would give an analog of Schreiber's theorem.)

(III) When is Q mixing of "higher order", at least when restricted to sets of Σ?

(IV) Can we formulate an analogous statement for martingales?

A Method for the Study of Positive Operators

GIAN-CARLO ROTA

The notions of dilation and compression of linear operators, introduced by Halmos [3], have been investigated in the case of Hilbert space (see [8] for a summary of results). Here we investigate a special type of dilation of positive (order preserving) operators acting on $L_\infty = L_\infty(S, \Sigma, \mu)$, where $\mu(S) = 1$. Such a dilation will be called a *strict* dilation, and is defined as follows. Let V be a linear operator on L_∞, let T be a linear operator in $L_\infty(S, \Sigma', \mu)$, where Σ' is a sub-σ-field of Σ, and let E be the *conditional expectation* of $L_\infty(S, \Sigma, \mu)$ onto $L_\infty(S, \Sigma', \mu)$. We say that V is a *strict dilation* of T when $EVf = Tf$ for all f in $L_\infty(S, \Sigma', \mu)$. The term is suggested by the familiar duality between "strict sense" and "wide sense" theorems in probability theory. (For a study of conditional expectation, see [7] or [9]). In the following we shall assume that the probability space (S, Σ', μ) is separable.

A particular class of linear operators in L_∞ is that of *measurable transformations*, namely, linear operators of the form $Vf(s) = f(\phi(s))$, where $\phi : S \to S$ is a transformation of the set S transforming measurable sets into measurable sets, and such that $\mu(\phi(E)) = 0$ if $\mu(E) = 0$ for a set $E \in \Sigma$. There will be no confusion in denoting by the term "measurable transformation" both ϕ and V.

We are interested in the following question: which positive operators in L_∞ have strict dilations? The answer is given by the following.

THEOREM 1. Let T be a positive operator in the probability space $L_\infty(S, \Sigma', \mu)$ such that $T1 = 1$, where 1 is the function identically equal to one. Then there exists a strict dilation V of T which is a measurable transformation, and such that

$$EV^n f = T^n f, \quad n = 0, 1, 2, \ldots, \quad f \in L_\infty(S, \Sigma', \mu).$$

Sketch of proof. One begins by proving the following lemma, which is closely related to classical Daniell-Kolmogorov type consistency theorems and to the moment problem, but whose uniqueness assertion will be needed.

Let q be a real-valued function defined for each n on ordered n-tuples of pairs (g, i), where $g \in G$ and i is a non-negative integer, thus:

$$q\begin{pmatrix} g_1 & g_2 & \cdots & g_n \\ i_1 & i_2 & \cdots & i_n \end{pmatrix}.$$

Here G can be an arbitrary set.

Unpublished manuscript, Massachusetts Institute of Technology, Spring, 1961.

For every polynomial

$$p(x_1, \ldots, x_n) = \sum a_{i_1 \ldots i_n} x_1^{i_1} \cdots x_n^{i_n},$$

we set

$$p(q; g_1, \ldots, g_n) = \sum_{i_1=1}^{N_1} \cdots \sum_{i_n=1}^{N_n} a_{i_1 \ldots i_n} q \begin{pmatrix} g_1 & g_2 & \cdots & g_n \\ i_1 & i_2 & \ldots & i_n \end{pmatrix}.$$

LEMMA. Suppose that

(a) For fixed n, q is invariant under any permutation of the pairs $\begin{pmatrix} g_k \\ n_k \end{pmatrix}$.

(b) For all n,

$$q \begin{pmatrix} g_1 & \cdots & g_{n-1} & g_n \\ i_1 & \ldots & i_{n-1} & 0 \end{pmatrix} = q \begin{pmatrix} g_1 & \cdots & g_{n-1} \\ i_1 & \ldots & i_{n-1} \end{pmatrix}.$$

(c) If $p \geq 0$ is any non-negative-valued polynomial, then $p(q; g_1, \ldots, g_n) \geq 0$.

(d) $\limsup_{n \to \infty} \begin{pmatrix} g \\ 2n \end{pmatrix}^{1/(2n)} < \infty$ for every $g \in G$.

(e) $q \begin{pmatrix} g_1 & g_2 & \cdots & g_n \\ 0 & 0 & \ldots & 0 \end{pmatrix} = 1$ for all n and all g_k.

Then there exists a probability space (S, Σ, μ), and for each $g \in G$ a bounded real-valued measurable function $g' \in L_\infty(S, \Sigma, \mu)$ such that

$$q \begin{pmatrix} g_1 & g_2 & \cdots & g_n \\ i_1 & i_2 & \ldots & i_n \end{pmatrix} = \int_S g_1'(s)^{i_1} g_2'(s)^{i_2} \cdots g_n'(s)^{i_n} \mu(ds).$$

The structure consisting of the space (S, Σ, μ), and the mapping $g \to g'$ is unique within isomorphism of measure spaces, provided it is "minimal" in the obvious sense.

The proof of this lemma will be omitted, as it can be carried out by standard techniques. The uniqueness assertion can be made to follow from a general uniqueness theorem for the moment problem, due to Haviland (cf. [11]).

Given the lemma, the theorem is obtained by choosing a non-negative function f such that the L_1-closure of the polynomials $\sum c_i f^i$ is the entire space $L_1(S, \Sigma', \mu)$ (such a function always exists), letting G be the set of non-negative integers, and setting

$$q\begin{pmatrix} n_1 & n_2 & .. & .. & n_k \\ i_1 & i_2 & .. & .. & i_k \end{pmatrix} = \int_S T^{n_1}(f^{i_1})T^{n_2}(f^{i_2})\cdots T^{n_k}(f^{i_k})\mu(ds).$$

It is then shown that q, thus defined, satisfies the hypotheses of the lemma. Let g_n $(n = 0, 1, 2, \ldots)$ be the function corresponding to $q\begin{pmatrix} n \\ 1 \end{pmatrix}$ as in the lemma. If $p = p(g_0, g_1, \ldots, g_n)$ is a polynomial in $g_0, g_1, \ldots, g_n$, define V_p to be the function $p(g_1, g_2 \ldots, g_{n+1})$. Clearly V is multiplicative $(V(fg) = VfVg)$ and $V1 = 1$, hence V is positive, and can be extended to a measurable transformation of $L_\infty(S, \Sigma, \mu)$. Let Σ' be the smallest σ-field in which the function g_0 is measurable. Then by mapping $p(f)$ into $p(g_0)$ and closing in L_1, we obtain an isomorphism of the given probability space with $L_\infty(S, \Sigma', \mu)$. We shall consider this identification as made. To see that $EV^n f = T^n f$ for $f \in L_\infty(S, \Sigma', \mu)$, it suffices to verify that

$$\int gV^n f = \int gT^n f$$

for all $g \in L_\infty(S, \Sigma', \mu)$. But this identity follows immediately from the definition of V. □

The result has several applications to the spectral theory of positive operators. One type of application consists in exploiting the fact that bounded almost everywhere convergence is preserved under conditional expectation. In view of this, it is to be expected that theorems related to the almost everywhere convergence of ergodic averages of $n^{-1}(I + T + T^2 + \ldots + T^{n-1})f$ can be inferred from corresponding convergence statements for $n^{-1}(I + V + V^2 + \ldots + V^{n-1})f$. This leads to simplifications of some existing ergodic theorems and to some novel convergence results. These results will be given in the detailed exposition of the present work.

A second type of application of Theorem 1 consists in the possibility it affords of extending the classical Frobenius-Perron theory of matrices (see e.g. Gantmacher, [2]) with non-negative elements to arbitrary order-preserving operators. It has been conjectured for a long time that many of these properties should hold irrespective of dimensions; however, the methods employed so far (see for an excellent survey, Karlin [5]) seem too restrictive. Theorem 1 gives a strong technique for extending matrix methods to arbitrary operators. To illustrate this technique, we give below the proofs of a result conjectured by Karlin [5].

PROPOSITION. Assume that $Tf \geq 0$ if $f \geq 0$, $T1 = 1$ and $\int |Tf| \leq \int |f|$ for $f \in L_\infty(S, \Sigma', \mu)$. Then, if $Tf = f$ for some f, and Σ'' is the smallest σ-field in which f is measurable, every Σ''-measurable function g is also fixed under T. In particular, to every f such that $Tf = f$ there corresponds a non-negative g whose support coincides with the support of f, for which $Tg = g$.

Proof. We prefix the following remark on conditional expectation. It is well known (cf. Loève, [7]) that a conditional expectation satisfies all the convexity inequalities satisfied by ordinary definite integrals, or "expectations". In particular, we have Schwarz's inequality, namely, $|E(fg)|^2 \leq E|f|^2 E|g|^2$. It can be easily shown (cf. Loève, [7]) that if $g \neq 0$, *equality* holds in Schwarz's inequality if and only if $f = \alpha g$, where α is a non-zero function such that $E\alpha = \alpha$. In particular, if $|Ef|^2 = E|f|^2$, then $f = Ef$ a.e.

We first show that from $T \geq 0$ and $Tf = f$ there follows $Vf = f$. Indeed, $T(f^2) = EV(f^2) = E(Vf)^2$, since V is multiplicative, and, since conditional expectation satisfies the analogue of Schwarz's inequality, $E\,(Vf)^2 \geq (EVf)^2$. We therefore obtain $T(f^2) \geq (Tf)^2 = f^2$. But $\int T(f^2) \leq \int f^2$; hence $Tf^2 = f^2$. Since equality obtains in Schwarz's inequality, it also follows that $EVf = Vf$. This in turn implies that if $Tf = f$, then $Vf = f$. Polarizing, we infer that if $Tf = f$, $Tg = g$, and f, g are non-negative, then $T(fg) = fg$. Next, we show that, if $Tf = f$, then $T|f| = |f|$. In fact, we prove the following more general statement, which will be used in the proof of the following theorem: if $Tf = \alpha f$ with $|\alpha| = 1$, then $T|f| = |f|$.

Indeed, from $|EVf| \leq E|Vf| = EV|f|$ (since E is a conditional expectation) it follows that $EV|f| \geq |f|$, or $T|f| \geq |f|$. But $\int T|f| \leq \int |f|$; hence as before, $T|f| = |f|$. From this it follows easily that if $Tf = f$ and $Tg = g$ (f and g not necessarily positive) then $T(fg) = fg$.

The statement now follows by noticing that T is weak $*$-continuous, and hence any function in the weak $*$-closure of the algebra of fixed functions is again left fixed by T. But, by a result due to Segal [10], such a weak $*$-closure is exactly a set of the form $L_\infty(S, \Sigma'', \mu)$, for some σ-field $\Sigma'' \subset \Sigma'$. This completes the proof.

As an application of the preceding proposition, we obtain the following extension of a classical theorem proved by Frobenius for matrices with non-negative elements.

THEOREM 2. Let T satisfy the hypotheses of the preceding proposition. If $Tf = \alpha f$ for $|\alpha| = 1$, then $\alpha^2, \alpha^3, \ldots$ are eigenvalues of T. If $f = |f|g$, with $|g| = 1$, then an eigenfunction belonging to the eigenvalue α^n is $|f|g^n$.

Sketch of proof. It has been shown above that if $Tf = f$ then $Vf = f$. In view of the above, we can now show that, if $Tf = f$, then $T(fg) = fTg$. Indeed, $T(fg) = EV(fg) = E(VfVg) = E(fVg) = fEVg = fTg$. Therefore we can express T as a direct integral of operators satisfying the same conditions and the additional condition that if $Tf = f$, then $f = 1$, and it suffices to prove the theorem with this additional assumption.

Now let $Tf = \alpha f$, $|\alpha| = 1$. Then $T|f| = |f|$, hence $|f| = 1$. Consequently, $|Vf|^2 = Vf V\bar{f} = V(f\bar{f}) = V1 = 1$. The theorem will be proved if we can show that $Vf = \alpha f$, for then

$$Tf^2 = EVf^2 = E(Vf)^2 = E\alpha^2 f^2 = \alpha^2 f^2.$$

To show $Vf = \alpha f$, we note that $|Vf|^2 = 1$, and $1 = E|Vf|^2 \geq |EVf|^2 = |\alpha f|^2 = 1$; hence, equality holding in Schwarz's inequality, $Vf = EVf$, whence $Vf = \alpha f$, q.e.d.

Finally, we remark that Theorem 1 admits the following generalization. Let G be a semigroup, and let T_g, for $g \in G$, be any family of positive operators in

$L_\infty(S, \Sigma', \mu)$ such that $T_g 1 = 1$ for all g. Then there exists a *representation* $g \to V_g$ of G by measurable transformations V_g in an extended space (S, Σ, μ) such that $EV_g = T_g$ for all g, where E is again conditional expectation.

I wish to thank Professors Furstenberg, McCarthy, Segal and Singer for patiently listening to expositions of the present work, and for offering valuable suggestions and criticisms.

Bibliography

1. Dunford and Schwartz, *Linear Operators,* Vol. I, Interscience, New York, 1958.

2. Gantmacher, *The Theory of Matrices,* Vol. 2, New York, 1959.

3. Halmos, *Lectures on Ergodic Theory,* Tokyo, 1956.

4. Halmos, *Normal dilations and extensions of linear operators,* Summa Brasiliensis Math. **2**(1950), 125–134.

5. Karlin, *Positive operators,* J. Math. Mech., **8**(1959), 907–938.

6. Kelley, *Averaging operators in* $C_\infty(X)$, Illinois J. Math. **2**(1958), 214–223.

7. Loève, *Probability Theory,* New York, 1955.

8. Sz.-Nagy, *Prolongements des transformations de l'espace de Hilbert qui sortent de cet espace,* Appendix to Riesz and Sz.-Nagy, *Leçons d'analyse fonctionnelle,* Budapest, 1955.

9. Rota, *On the representation of averaging operators,* Rend. Sem. Mat. Univ. Padova, **30** (1960), 52–64.

10. Segal, *Abstract probability spaces and a theorem of Kolmogoroff,* Amer. J. Math. **76** (1954), 721–732.

11. Shohat and Tamarkin, *The Problem of Moments,* New York, 1943.

Bulletin of the American Mathematical Society, **67**:6 (1961), 556–558

ON THE EIGENVALUES OF POSITIVE OPERATORS[1,2]

BY GIAN-CARLO ROTA

Communicated by I. M. Singer, August 12, 1961

The classical theory of Frobenius-Perron concerning the distribution of eigenvalues of a matrix with non-negative elements has been variously extended to positive operators, that is, linear operators on function spaces transforming non-negative functions into non-negative functions. Since the classical work of Jentzsch (see bibliography), there have been two kinds of extensions: (a) it has been established under various conditions that a positive operator shall have a positive eigenvector with positive eigenfunction (see e.g. Birkhoff [**1**], Karlin [**4**], Samelson [**7**], Schaefer [**8**]), and (b) attempts have been made to extend the Frobenius theorem stating that, for a non-negative matrix with spectral radius one, the eigenvalues on the unit circle are roots of unity. This result has been extended to positive operators under strong additional assumptions, all of them guaranteeing that the intersection of the spectrum with the unit circle shall consist of isolated points only. Results of this kind are summarized in Karlin. It is our present purpose to establish the second result under fairly general conditions. Our result is the following:

THEOREM. *Let P be a linear operator of norm at most one (a "contraction") in $L_1(S, \Sigma, \mu)$, where (S, Σ, μ) is a measure space of finite measure. Suppose that P is bounded, with norm at most one, in $L_\infty(S, \Sigma, \mu)$. Then*:

If α is an eigenvalue of P, and $Pf = \alpha f$ for some nonzero integrable f, then α^2 is also an eigenvalue of P.

If f is written in the form $f = |f| g$, where the function g is of absolute value one at every point, then the eigenfunction belonging to α^2 is $|f| g^2$.

Operators of the type described in this theorem, the so-called dissipative operators, are commonly encountered in various circumstances, for instance in the study of semigroups generated by second-order differential operators. The property of contractivity in L_1 is also crucially assumed in the pointwise ergodic theorems.

The proof of this theorem can be based upon elementary functional-analytic techniques, and will be given in full here. We begin by making the following two simplifications: (a) It can be assumed

[1] Work done by contract with the Office of Naval Research.

[2] I wish to thank Professor Jack Feldman of Berkeley for several enlightening conversations on this subject.

that $\mu(S)=1$; and (b) by going over to the Stone-Čech compactification of $L_\infty(S, \Sigma, \mu)$, we can assume that all the bounded functions treated are continuous on a compact Hausdorff space. This second simplification is not strictly necessary, but it is convenient to talk about values of a function at a point. The proof results from the following seven steps.

(1) For a continuous non-negative function f, suppose that $Pf=f$, then $P(f^2)=f^2$.

Indeed, for a fixed point s in the compact Hausdorff space S, the linear functional $f\to Pf(s)$ is positive and continuous. Hence, by the Riesz representation theorem, there exists a non-negative measure $m(s, dt)$ such that $Pf(s)=\int f(t)m(s, dt)$ (integrations will always be carried out over the whole space). By Schwarz's inequality we have $(\int f(t)m(s, dt))^2 \leqq \int 1m(s, dt)\int f(t)^2m(s, dt)$, where 1 is the function identically equal to one. Letting $P1=h$, where h is a non-negative function taking values between zero and one, the last inequality can be rewritten as $(Pf(s))^2 \leqq h(s)(Pf^2)(s)$. But $Pf=f$, hence this gives $f^2 \leqq P(f^2)$. In view of the assumption that P decreases the integrals of positive functions, this can only be true if $P(f^2)=f^2$.

(2) If $|Pf|=|f|$, then $P|f|=|f|$, where $|f|$ is the function everywhere equal to $|f(s)|$. This is shown in the same way as (1), using the inequality $|\int f(t)m(s, dt)| \leqq \int |f(s)| m(s, dt)$.

(3) From (1) and (2) it follows easily that, if $Pf=f$ and $Pg=g$ for any two functions f and g, then $P(fg)=fg$. Therefore, the fix-points of P form a *subalgebra* which, in view of the assumption, will be isomorphic to one of the form $L_\infty(E, \Sigma_0, \mu)$ on the original measure space, where Σ_0 is a σ-subfield of Σ, and E a subset of S. In particular, the characteristic function (also known as the indicator) χ_E of the subset E will belong to the subalgebra, that is $P\chi_E=\chi_E$.

(4) Now let F be the complement of E in the set S, and let $P\chi_F=q$. Then the function q lives on F; for otherwise, $P1=P(\chi_E+\chi_F)>1$ at some point of E, contradicting the assumption that P is a contraction in L_∞. Hence the subspace $L_1(E, \Sigma, \mu)$, (where we denote by Σ the σ-field obtained by intersecting the sets of Σ with the set E) is invariant under P. Now, if $Pf=\alpha f$ for some α of absolute value one, then, by (2), $P|f|=|f|$, hence f lives on E. We conclude that it suffices to prove the theorem under the additional assumption that $P1=1$, an assumption which we shall make from now on.

(5) If $Pf=f$ and g is arbitrary, then $P(fg)=fPg$. It suffices to prove this when $f\geqq 0$. Using the device of (1), we have $\int f(t)m(s, dt)=f(s)$ for each s. By Schwarz's inequality $(\int f(t)m(s,dt))^2$

$\leqq \int f(t)^2 m(s, dt)$. But by (1), equality holds in Schwarz's inequality. Hence $f(t)$ is, for fixed s, constant almost everywhere relative to the measure $m(s, dt)$. The assertion now becomes obvious when rewritten in terms of integrals relative to the measure $m(s, dt)$.

(6) Now let $Pf=\alpha f$ for some α such that $|\alpha|=1$. Write $f=|f|g$, where $|g(s)|=1$ for all s in its support, and let G be the support of f. By (5), to complete the proof of the theorem, it suffices to prove that $P(g^2)=\alpha^2 g^2$. Since the characteristic function of G is invariant under P, we can restrict ourselves to functions living on G. Hence we shall assume that $|g(s)|=1$ for all s.

(7) The proof is now completed as follows. By Schwarz's inequality, $|\int g(t)m(s, dt)|^2 \leqq \int |g(t)|^2 m(s, dt)=1$. But, since $|Pg|=|g|$, equality holds in the inequality, and this can only happen if $g(t)$ is constant almost everywhere relative to the measure $m(s, dt)$. In view of the identity $\int g(t)m(s, dt)=\alpha g(s)$, it follows that, for fixed s, $g(t)=\alpha g(s)$ for a set of points t of $m(s, dt)$-measure one. But if this is the case, then also $g(t)^2=\alpha^2 g(s)^2$ for a set of points of $m(s, dt)$-measure one. Integrating, we obtain $\int g(t)^2 m(s, dt)=\alpha^2 g(s)$, that is, changing notation, $P(g^2)=\alpha^2 g^2$, q.e.d.

We conjecture that the assumptions of the above theorem are essentially sharp, that is, any significant class of operators for which the assertion holds is equivalent, after a change of measure, to the one considered above. For example, it is easy to see that those quasi-compact operators for which the assertion is true are equivalent, after a change of measure, to contractions.

Bibliography

1. Birkhoff, *Extensions of Jentzsch's theorem*, Trans. Amer. Math. Soc. vol. 85 (1957) pp. 219–227.

2. Frobenius, *Ueber Matrizen aus positiven Elementen*, S.-B. Preuss. Akad. Wiss. Berlin (1908) pp. 471–476, (1909) pp. 514–518, (1912) pp. 456–477.

3. Jentzsch, *Ueber Integralgleichungen mit positiven Kern*, Crelle's Journal vol. 141 (1912) pp. 235–244.

4. Karlin, *Positive operators*, J. Math. Mech. vol. 8 (1959) pp. 907–938.

5. Perron, *Zur Theorie der Matrizen*, Math. Ann. vol. 64 (1907) pp. 248–263.

6. Putnam, *On bounded matrices with non-negative elements*, Canad. J. Math. vol. 10 (1958) pp. 587–591.

7. Samelson, *On the Perron-Frobenius theorem*, Michigan Math. J. (1957) pp. 57–59.

8. H. Schaefer, *Some spectral properties of positive linear operators*, Pacific J. Math. vol. 10 (1960) pp. 1009–1019.

9. J. T. Schwartz, *Eigenvalues of positive compact operators*, to appear in Communications PAM.

Massachusetts Institute of Technology

ADDENDUM TO
ON THE EIGENVALUE OF POSITIVE OPERATORS[1]

BY GIAN-CARLO ROTA

Two important assumptions on the operator P in the Theorem stated in this note were omitted by a clerical error. They are

(a) the operator P is *positive* ($Pf \geqq 0$ if $f \geqq 0$).

(b) the number α is of absolute value *one*.

These assumptions are stated in the informal discussion at the beginning of the note, and are used throughout in the proof.

[1] See volume 67, no. 6 (November 1961) pp. 556–558 of this Bulletin.

Linear operators. Part II. *Spectral theory*. By Nelson Dunford and Jacob T. Schwartz, with the assistance of William G. Bade and Robert G. Bartle. Interscience, New York, 1963. Pp. 859–1923. $35.

If a mathematical visitor from another century were to visit these shores and inquire about the present state of mathematical analysis, he would not miss much if he limited himself to leafing through the pages of this book. With the long awaited appearance of the second volume, we now see before our eyes a panoramic view, rich in colorful detail, of the whole output of a school of mathematical analysis that started with the work of Volterra and Fréchet near the turn of the century, and reached its peak in Poland, Hungary and the Soviet Union as well as at Chicago and Yale in the thirties and forties.

It is impossible to do justice to this volume by merely listing the contents, as has become customary in a review; we shall instead highlight some of the most valuable and typical parts, referring to Halmos's detailed review of the first volume for a description of the techniques of presentation of the material.

The guiding idea of the entire work is the spectral theory of a single linear operator, and its varied applications. Algebras of operators, specifically B^*-algebras, enter only in an ancillary function as aids in the proof of the spectral theorem, or as convolution algebras. This is all to the good, in view of the emphasis on concrete applications and special linear operators. It is rumored that von Neumann algebras and group representations are to be treated in detail in a separate volume by J. T. Schwartz anyway.

Chapter X, containing a proof of the spectral theorem, and probably written at an early stage, should be read by whoever wants to get the flavor of the exposition. The style is matter-of-fact in a pleasant no-nonsense way; it is a great improvement over certain excesses of rationalistic presentation with which we have become all too familiar. The overwhelming use of intricate notation, also an endemic vice of the day, is carefully avoided; one can open the book at almost any page and easily figure out the meaning of all symbols without having to trace it backwards ad infinitum. Theorems are seldom stated in their full stratospheric generality, but rather at that level where the typical application is to be found. A happy instance of this wise limitation is the treatment of multiplicity theory and spectral representation, in both Chapter X and Chapter XII; here the assumption of separability is a welcome simplification; the proofs, here as elsewhere, are based upon the natural, at times even pedestrian approaches, avoiding clever mystification and elegance at all costs, in favor of sound motivation.

In these days of extreme fragmentation of fields, when some mathematicians take pride in not knowing the applications of their own work, the authors' constant concern with the "Zusammenhang" between theory and application, and between distinct branches of mathematics, accounts for the unusual length of this treatise. This concern is quite apparent in the awesome Chapter XI, bearing the understating title "Miscellaneous Applications." A partial list of the topics treated here might make a good list of headings for as many courses in analysis, as follows: compact groups (including Haar measure and the Peter-Weyl theorem); almost periodic functions; convolution algebras (Plancherel's theorem and neighboring results); closure theorems of Wiener type; spectral synthesis (several detailed results); Hilbert-Schmidt operators, including a beautiful and little-known inequality of Carleman bounding the growth of the resolvent of a Hilbert-Schmidt operator, and some profound and largely original results on the completeness of the eigenfunctions of non-self-adjoint Hilbert-Schmidt operators; the Hilbert-transform theory of Calderón and Zygmund; a detailed previously unpublished study of compact operators, classified according to their trace classes; subdiagonalization of compact operators, including some of the deep results on invariant subspaces initiated by the Russian school. The section on "Notes and Remarks" at the end of the chapter gives, among other tidbits, proofs of the Pontrjagin duality theorem, the Marcinkiewicz interpolation theorem, with recent refinements due to Hörmander, the Paley-Littlewood inequality, as well as thorough references to the literature for topics touched upon but not fully treated in the text, such as eigenvalue distribution for compact operators and various results on topological groups. Finally, two lengthy sections of exercises, written, as elsewhere throughout the book, in the interlocking Pólya-Szegö style, provide drill in special Tauberian theorems, various results on Fourier integrals and singular integrals, location of eigenvalues for finite matrices (work of A. Brauer, inequalities of H. Weyl, Perron-Frobenius theory, Fredholm theory, etc.). The conscientious student may supplement these exercises with the set of 53 given in the following chapter, "Unbounded Selfadjoint Operators" (including unitary semigroups and moment problems); here he will find all the folklore on unbounded operators, dilations of operators, von Neumann's spectral sets, various versions of the Heisenberg uncertainty principle, comparison theorems for completeness of Paley-Wiener type, etc., etc.

Chapter XIII is the central part of the book; it is a complete and definitive treatise on the spectral theory of ordinary selfadjoint differential operators, centering upon the Coddington-Kodaira-Krein-Levinson-Titchmarsh-Weyl theory of the spectral representation. The results on location of the continuous spectrum and the limit point-limit circle dichotomy are a complete survey of work done in the field up to 1958, *literally*. Those results that are not proved in the text are summarized in a lengthy section on notes and remarks, and in a set of 128 exercises which will tax the most conscientious student. A final section of this chapter, 25 pages long, contains specific techniques for computation of spectral resolutions of hypergeometric differential operators. The results derived here are a great improvement over the only treatment previously available on the subject by the late E. C. Titchmarsh, and will be invaluable to the worker in group representations, as well as to any physicist with no more than the average physicist's deficiencies in mathematics.

The final Chapter XIV, on linear partial differential operators, displays, in the words of the authors, "a bouquet of applications of functional analysis." It would be hard to find a bouquet of similar fragrance in the literature: Sobolev's theorem, the theory of distributions (admirably done without the largely irrelevant locally convex frills), the elliptic boundary value problem, including the remarkable Browder completeness theorem, not previously included in a textbook to the reviewer's knowledge, and finally the Friedrichs-Lax-Phillips theory of symmetric hyperbolic systems.

Whoever has followed this book since its beginning, and has derived from it much of his mathematical education and inspiration, as the reviewer has and as a whole generation of analysts will, cannot forget the immense amount of human effort and loving care that went into its composition. As one reads through these pages, the entire world of present-day mathematical analysis springs to life. Since Picard and Goursat, there has not been a textbook on such a gigantic scale. The authors have successfully accomplished a task that many would have regarded as impossible.

To have a nearly complete view of an entire field, that of functional analysis, available in a coherent exposition (which, we hope, will eventually reach several thousand pages with the publication of the remaining $n-2$ volumes), is a benefit that few mathematical disciplines have ever enjoyed—one thinks of Fricke-Klein or Russell-

Whitehead as feeble precedents; inevitably it will have a decisive influence upon the future development of all analysis, as well as of much theoretical physics and other neighboring disciplines.

No serious student of analysis can afford to ignore this treatise. More than our thanks, the authors deserve to be read by anyone who is interested in learning the results of a whole age of mathematics.

GIAN-CARLO ROTA

Invariant Subspaces and Models for Linear Operators

James Rovnyak

These papers announce a remarkable property of shift operators which today is known to every operator theorist. Let $\mathfrak{H}$ be the Hilbert space of square summable sequences with entries in a Hilbert space $\mathfrak{K}$. The shift operator

$$S\colon (k_0, k_1, k_2, \ldots) \to (0, k_0, k_1, k_2, \ldots)$$

is an isometry on $\mathfrak{H}$ with adjoint

$$S^*\colon (k_0, k_1, k_2, \ldots) \to (k_1, k_2, k_3, \ldots).$$

The dimension of $\mathfrak{K}$ is called the multiplicity of S. Rota showed that S^* is a universal operator in the sense that its restrictions to closed invariant subspaces in some sense include all operators: if $\mathfrak{K}$ has countably infinite dimension, every bounded linear operator on a separable Hilbert space is similar to a constant multiple of the restriction of S^* to some closed invariant subspace. In "Note on the invariant subspaces of linear operators", this is proved by showing that every Hilbert space operator T satisfying $\|T\| < 1$ is similar to the restriction of S^* to a closed invariant subspace. In "On models for linear operators", the same conclusion is obtained more generally for an operator whose spectrum is contained in the open unit disk. It is the latter form of the result that is known today as the *Rota model* [3,13,17]. A variant of the Rota model replaces similarity by unitary equivalence:

> *Let T be a bounded linear operator on a Hilbert space $\mathfrak{G}$ such that $\|T\| \leq 1$ and $\|T^n f\| \to 0$ as $n \to \infty$ for each f in $\mathfrak{G}$. Let S be a shift operator of multiplicity at least as large as the dimension of the closure of the range of $1 - T^*T$. Then T is unitarily equivalent to the restriction of S^* to some closed invariant subspace.*

For a proof and history of this version, see Rosenblum and Rovnyak [25].

One of the original hopes of the Rota model was to prove the existence of nontrivial invariant subspaces for bounded operators on a Hilbert space of dimension greater than one. Since the Beurling-Lax theory represents invariant subspaces of the shift operator in terms of operator-valued inner functions on the unit disk, the Rota model implies that the problem of finding invariant subspaces is equivalent to the problem of factoring inner functions. Helson [15] attributes this observation to D. Lowdenslager and shows that it can be used to prove the existence of invariant subspaces when the shift operator has finite multiplicity. Despite other efforts, such as the failed attempt by de Branges and Rovnyak [8], the hope to resolve the invariant subspace problem for Hilbert space operators by this (or any other) method has not been realized to this day.

The notion of an operator model has survived, however, and a host of models are in use in operator theory today. Models provide a means to study operators whose structure cannot be exactly described. They do this by identifying cogent examples and by providing a setting for calculations and often interaction with function theory. The Rota model was the first and is still one of the most striking examples of the concept. Some of the subsequent models are directly related to the Rota model, while others are at a considerable distance or have no connection with the Rota model. The genre is important in modern operator theory, and we select a few illustrations with no pretense of completeness.

The canonical models of de Branges and Rovnyak [9] and Sz.-Nagy and Foiaş [28] include a universal model and also make connections with dilation theory and the theory of characteristic operator functions. Such models provide a natural framework for other results, such as the generalized interpolation theorem of Sarason [26], its extension in the commutant lifting theorem of Sz.-Nagy and Foiaş [28], and applications to interpolation theory as in Foiaş and Frazho [14]. The model spaces are related to function theory in the unit disk by Sarason [27]. See also Helton [16] and Ball and Cohen [4] for the use of models in linear systems theory. A coordinate-free generalization and other applications are described in Nikolski and Vasyunin [19].

Herrero [17] and Voiculescu [29] generalized the Rota model to operators with spectra in general domains, and these results are applied in characterizations of closures of similarity orbits of operators in Fialkow [12] and Herrero [18]. Generalizations to sets of commuting operators are given by Ball [3], Curto and Herrero [5], and Curto and Vasilescu [6].

If we go farther afield, many other examples can be mentioned. Agler [1] describes an axiomatic approach to operator models, recent examples of which are given by Dritschel and McCullough [10] for hyponormal operators and by Dritschel and Woerdeman [11] for operators with numerical radius at most one. Models also play an important role in the theory of n-isometries of Agler and Stankus [2] and Richter [24]. Functional models for hyponormal and subnormal operators have been given by Pincus, D. X. Xia and J. B. Xia [21] and D. X. Xia [30]. A subscalar model for hyponormal operators is constructed by Putinar [23].

Finally, it should be noted that the Rota model is an early instance of a theorem that gives conditions for a Hilbert space operator T to be similar to a contraction. The problem of characterizing such operators is a difficult one and has been intensively studied. Paulsen [20] showed that a Hilbert space operator T is similar to a contraction if and only if it is completely polynomially bounded. A long-standing open problem, to determine if a polynomially bounded operator is always similar to a contraction, was recently settled in the negative by Pisier [22]. A survey of this field may be found in the featured review of Pisier's paper by Davidson [7].

References

[1] J. Agler, An abstract approach to model theory. In *Surveys of some recent results in operator theory, Vol. II*, Longman Sci. Tech., Harlow, 1988, pp. 1–23.

[2] J. Agler and M. Stankus, m-isometric transformations of Hilbert space. I, II, and III. *Integral Equations Operator Theory* **21** (1995), 383–429; **23** (1995), 1–48; **24**, (1996), 379–421.

[3] J. A. Ball, Rota's theorem for general functional Hilbert spaces. *Proc. Amer. Math. Soc.* **64** (1977), 55–61.

[4] J. A. Ball and N. Cohen, de Branges-Rovnyak operator models and systems theory: a survey. In *Topics in matrix and operator theory, Proceedings of the Fifth Workshop held at Erasmus University, Rotterdam, 1989*, H. Bart, I. Gohberg and M. A. Kaashoek, eds., Birkhäuser, Basel, 1991, pp. 93–136.

[5] R. E. Curto and D. A. Herrero, On closures of joint similarity orbits. *Integral Equations Operator Theory* **8** (1985), 489–556.

[6] R. E. Curto and F.-H. Vasilescu, Standard operator models in the polydisc. I and II. *Indiana Univ. Math. J.* **42** (1993), 791–810; *ibid*, **44** (1995), 727–746.

[7] K. R. Davidson, Featured review of G. Pisier, "A polynomially bounded operator on Hilbert space which is not similar to a contraction," *Mathematical Reviews*, 97f:47002, 1997.

[8] L. de Branges and J. Rovnyak, The existence of invariant subspaces. *Bull. Amer. Math. Soc.* **70** (1964), 718–721; correction, *ibid.* **71** (1965), 396,

[9] L. de Branges and J. Rovnyak, Canonical models in quantum scattering theory. In *Perturbation theory and its applications in quantum mechanics (Proc. Adv. Sem. Math. Res. Center, U.S. Army, Theoret. Chem. Inst., Univ. of Wisconsin, Madison, WI, 1965)*, Wiley, New York, 1966, pp. 295–392.

[10] M. A. Dritschel and S. McCullough, Model theory for hyponormal contractions. *Integral Equations Operator Theory* **36** (2000), 182–192.

[11] M. A. Dritschel and H. J. Woerdeman, Model theory and linear extreme points in the numerical radius unit ball, *Mem. Amer. Math. Soc.* **129** (1997).

[12] L. A. Fialkow, Closures of similarity orbits of Hilbert space operators. In *Surveys of some recent results in operator theory, Vol. I*, Longman Sci. Tech., Harlow, 1988, pp. 169–219.

[13] C. Foiaş, A remark on the universal model for contractions of G. C. Rota. *Com. Acad. R. P. Romîne* **13** (1963), 349–352.

[14] C. Foiaş and A. E. Frazho, *The commutant lifting approach to interpolation problems.* Oper. Theory: Adv. Appl., Vol. 44, Birkhäuser, Basel, 1990.

[15] H. Helson, *Lectures on invariant subspaces.* Academic Press, New York, 1964.

[16] J. W. Helton, Discrete time systems, operator models, and scattering theory. *J. Funct. Analysis* **16** (1974), 15–38.

[17] D. A. Herrero, A Rota universal model for operators with multiply connected spectrum. *Rev. Roumaine Math. Pures Appl.* **21** (1976), 15–23.

[18] D. A. Herrero, *Approximation of Hilbert space operators. Vol. 1*, 2nd ed., Longman, Harlow and Wiley, New York, 1989.

[19] N. Nikolski and V. Vasyunin, Elements of spectral theory in terms of the free function model. I. Basic constructions. In *Holomorphic spaces, (Berkeley, CA, 1995)*, Math. Sci. Res. Inst. Publ. 33, Cambridge Univ. Press, Cambridge, 1998, pp. 211–302.

[20] V. I. Paulsen, Every completely polynomially bounded operator is similar to a contraction. *J. Funct. Anal.* **55** (1984), 1–17.

[21] J. D. Pincus, D. Xia, and J. B. Xia, The analytic model of a hyponormal operator with rank one self-commutator. *Integral Equations Operator Theory* **7** (1984), 516–535; Note, *ibid.* **7**(1984), 516–535.

[22] G. Pisier, A polynomially bounded operator on Hilbert space which is not similar to a contraction. *J. Amer. Math. Soc.*, **10** (1997), 351–369.

[23] M. Putinar, Hyponormal operators are subscalar. *J. Operator Theory* **12** (1984), 385–395.

[24] S. Richter, A representation theorem for cyclic analytic two-isometries. *Trans. Amer. Math. Soc.* **328** (1991), 325–349.

[25] M. Rosenblum and J. Rovnyak, *Hardy classes and operator theory.* Oxford Univ. Press, New York, 1985; reprinted Dover, Mineola, NY, 1997.

[26] D. Sarason, Generalized interpolation in H^∞. *Trans. Amer. Math. Soc.* **127** (1967), 179–203.

[27] D. Sarason, *Sub-Hardy Hilbert spaces in the unit disk*, Wiley-Interscience, New York, 1994.

[28] B. Sz.-Nagy and C. Foiaş, *Harmonic analysis of operators on Hilbert space.* North-Holland, Amsterdam, 1970.

[29] D. Voiculescu, Norm-limits of algebraic operators. *Rev. Roumaine Math. Pures Appl.* **19** (1974), 371–378.

[30] D. Xia, The analytic model of a subnormal operator. *Integral Equations Operator Theory* **10** (1987), 258–289.

Gian-Carlo Rota's Work on Positive Operators

H. H. Schaefer and X.-D. Zhang

The classical theory of Perron-Frobenius deals with the distribution of eigenvalues of matrices with non-negative entries. One of the main results of this theory is that the peripheral spectrum (the set of eigenvalues with maximum modulus) of such matrices is cyclic, that is, if $a = re^{i\theta}$ is an eigenvalue, then $re^{in\theta}$ is also an eigenvalue for every integer n. In the early 1960's Gian-Carlo Rota introduced various new methods for studying positive contractions on L_p-spaces. His paper "On the Eigenvalues of Positive Operators" was essentially the only paper of Rota's concerned with the spectral theory of positive linear operators, but it was indeed a very influential one; in fact, more than thirty references to this paper - extending into the seventies - can be found in the standard literature. The essence of his result is this: Let (X, Σ, μ) be a finite measure space, and P a positive contraction of $L_1(\mu)$ such that the restriction of P to $L_\infty(\mu)$ is a contraction of $L_\infty(\mu)$. If α with $|\alpha| = 1$ is an eigenvalue of P with eigenfunction $f = |f|g$, then α^n is also an eigenvalue of P with eigenfunction $|f|g^n$ for every integer $n \geq 0$. The conclusion is obviously strong and surprising, as compactness is not behind it. On the other hand, both in the finite and infinite dimensional situations of a positive linear operator on some Banach lattice E, the result is very special - it draws heavily on a measure theoretic background. Still, it was a result that heralded a rather elaborate extension of the classical Perron-Frobenius theory to infinite dimensions and thus, Rota occupies an historic place at the beginning of a period (roughly, from 1960 to 1975) in which the classical theory, dating to the beginning of the last century, found what appears to be its appropriate and deserved extension. For a reasonably concise report on this, the reader is referred to *Banach Lattices and Positive Operators* ([1], Chapter V, Sections 4-6). During the same period (the early sixties), Rota also investigated the problem of dilations of positive contractions on L_p-spaces and obtained several new results. One of the main results is: If T is a positive contraction on $L_\infty(X, \Sigma', \mu)$, where (X, Σ', μ) is a probability space, such that $T\mathbf{1} = \mathbf{1}$, where $\mathbf{1}$ is the constant function always taking value 1, then there exist a larger probability space (X, Σ, μ) and a linear transformation V on $L_\infty(X, \Sigma, \mu)$ (induced by a measurable transformation from (X, Σ, μ) into itself) such that $EV^n f = T^n f$ for all f in $L_\infty(X, \Sigma', \mu)$ and all integers $n \geq 0$, where E is the conditional expectation operator from $L_\infty(X, \Sigma, \mu)$ onto $L_\infty(X, \Sigma', \mu)$. These results find wide applications in the theory of probability and stochastic processes.

References

[1] H. H. Schaefer, *Banach Lattices and Positive Operators.* Springer-Verlag, Berlin and New York, 1974.

The Joint Spectral Radius

Gilbert Strang

The paper "A note on the joint spectral radius" gives a natural definition of the joint spectral radius of two matrices A and B :

$$\rho(A,B) = \liminf_{N \to \infty} \quad (\text{largest norm of any product with } N \text{ factors})^{1/N}$$

The product can have A's and B's in any order (and the lim inf is actually a limit). I was very proud to be a joint author with Gian-Carlo, but I am now a little ashamed that we never thought seriously about how to compute this number ρ. For a single matrix it equals the largest magnitude of the eigenvalues $\lambda(A)$. But products of A and B can produce norms and eigenvalues that are very hard to estimate (as N increases) from the two matrices. The Lyapunov exponent is a similar number, using averages over products of length N instead of maxima, and it suffers from the same difficulty (impossibility?) in actual computation. An equivalent definition of ρ is the infimum over all matrix norms of $\max(\|A\|, \|B\|)$. The definitions extend directly to sets of more than two matrices, and an l_p norm joint spectral radius has also proved useful [7,8].

Every few years, Gian-Carlo would ask me whether anyone ever read our paper. After I had tenure, I could tell him the truth: "not often". Part of the reason may have been the relatively unfamiliar journal, and the nonexistence of the Internet, and even the atypically obscure language that he had chosen to express our (very simple) idea. In recent years I could change my answer! The joint spectral radius suddenly found application in wavelet theory, especially in the work of Ingrid Daubechies and Jeff Lagarias [5]. Of special interest is the question whether $\rho < 1$, so that products with more and more factors of A and B all go to zero.

The application comes in solving the key equation in wavelet theory—the refinement equation (or dilation equation) for the scaling function. Its special feature is the presence of two time scales t and $2t$. Written as a system for a vector unknown as in [9], it involves two matrices $m(0)$ and $m(1)$:

$$\phi(t) = m(0)\phi(2t) + m(1)\phi(2t-1) \qquad \text{on} \quad (0,1).$$

At the point t with binary expansion $.t_1t_2t_3t_4\ldots$, either $m(0)$ or $m(1)$ enters on the right side according to whether t_1 is 0 or 1. And the recursion continues, involving all the bits in the expansion of t:

$$\phi(t) = m(t_1)\phi(.t_2t_3t_4\ldots) = m(t_1)m(t_2)\phi(.t_3t_4t_5\ldots) = \ldots$$

A nearby point T begins with the same bits as t. To prove continuity of the solution is to show that $\phi(T)$ is close to $\phi(t)$. Both $m(0)$ and $m(1)$ have an eigenvalue at 1, but its effect cancels in the difference $\phi(T) - \phi(t)$. What matters is the restrictions

A and B of $m(0)$ and $m(1)$ to their common invariant subspace, orthogonal to $(1,\ldots,1)$. Continuity will hold if the product of A and B in every order is small—thus if $\rho(A,B) < 1$. The exact value of ρ determines the exponent of Hölder continuity of $\phi(t)$.

A beautiful result of Berger and Wang [1] allows the option of estimating ρ from the eigenvalues of the products, instead of their norms. The eigenvalues (we take their absolute values) approach from below and the norms from above. Thus

$$\rho(A,B) = \limsup_{N\to\infty} \quad (\text{largest }|\text{eigenvalue}|\text{ of any product with } N \text{ factors})^{1/N}.$$

Lagarias and Wang [8], and also Gurvits [6], asked whether this limit is necessarily attained for a finite product. This "finiteness conjecture" has just been disproved in 2000 by Bousch and Mairesse [4], using their deep results for dynamical systems. They are able to give the form of a 2 by 2 counterexample, but not explicit matrices A and B. Maesumi (private communication) now conjectures that such counterexamples form a set of measure zero. Blondel, Theys, and Vladimirov [2] have just established that

$$A = \begin{bmatrix} 1 & 1 \\ 0 & 1 \end{bmatrix} \quad \text{and} \quad B = \begin{bmatrix} b & 0 \\ b & b \end{bmatrix}$$

yield counterexamples for infinitely many b. A specific b remains unknown in November 2001.

The logical decidability of $\rho \leq 1$ has also been an active question. Blondel and Tsitsiklis [3] have now given a negative answer. Perhaps we will never have an explicit counterexample to finiteness? Gian-Carlo would have been quite happy, if he knew that our simple definition led to such profound questions (and answers).

A much more complete list of references to the joint spectral radius is in the version of this commentary on the web page `math.mit.edu/~gs`. *It is* #146 *in the list of Gilbert Strang's publications.*

References

[1] M. A. Berger and Y. Wang, Bounded semigroups of matrices. *Linear Algebra Appl.* **166** (1992), 21–27.

[2] V. D. Blondel, J. Theys, and A. A. Vladimirov, An elementary counterexample to the finiteness conjecture. Preprint, 2001, `www.inma.ucl.ac.be/~blondel`.

[3] V. D. Blondel and J. N. Tsitsiklis, The boundedness of all products of a pair of matrices is undecidable. *Systems & Control Letters* **41(2)** (2000), 135–140.

[4] T. Bousch and J. Mairesse, Asymptotic height optimization for topical IFS, Tetris heaps and the finiteness conjecture. *J. Amer. Math. Soc.*, to appear.

[5] I. Daubechies and J. C. Lagarias, Sets of matrices all infinite products of which converge. *Linear Algebra Appl.*, **161** (1992), 227–263.

[6] L. Gurvits, Stability of discrete linear inclusions. *Linear Algebra Appl.* **231** (1995), 47–85.

[7] R. Q. Jia, Subdivision schemes in L_p spaces. *Adv. Comput. Math.* **3** (1995), 309–341.

[8] J. C. Lagarias and Y. Wang, The finiteness conjecture for the generalized spectral radius of a set of matrices. *Linear Algebra Appl.*, **214** (1995), 17–42.

[9] G. Strang and T. Nguyen, *Wavelets and Filter Banks.* Wellesley-Cambridge Press, Wellesley MA, 1996.

CHAPTER 3

Reynolds Operators

Estratto dal volume XXX (1960)
dei « Rendiconti del Seminario Matematico dell' Università di Padova »

ON THE REPRESENTATION OF AVERAGING OPERATORS

Nota (*) *di* G. C. Rota (*a Cambridge, Mass.*) (1)

1. Introduction. - Recent developments in the theory of turbulence (see (7), (8) and especially (6), where further references will be found) have led to a systematic study of some concepts which were formerly only implicitly used. One of these is that of a Reynolds operator, studied by Kampé de Fériet, G. Birkhoff and M.me Dubreil-Jacotin; a related notion is that of an *averaging operator,* whose importance was first pointed our in (6) and (7). The history of the subject is excellently explained in (6) and (7).

The present work intends to fill a gap in the theory of averaging operators. Our aim is to give a complete representation theory of averaging operators in some of the function spaces of most frequent occurrence. This is done by starting from a « minimal » system of postulates for averaging operators which are independent from any order structure of the function space. The thesis of Mrs. Moy (12) gives a result related to ours, starting however from a less stringent system of postulates than those set forth below.

The exposition is self-contained. Lemma 1 is stated in slightly different form by Mrs. Moy.

(*) Pervenuta in Redazione il 17 ottobre 1959.
Indirizzo dell:A.: Massachussetts Institute of Technology, Cambridge Mass. (U.S.A.)
1) Research supported by contract 7667 with the Office of Naval Research.

2. Averaging Operators. - In this and the following section we shall only consider probability spaces (S, Σ, μ), that is, measure spaces [2]) for which $\mu(S) = 1$. By « functions » we mean as usual equivalence classes modulo sets of measure zero. If Σ' is a subfield of the σ-field Σ, then we denote by (S, Σ', μ) the measure space obtained by restricting the measure μ to Σ'. All function spaces are tacitly taken over the real field endowed with its natural Borel structure.

DEFINITION. An *averaging operator* (or *average*) A in $L_p(S, \Sigma, \mu)$ (where p is a *fixed* real number, $1 \leq p \leq \infty$) is a linear operator in $L_p(S, \Sigma, \mu)$ with the following three properties:

(1) A is a contraction operator:

$$\int_S |(Af)(s)|^p \mu(ds) \leq \int_S |f(s)|^p \mu(ds),$$

for f in $L_p(S, \Sigma, \mu)$.

(2) If f is of class $L_p(S, \Sigma, \mu)$ and g is an essentially bounded function on (S, Σ, μ), then the function $(Af)(s)(Ag)(s)$ is of class $L_p(S, \Sigma, \mu)$ and

$$A(g\,Af) = (Ag)(Af).$$

(3) If $I(s)$ is the function identically equal to one on S, then $AI = I$ [3]).

The following examples may be found enlightening by the reader who is puzzled by this definition. Consider first the case when the range of the operator A is a one-dimensional subspace. Then our main theorem below will show that A is the operator

$$Af = \left(\int_S f(s)\mu(ds)\right) I,$$

---

[2]) We follow the notation in Dunford and Schwartz, (1).

[3]) It follows from the main theorem proved below that every averaging operator in this sense is also an averaging operator in the sense of Birkhoff (6) and Sopka (10).

that is, the ordinary « average » or mean of the function f. A slightly more general example is obtained when the range of A is a finite-dimensional subspace. Again, it is a consequence of the main theorem that the operator A can be described as follows. There is a partition of S into a finite number of disjoint measurable sets $E_1, E_2, \ldots, E_n$ of positive measure, and for every function f in the given space,

$$(3) \qquad (Af)(s) = [\mu(E_k)]^{-1} \int_S f(s)\mu(ds) \qquad \text{for } s \text{ in } E_k.$$

The second example is immediately extended to a countable partition of S. However, any further non-trivial example of averaging operator lies beyond this « naive » approach, and requires the following notion from measure theory.

Let Σ' be a σ-subfield of Σ, and let f be of class $L_p(S, \Sigma, \mu)$ (again for fixed p). Then the countably additive set function on Σ'

$$\mu_f(E) = \int_E f(s)\mu(dsZ), \qquad E \in \Sigma'$$

is μ-continuous. By the Radon-Nikodym Theorem there exists a *unique* Σ'*-measurable* function f' such that

$$\int_E f'(s)\mu(ds) = \int_E f(s)\mu(ds), \qquad E \in \Sigma'.$$

The function f' is the Radon-Nikodym derivative of f relative to the σ-subfield Σ'. It is evident that the operator $Af = f'$ is linear and bounded; in fact, it is easily verified that it is an averaging operator. In probability theory f' is the conditional expectation of f relative to the σ'-field Σ'.

The first of the examples given above is obtained by taking the Radon-Nikodym derivative of f relative to the minimal subfield consisting of the two sets ($\emptyset$ and S); the second is obtained by considering the σ-subfield Σ' generated by a partition, and again taking the Radon-Nikodym derivative. Thus the operator $f \to f'$ is a natural generalization of the elementary mean taken over a partition. Our main result states that every averaging operator in $L_p(S, \Sigma, \mu)$ is obtained by taking a Radon-Nikodym derivative relative to a suitably defined subfield.

Before embarking upon a proof of this result, consider the intuitive meaning of the crucial property (2) of averages. This property was obtained inductively from the study of certain operators introduced by Osborne Reynolds in turbulence theory; it also appears in Tarski's quantification axiom (see (6)). Algebraically it can be considered as a strenghthening of the property of linearity of the operator A, for it states that A remains linear when the vector space is endowed with the richer structure of module over a certain subring of the ring of essentially bounded functions (cf. Lemma 2 below).

3. Main Representation Theorem. We begin by establihing some unavoidable technical lemmas, some of them of independent interest. It is always tacitly assumed that p is a real number greater than one.

LEMMA 1. Let g be an essentially bounded function on (S, Σ, μ), and let $\mathcal{C}$ be the collection of all functions of the form $p(g)$, where p is a real polynomial. Let $\Sigma(g)$ be the σ-field generated by sets of the form $\{s \mid g(s) \in F\}$, where F is a Borel set on the line. Then the closure of $\mathcal{C}$ in $L_p(S, \Sigma, \mu)$ is the subspace $L_p(S, \Sigma(g), \mu)$.

Proof. A measure μ_g on the line (the « distribution function » of g) is defined by the formula

$$\mu_g(F) = (\{ s \mid g(s) \in F \})$$

where F is any Borel set. Since g is bounded, the measure μ_g has compact carrier, say the interval $[r, s]$. For any real-valued Borel function h on the line we have the identity

$$\int_S h(g(s))\mu(d\ ) = \int_r h(t)\mu_g(dt).$$

In particular, if E is a set in $\Sigma(g)$, and $g(E) = F$, then F

is a Borel set, and

$$\int_S \chi_E(s)\mu(ds) = \int_S \chi_F(g(s)\mu(ds) =$$

$$= \int_S \chi_F(t)\mu_g(dt).$$

Since the characteristic function χ_F is a pointwise limit everywhere on $[r, s]$ of a sequence of polynomials in the real variable t, it follows from the bounded convergence theorem that χ_F is a limit of polynomials in $L_p([r, s], \Sigma, \mu)$ and hence, that χ_E is a limit in norm of a sequence of polynomials in the function g in $L_p(S, \Sigma, (g), \mu)$. Thus the characte ristic function χ_E lies in the closure of $\mathfrak{A}$. The statement then follows from the fact that simple functions are dense in $L_p(S, \Sigma, \mu)$.

PROPOSITION 1. (Birkhoff). Every average is a projection (that is, $A^2 = A$).

Proof. By (2) and (3) we have

$$A^2 f = A(IAf) = (AI)(Af) = Af.$$

LEMMA 2. Let g be an essentially bounded function for which $Ag = g$, and let $\Sigma(g)$ be the σ-field generated by g, as in Lemma 1. Then for every measurable set E in $\Sigma(g)$ we have

$$A\chi_E = \chi_E.$$

Proof. If g is fixed under A and bounded, then g^2 is also bounded and fixed under A, for by (2) and Prop. 1 we have

$$A(g^2) = A(gA(g)) = (Ag)^2 = g^2.$$

It follows by induction that $A(g^n) = g^n$ for any positive integer n, and hence that $A(p(g)) = p(g)$ for any real polynomial p. Lemma 1 and the uniform continuity of A entail that $Af = f$

for any f in $L_p(S, \Sigma(g), \mu)$. The assertion of the Lemma is a special case of this conclusion.

PROPOSITION 2. Every averaging operator in $L_p(S, \Sigma, \mu)$ maps essentially bounded functions into essentially bounded functions.

Proof. Let f be essentially bounded. From the knowledge that A is bounded in L_p one can only infer that the function $g = Af$ is of class L_p. However, the « averaging identity » (2) shows that $A(fg) = g^2$ is also of class L_p, and by induction that all powers $g^n = A(fg^{n-1})$ are of class L_p for any positive integer n. Applying (1) and Hölder's inequality we find that

$$\int_S |g(s)|^{np}\mu(ds) \leq \int_S |(Afg^{n-1})(s)|^p \mu(ds) \leq$$

$$\leq \int_S |f(s)|^p |g(s)|^{(n-1)p}\mu(ds) \leq$$

$$\leq \left(\int_S |f(s)|^{np}\mu(ds)\right)^{1/n}\left(\int_S |g(s)|^{np}\mu(ds)\right)^{1-(1/n)}$$

and hence that

$$\left(\int_S |g(s)|^{np}\mu(ds)\right)^{1/n} \leq \left(\int_S |f(s)|^{np}\mu(ds)\right)^{1/n} \leq |f|_\infty^p$$

for all positive n. Letting n tend to infinity we find (cf. Loomis [3], p. 39, Thm. 14F) that the left hand side tends to $|g|_\infty^p$, which is thus seen to be finite, q. e. d.

LEMMA 3. The equality

$$\int_S (Af)(s)\mu(ds) = \int_S f(s)\mu(ds)$$

holds for every function f in $L_p(S, \Sigma, \mu)$.

Proof. Let A^* be the adjoint operator of A in $L_q(S, \Sigma, \mu)$ $(p^{-1} + q^{-1} = 1)$. We shall first prove that $A^*I = I$. Note that for $g(s) = (A^*I)(s)$ we have by (3)

$$\int_S g(s)\mu(ds) = \int_S I(s)(AI)(s)\mu(ds) = 1.$$

Secondly, from

$$|g|_q = \sup_{|f|_p \leq 1} \left| \int_S g(s)f(s)\mu(ds) \right| =$$
$$= \sup_{|f|_p \leq 1} \left| \int_S (Af)(s)\mu(ds) \right| = 1$$

we infer that

$$1 = \int_S g(s)\mu(ds) \leq |g|_q |I|_p = 1.$$

Since equality obtains in Hölder's inequality, we conclude ((1), III.9.42) that $g = I$ almost everywhere.

Now to the proof of the Lemma. For f of class L_p we obtain

$$\int_S (Af)(s)\mu(ds) = \int_S g(s)f(s)\mu(ds) = \int_S f(s)\mu(ds),$$

q.e.d. [4]).

The proof of the main Theorem is now almost concluded.

THEOREM 1. Let A be an averaging operator in $L_p(S, \Sigma, \mu)$ (for fixed p). Then there exists a unique σ-subfield Σ' of Σ such that $Af = f'$ for f in L_p, where f' is the Radon-Nikodym derivative of f relative to the σ-field Σ'.

Proof. Let Σ' be the smallest σ-field containing all σ-fields $\Sigma(g)$, where g is any bounded function fixed under A. It is immediate from Lemma 2 and assumption (2) that for every E in Σ', $A\chi_E = \chi_E$. Therefore, by Lemma 3 and (2), for all E in Σ',

$$\int_S (Af)(s)\mu(ds) = \int_S \chi_E(s)(Af)(s)\mu(ds) =$$
$$= \int_S (A(\chi_E f))(s)\mu(ds) = \int_E f(s)\mu(ds).$$

[4]) The preceding argument does not include the case $p = 1$, but this case is even simpler.

If the function f is bounded, then Af is Σ' measurable. Therefore the first and last members in the above equalities now show that Af is the Radon-Nikodym derivative f' of f relative to the σ-field Σ'. Thus $Af = f'$ whenever f is essentially bounded. Since A has a unique extension to the entire space, the proof is complete.

4. Averages of Essentially Bounded Functions. - Two properties of averaging operators in $L_p(S, \Sigma, \mu)$ for finite p which follow at once from the preceding Theorem are the following:

(A) An averaging operator is uniquely determined by its range.

(B) Every averaging operator is selfadjoint. That is, if the operator A, as well as its adjoint A^*, are restricted to the dense subspace of essentially bounded functions, then $A = A^*$. This follows from the identity for Radon-Nikodym derivatives, for bounded f and g,

$$\int fg' = \int f'g' = \int f'g.$$

Neither (A) nor (B) hold for averaging operators in $L_\infty(S, \Sigma, \mu)$. This happens because there is a far wider variety of averages of essentially bounded functions. We derive below the structure of such operators, as far as it can be described in full generality.

PROPOSITION 3. Every averaging operator in $L_\infty(S, \Sigma, \mu)$ is an order-preserving transformation.

Proof. We can find an order-preserving isometric isomorphism F of the algebra $L_\infty(S, \Sigma, \mu)$ onto $C(S_1)$, the B-algebra of all continuous functions on a compact Hausdorff space S_1 (for a proof, see (1), V.8.11). It follows that if A is an averaging operator in L_∞, then the operator $A' = FAF^{-1}$ is a bounded operator in the B-space $C(S^1)$ with the following three properties:

(1′) A' is a contraction,

(2′) $A'(fAg) = A'f\ A'g$

for any tho continuous functions f and g,

(3′) $A'I = I$.

Since the isomorphism F is order-preserving, the assertion will be established if it is shown that the operator A' preserves order. If this were not the case, then we could find a real function f in $C(S_1)$, which without loss of generality we can assume to take values between 0 and 1, such that $(A'f)(s_0) = -\varepsilon < 0$ at some point s_0 in S_1. Therefore $(A'(I-f))(s_0) = 1 + \varepsilon > 1$. Since the function $1 - f$, lies between 0 and 1, this contradicts the assumption that A′ is a contraction, q.e.d.

From the preceding proof it is seen that the family of averaging operators in $L_\infty(S, \Sigma, \mu)$ is «isomorphic» to the family of all averaging operators in $C(S_1)$, as defined by (1′) — (3′). The latter operators have been studied by G. Birkhoff in (7), J. Sopka in (9) and J. L. Kelley in (10). Using the results of these authors together with the representation of L_∞ as a C-space, one can obtain some information about general averaging operators in L_∞. The following discussion is therefore limited to those results which do not follow from such a representation technique.

Let $\Sigma(A)$ be the field of all measurable sets whose characteristic functions are invariant under A. We call $\Sigma(A)$ the field *associated with* the averaging operator A. In general, $\Sigma(A)$ is not a σ-field. One can characterize the range of the operator A as follows.

PROPOSITION 4. Let A be an averaging operator in $L_\infty(S, \Sigma, \mu)$, and let $\Sigma(A)$ be the associated field. Then the range of A is the subspace of all essentially bounded functions f such that, for any real a and b (possibly infinite) the set

$$[s \mid a < f(s) \leq b] \qquad [4]$$

belongs to the field $\Sigma(A)$.

Proof. By Prop. 1 the range of A consists of all functions invariant under A. Let f be one such function, and let E be the set [4]. Since the product of two invariant characteristic functions is again invariant, we can assume without loss of generality that $a = 0$ and b is infinite. From the identity $A(\chi_E f) = fA\chi_E$ we infer that the function $fA\chi_E$ is non-negative; therefore $A\chi_E$ is supported on E, that is, $A\chi_E = \chi_E A\chi_E$. Let $E' = S - E$; then $I = A\chi_E + A\chi_{E'} = \chi_E A\chi_E + \chi_{E'} A\chi_{E'}$ by (3). Multiplying both sides by χ_E we obtain $\chi_E = A\chi_E$, so that χ_E is invariant, as we wanted to show.

Conversely, suppose that f safisfies the condition of the Proposition. We show that f is invariant under A. If $Af \neq f$ on a set of positive measure, then there exist real numbers a and b such that the sets in [4] for f and Af do not coincide almost everywhere. We can assume without loss of generality that $a = 0$ and b is infinite, and that the set where Af is non-negative does not contain the set E in [4], where f is non-negative (otherwise take $-f$). Thus χ_E is invariant under f, by hypothesis, but $\chi_E Af$ is negative on a set of positive measure. This contradicts Prop. 3.

In virtue of Prop. 4, every function f invariant under A is measurable relative to the σ-field $\Sigma'(A)$ generated by the associated field $\Sigma(A)$. The converse of this statement is false in general; and this indicates that the introduction of purely finitely additive measures is essential to the problem.

Let ν be a finitely additive real measure on the σ-field Σ and vanishing on all μ-null sets. It is well-known that the indefinite integral

$$\nu(f;\, E) = \int_E f(s)\nu(ds), \qquad E \text{ in } \Sigma$$

is well-defined for all essentially bounded functions f. Now suppose that the restriction of the measure ν to a given sub-σ-field Σ' is countably additive. Then the set function $\nu\,(f;\, E)$ restricted to Σ' is absolutely continuous relative to μ, and the Radon-Nikodym derivative of $\nu(f;\, E)$ is well-defined; we call it the Radon-Nikodym *derivative of the pair* $(f,\, \nu)$ relative to the σ-field Σ'.

THEOREM 2. Let A be an averaging operator in $L_\infty(S, \Sigma, \mu)$ with associated field $\Sigma(A)$, and let $\Sigma'(A)$ be the σ-field generated by $\Sigma(A)$. Then there exists a unique positive finitely additive measure ν_A defined on Σ and vanishing on μ-null sets, such that:

(a) for every E in $\Sigma'(A)$, $\nu_A(E) = \mu(E)$,

(b) for every essentially bounded function f, the function Af is the Radon-Nikodym derivative of the pair (f, ν_A) relative to the σ-field $\Sigma'(A)$.

Proof. We recall that the conjugate space of $L_\infty(S, \Sigma, \mu)$ is the B-space $ba(S, \Sigma, \mu)$ of all finitely additive set functions which vanish on μ-mull sets (cf. (1), IV.8.16). Denoting again by A^* the adjoint operator of A, let $\nu_A = A^*\mu$. We have then the identity

$$(5) \qquad \int_E (Af)(s)\mu(ds) = \int_S A(\chi_E f)(s)\mu(ds) = \\ = \int_S \chi_E(s)f(s)\nu_A(ds) = \int_E f(s)\nu_A(ds)$$

for E in $\Sigma(A)$. In particular for $f = I$ we obtain $\nu_A(E) = \mu(E)$; therefore ν_A is countably additive on $\Sigma(A)$, and by a well-known theorem of Carathéodory also countably additive on the σ-field $\Sigma'(A)$. Thus the Radon-Nikodym derivative of the pair (f, ν_A) is well-defined, and [5] holds for all E in $\Sigma'(A)$. The first and last members of [5] show that Af is the Radon-Nikodym derivative of the pair (f, ν_A) relative to $\Sigma'(A)$. The uniqueness and positivity of the measure ν_A are also immediately inferred from [5].

The converse of Thm. 2 is false, as may be expected from its counterpart in $C(S)$, as shown in (7). This raises the question of whether there is a natural characterization among all averaging operators with the same range of the one which is obtained by taking the Radon-Nikodym derivative relative to the σ-field $\Sigma'(A)$, as in Thm. 1. Such a characterization is derived below. We say that a bounded operator A in L_∞ is *weakly continuous* when it is continuous in the weak L_1-topology of L_∞.

PROPOSITION 5. Let A be an averaging operator in $L_\infty(S, \Sigma, \mu)$. The associated field $\Sigma(A)$ is a σ-field and $Af = f'$ (the Radon-Nikodym derivative of f relative to the σ-field $\Sigma(A)$) if and only if $A^*\mu = \mu$.

Proof. The necessity of the condition is clear, in virtue of Remark (B). To establish the sufficiency, we remark that the assumption gives the identity

$$\int_S f(s)\mu(ds) = \int_S (Af)(s)\mu(ds)$$

for all essentially bounded f. Writing $f = f^+ - f^-$, where f^+ and f^- are non-negative, we get by Prop. 3 that $(Af^+)(s) + (Af^-)(s) \geq |(Af)(s)|$, and hence

$$\int_S |f(s)|\mu(ds) = \int_S [f^+(s) + f^-(s)]\mu(ds) \geq \int_S |(Af)|(s)\mu(ds)$$

Therefore A can be extended to an averaging operator in $L_1(S, \Sigma, \mu)$, and the result follows from Thm. 1 (*).

Remark. The assumption of weak continuity of A is not strong enough to give the conclusion of Prop. 4. To see this, it suffices to consider the case when S is a finite atomic measure space, and apply the Theorem of Birkhoff (6) for continuous functions, noting that in this simple case the conditions given there for an averaging operator in $C(S)$ are also sufficient. Such an example also shows that in general the adjoint operator of an averaging operator is not an averaging operator.

(*) Dr. P. C. Shields of M. I. T. was kind enough to point out an oversight in the first draft of this proof, and to suggest an improvement.

BIBLIOGRAPHY

(1) DUNFORD and SCHWARTZ: *Linear Operators*, Vol. I, New York, 1959.

(2) LOÈVE: *Probability Theory*, New York, 1955.

(3) LOOMIS: *Abstract Harmonic Analysis*, New York, 1953.

(4) Mme. M.-L. DUBREIL - JACOTIN: *Étude Algébrique des Trasformations de Reynolds*, Colloque d'Algèbre superieure, CBRS, Louvain, 1957.

(5) J. KAMPÉ DE FÉRIET: *La notion de moyenne dans la théorie de la turbulence*, Rendiconti Seminario Matematico e Fisico di Milano (27), (1955-56), pp. 1-43.

(6) G. BIRKHOFF: *Moyennes des Fonctions Bornées*, Colloque Internationale No. 24 du CNRS, Paris 1950.

(7) G. BIRKHOFF: *Averaging Operators*, Symposium in Lattice Theory, AMS, 1960.

(8) SHU-TEH-CHEN-MOY: Pacific Journal of Mathematics, 4 (1954), pp. 47-63.

(9) J. SOPKA: Harvard Doctoral Dissertation, 1950.

(10) J. L. KELLEY: *Averaging Operators in* $C_\infty(X)$, Illinois J. M. 2 (1958), pp. 214-223.

Reprinted from the Proceedings of the NATIONAL ACADEMY OF SCIENCES
Vol. 46, No. 6, pp. 863–868. June, 1960.

SPECTRAL THEORY OF SMOOTHING OPERATORS

BY GIAN-CARLO ROTA*

DEPARTMENT OF MATHEMATICS, MASSACHUSETTS INSTITUTE OF TECHNOLOGY

Communicated by J. L. Walsh, April 27, 1960

1. *Introduction.*—Linear operators on spaces of differentiable, continuous or measurable functions which enjoy smoothing properties of various sorts have been studied from several points of view.[1] Only recently, however, has an algebraic operator identity been discovered which expresses a smoothing property of the operator. This is the Reynolds[2] Identity $R(fg) = RfRg + R[(f - Rf)(g - Rg)]$, (denoted below as *R.I.*), where R is a linear operator acting on a function algebra, or more generally on an arbitrary commutative algebra. This identity originated from the study of stochastic solutions of the Navier-Stokes equations;[3] however, its usefulness in other contexts, of which we mention below the most important, warrants an independent study. The simplest, and in many respects the typical, example of an operator satisfying the *R.I.*, is the operation of conditioning in probability theory.[4] If X is a random variable and B a Borel subfield, then the operator $X \rightarrow E^B X$, the *conditional expectation* of X relative to B, satisfies the Reynolds Identity. If the probability space is a finite set, then every order-preserving operator satisfying the *R.I.* is essentially a conditioning.[5] This is not the case in general: the operator $f \rightarrow \int_{-\infty}^{0} f(t + x)e^t dt$ is easily seen to satisfy the Reynolds Identity. This operator performs a weighted average of the function f over "its past," the "remote past" being counted less heavily than the immediate past; it can be interpreted as a generalization of conditional expectation. The *R.I.* has an important connection with ergodic theory. Let $s \rightarrow \varphi(s)$ be a *measurable transformation* on a finite measure[6] space (S, Σ, μ). Assume that for the linear operator $f(s) \rightarrow Tf(s) = f(\varphi(s))$ the operator $\log T$ can be defined by standard operational calculus. It can be shown that for any complex number λ not in the spectrum of $\log T$ the operator $R = \lambda(\lambda I - \log T)^{-1}$ satisfies the *R.I.* It turns out that some classical problems of ergodic theory concerning measurable transformations can be interpreted in terms of operators satisfying the *R.I.* Paramount among these is the problem of the existence of a finite measure equivalent to μ under which the transformation φ is invariant. The restatement of this problem in terms of operators R is solved below for some classes of measurable transformations.

A further motivation for the study of operators satisfying the *R.I.* is that these include as special cases both conditional expectation operators and operators originating from measurable transformations; this provides a link between the convergence theory of *martingales* and classical *ergodic theory.*

A problem studied below is how far the Reynolds Identity goes toward determining the operator as belonging to one of these two classes. The question is solved for operators with pure point spectra which are weak*-continuous in $L_\infty(S, \Sigma, \mu)$; this class of operators is wide enough to include all operators bounded in $L_p(S, \Sigma, \mu)$ and satisfying the Reynolds Identity on the subalgebra of essentially bounded functions. The theory summarized below is in many ways an extension of ergodic theory.

2. *Reduction Theory.*—Little can be obtained from a purely algebraic study of

the Reynolds Identity. It is easily seen to be equivalent to the identities $R(fRg + gRf) = RfRg + R(RfRg)$ and $2R(fRf) = (Rf)^2 + R(Rf)^2$. If the operator R is idempotent, then the simpler identity $R(fRg) = RfRg$ follows from the *R.I.*; under very mild conditions this identity characterizes conditional expectation.[7] Another consequence of the Reynolds Identity is the identity $R(fRg) = fRg$ for all f such that $Rf = f$. This implies that the fixpoints of R form a subalgebra. A deeper consequence of the *R.I.*, valid in every algebra without nilpotent elements, is the following: if $R^n g = 0$ for some $n \geq 1$, then $R^2 g = 0$.[8]

DEFINITION. *A weak*-continuous operator R in $L_\infty(S, \Sigma, \mu)$ is a smoothing operator (S.O.) when (a) $Re = e$, where e is the identity operator, (b) R satisfies the Reynolds Identity, (c) R is strongly positive, that is $Rf \geq 0$ if $f \geq 0$ and $Rf \neq 0$ if $f \geq 0, f \neq 0$.*

Operators of the form $\lambda(\lambda I - \log T)^{-1}$ for real λ, where T is defined as above in terms of a measurable transformation, are *S.O.s*. Every bounded operator in $L_p(S, \Sigma, \mu)$ which satisfies (*a*), (*b*), and (*c*) for all essentially bounded f and g, can be shown to be a *S.O.* Every *S.O.* on a real vector space can be trivially extended to a *S.O.* on the corresponding complexification. A *simple* smoothing operator is one for which the identity function is the only fixpoint. Simple *S.O.s.* extend the notion of an *ergodic* transformation, to which they are equivalent in special cases. For a general *S.O.* in $L_p(S, \Sigma, \mu)$ it can be shown that the set of all f in $L_p(S, \Sigma, \mu)$ fixed under R is a closed subspace of the form $L_p(S, \Sigma', u)$, where Σ' is a sub-σ-field of the σ-field Σ. Using this result one can prove the following reduction theorem, which extends a similar result for measure-preserving transformations.

THEOREM 1. *Every smoothing operator on a finite measure space $L_2(S, \Sigma, u)$ is the direct integral of simple smoothing operators.*

The proof relies strongly on the classical reduction theory for von Neumann algebras;[9] the main difficulty in the direct integral decomposition of $L_2(S, \Sigma, \mu)$ as $\int^{\oplus} H a d\alpha$ is to choose the integral H_α as $L_2(S_\alpha, \Sigma_\alpha, \mu_\alpha)$ so that the "differential" R_α satisfies the *R.I.* for almost all α. In the trivial case when R is a conditional expectation operator, one obtains a representation of R as a direct integral of expectation operators, that is, operators of the form $f \rightarrow e \int_S f(s)\mu(ds)$, where e is the identity function.[10] Theorem 1 is not used in the sequel: it is preferable to classify *S.O.s* directly, without previous reduction.

3. *Invariant Measures.*—The spectrum $\sigma(R)$ of a *S.O.* in $L_\infty(S, \Sigma, \mu)$ is contained in the unit disk $|\lambda| \leq 1$. If f and g are eigenfunctions of R belonging to eigenvalues λ and μ respectively, where $\lambda \neq \mu/(\mu - 1)$, then the function fg is an eigenfunction with eigenvalue $\lambda\mu/(\lambda + \mu - \lambda\mu)$.

A bounded weak*-continuous operator T in $L_\infty(S, \Sigma, \mu)$ is said to have *pure point spectrum* (*P.P.S.*) when (*a*) the spectrum of T is a totally disconnected (0-dimensional) set and (*b*) the eigenvectors of T span (in the weak*-topology) every weak*-closed subspace of $L_\infty(S, \Sigma, \mu)$ which reduces T. This notion extends the classical Hilbert space notion of operators with discrete spectra, introduced by Halmos and von Neumann.[11] The spectrum of the restriction of every operator with pure point spectrum to a weak*-closed reducing subspace can be shown to be a subset of the spectrum of T. The condition given below that a smoothing operator with *P.P.S.* be extendable to a normal operator in $L_2(S, \Sigma, u)$ amounts, in the case of an operator arising from a measurable transformation as indicated above, to a condition for the existence of an equivalent invariant measure. Specifically, let R be a smoothing operator; we say R is *invariant* when $R^*e = e$, where R^* is the adjoint of R in $L_1(S, \Sigma, \mu)$ and e is the identity function. When R is an operator of the form $\lambda(\lambda I\text{-log } T)^{-1}$, where $(Tf)(s) = f(\varphi(s))$ for a measurable transformation φ, then R is invariant whenever T is measure-preserving. Thus, whenever there is an equivalent measure μ_1 on (S, Σ) such that R is invariant in $L_\infty(S, \Sigma, \mu_1$ and $R = \lambda(\lambda I\text{-log } T)^{-1}$, then under the equivalent measure μ_1 the transformation φ is measure-preserving. The question whether a smoothing operator is invariant is thus seen to be closely related to—and in some cases to imply—the existence of an invariant measure for a measurable transformation. Surprisingly enough, the problem can be completely solved for smoothing operators with pure point spectra in $L_\infty(S, \Sigma, \mu)$. The study of these operators in the functions space $L_\infty(S, \Sigma, \mu)$, rather than in the usual Hilbert-space context, represents a notable extension of the allowable classes of measurable transformations, and contains the analogous theory in $L_p(S, \Sigma, \mu)$ $(1 \leq p < \infty)$.

THEOREM 2. *Let R be a smoothing operator with pure point spectrum in $L_\infty(S, \Sigma, \mu)$. Then there exists a finite measure μ_1 equivalent to μ such that R is invariant in $L_\infty(S, \Sigma, \mu_1)$.*

Only an indication of the proof of this result can be given here. One begins by establishing that the spectrum of R is contained in the curve $|\lambda - {}^1/_2| = {}^1/_2$, and that $\lambda = 1$ is an isolated point in $\sigma(R)$. Next it is shown that $\lambda = 1$ is a pole of the resolvent of R. This in turn, combined with results in the theory of positive operators,[12] gives the existence of integrable functions h such that $R^*h = h$. The difficulty lies in choosing h so that its support is the whole space; here the *R.I.* plays a crucial role. Note that a trivial condition guaranteeing that R is invariant is that R be a contraction operator in $L_p(S, \Sigma, \mu)$ for some finite $p \geq 1$; this condition also implies the weak*-continuity of R.

4. *Representation.*—If R is any invariant *S.O.*, then from the *R.I.* one easily obtains the identity $(R + R^*)f = 2R^*Rf$, valid for all essentially bounded f. This in turn implies that $I - 2R$ is isometric in $L_2(S, \Sigma, \mu)$, and hence can be extended to an everywhere defined bounded isometric operator. If in particular R has *P.P.S.*, then the operator R thus extended to $L_2(S, \Sigma, \mu)$ is *normal*. This implies that the operator R^* also satisfies the *R.I.* If R is invariant, a necessary and sufficient condition that R^* satisfy the *R.I.* is that the number ${}^1/_2$ does not belong to the spectrum of R. Whenever R^* satisfies the *R.I.* as well as R, and R is invariant, then R can be extended to a bounded operator in $L_p(S, \Sigma, \mu)$ $(1 \leq p \leq \infty)$, and the spectra of R in each of these spaces coincide. Thus, the structure of R is determined by its action on Hilbert space $L_2(S, \Sigma, \mu)$. Whenever R has *P.P.S.*, a complete representation for R is the following.

THEOREM 3. *An invariant smoothing operator R with pure point spectrum in $L_2(S, \Sigma, \mu)$ can be represented as follows. The space S is partitioned into uniquely determined subspaces $S_1 + S_2 + S_3$ such that (a) R acts as the zero operator on $L_2(S_1)$; (b) R coincides with a conditional expectation operator on S_2; (c) the space (S_3, Σ, μ) can be represented as a product space $I \times S_4$ of the unit interval (I, B, μ_e) with its natural Borel structure and a measure space (S_4, Σ_1, u_1). Let $f(t, s)$ be a real or complex-valued function on the product space $I \times S_4$, (where $0 \leq t < 1$ and $s \in S_4$), and* *extend f to be a periodic function on the entire real line, with period one. Then the action of the operator R on $L_2(S_3)$ is given by the Bochner integral*[13]

$$(*) \qquad (Rf)(t, s) = \frac{-r}{1 - e^r} \int_0^1 e^{ru} f(u + t, s) du, \qquad 0 \leq t < 1.$$

The real constant r is uniquely determined by the operator R.

The operator on the right side of (*), applied to complex-valued functions $f(t)$, can be considered as the Green's operator of the boundary value problem consisting[14] of the differential operator $r^{-1}f'(t) - f(t) = 0$ with the periodic boundary condition $f(0) = f(1)$. This operator is a finite realization of the "weighted average" discussed in section 1, generalizing conditional expectation.

In the proof of Theorem 3, the set S_1 is defined as the union of the supports of all eigenfunctions belonging to the eigenvalue $\lambda = 0$ of R; the set S_3 as the union of the supports of all eigenfunctions belonging to eigenvalue $\lambda \neq 0, 1$, and $S_2 = S - S_1 - S_3$. The representation of R in $L_2(S_2)$ follows from the remark that the restriction of R to this subspace is a projection. The representation of the restriction of R to $L_2(S_3)$ is obtained as follows. Let $\lambda_0 \neq 1$ be the eigenvalue of R nearest to one and lying in the upper half-plane. It can be shown that there exists a single eigenfunction f_0 belonging to the eigenvalue λ_0, such that $|f_0(s)| = 1$ almost everywhere on S_3. Assuming $\mu(S_3) = 1$ we have $\int_S f_0^m/(s) f_0^n(s) \mu(ds) = 0$, $m \neq n$. Furthermore $f_0(s) = f_0(s)^{-1}$ almost everywhere.

The decomposition of the space (S_3, Σ, μ)[15] is obtained by letting Σ' be the σ-field generated by f_0, and Σ'' the σ-field generated by all eigenfunctions belonging to the eigenvalue one. It can be shown that Σ' and Σ'' are independent σ-fields. To prove that (S_3, Σ', μ) is isomorphic to (I, B, μ_e) an isomorphism is established between the corresponding function spaces by setting $f_0 \rightarrow e^{irt}$, for suitably chosen r. This extends to an isomorphism of the algebra $L_\infty(S, \Sigma', \mu)$ onto $L_\infty(I, B, \mu_e)$. Finally, it is shown that the spectrum of the operator on the right side of (*) coincides with the spectrum of R.

5. *Sharp Conditions.*—The condition that R have pure point spectrum in Theorem 2, though the natural one in many respects, can be considerably relaxed. The resulting sharpened criteria are significant when viewed as conditions insuring that the operator be "spectral," and are in fact related to similar conditions on the resolvent studied in the theory of spectral operators.[16] If the spectrum of the smoothing operator R is totally disconnected, then necessary and sufficient conditions that R be equivalent to an invariant *S.O.* are that (*a*) the number one is an eigenvalue of R (that is, the range of the projection $E(1, R)$ is spanned by the functions f such that $(1 - R)^n f = 0$ for some n); (*b*) if for some essentially bounded f the only singularity of the vector-valued analytic function $(\lambda I\text{-}R)^{-1}f$ of λ in $L_\infty(S, \Sigma, \mu)$ lies at $\lambda = 0$, then in a neighborhood of $\lambda = 0$, $\|(\lambda I\text{-}R)^{-1}f\| = 0(|\lambda|^{-m})$ for some $m > 0$ as $|\lambda| \rightarrow 0$. These conditions imply in particular that R has pure point spectrum.

Conditions (*a*) and (*b*) can be considerably relaxed, and a complete representation theory can be obtained for weak*-continuous operators in L_∞ which satisfy the Reynolds identity and have pure point spectrum. It can be shown that any such operator can be represented by essentially trivial modifications of smoothing operators. The main device in arriving at such a representation is the existence of a partition of the space S into $S_1 + S_2$ such that the operator R_1: $f \rightarrow \chi_{S_1} Rf$ is a smoothing operator when restricted to $L_\infty(S_1)$, and $Rg = 0$ for all g supported on S_2. The difficulty is to prove that $R_1 e = e$ on S_1; this requires a previous study of the spectrum of R.

For *simple invariant* operators in $L_2(S, \Sigma, \mu)$ satisfying the *R.I.* and having *discrete* spectrum in the sense of Halmos and von Neumann, the following representation can be easily obtained. The space S is partitioned into $S_1 + S_2$, where R is the operator $f \rightarrow (\int f)e$ on $L_2(S_1)$ and $\exp(tR^{-1} - t)$ restricted to $L_2(S_2)$ is equivalent to a rotation on a compact abelian group for every real t. In particular if R is simple and has pure point spectrum, then R is isomorphic to a direct sum of the "definite integral" and $[1 + a(d/dt)]^{-1}$, where d/dt is the ordinary derivative on a compact interval of the line.

The main result, Theorem 2, can be applied to prove the existence of invariant measures for measurable transformations possessing an infinitesimal generator, or more generally for semigroups of maps in the sense considered by Hufford.[17] The determination of bounded eigenfunctions is often simpler than the study of the spectrum in other function spaces. The details of such applications will appear in a separate publication.

The algebraic theory of the Reynolds Identity is closely connected with the study of *derivations* in abstract algebras. It can be shown that $R^{-1} - I$ is a derivation whenever R^{-1} exists. It can be easily shown that R^{-1} is unbounded except in trivial cases; this leads to a proof of the Singer-Wermer Theorem.[18]

* This work was supported by the Office of Naval Research.

[1] Typically in I. J. Schoenberg's theory of variation-diminishing transforms; cf. Widder, D. V., *The Convolution Transform* (Princeton: Princeton Univ. Press, 1955). Note that the smoothing operators on the real line considered above are indeed variation-diminishing on continuous functions.

[2] After Reynolds, Osborne; cf. *Phil. Trans. Roy. Soc.*, **A186**, part I, 123–164 (1895). For a survey of the work up to 1957, cf. Dubreil-Jacotin, M.-L., *Etude Algébrique des transformations de Reynolds, Colloque d'Algèbre Supérieure* (Louvain: CBRS, 1957), pp. 9–27.

[3] Cf. Kampé, J., de Fériet, "Fonctions aléatories et théorie statistique de la turbulence," in *Theorie des Fonctions Aléatoires*, ed. A. Blanc-Lapierre et R. Fortet, (Paris: Masson, 1953). Cf. also Dubreil-Jacotin, M.-L., "Sur le Passage des Equations de Navier-Stokes aux Equations de Reynolds," *C. R. Acad. Sci., Paris*, **244**, 2887–2890 (1957).

[4] Cf. Loeve, *Probability Theory* (New York: Van Nostrand, 1955), p. 341 ff.

[5] Cf. Arbault, J., "*Transformations de Reynolds sur un ensemble fini*," *Corso sulla teoria della turbolenza* (Turin, 1957), pp. 115–121; Billik, M., and G.-C. Rota, "On Reynolds Operators in Finite-Dimensional Algebras," *J. Math. and Mechanics*, **9**, (1960); Molinaro, I., "Sur les Endomorphismes de Reynolds sur un Ensemble Fini," *Alger Mathématique*, 87–101 (1957).

[6] For notation and the explanation of terms referring to measure theory and operator theory, see Dunford, N., and J. T. Schwartz, *Linear Operators* (New York: Interscience, 1959), vol. 1.

[7] Cf. Moy, S. T. C., "Characterization of Conditional Expectation as a Transformation on Function Spaces," *Pacific J. Math.*, **4**, 47–63 (1954); Rota, G.-C., "On the Representation of Averaging Operators," *Padova Rendiconti*, 52–63 (1960).

[8] Cf. Billik and Rota, *loc. cit.* This result is not used in the proofs below.

[9] Cf. Dixmier, *Les Algèbres d'Opérateurs dans l'Espace Hilbertien* (Paris: Gauthier-Villars, 1956), chap. 3.

[10] For a similar representation of conditional expectation in terms of subdirect sums, see Birkhoff, G., *Moyennes des Fonctions Bornées, Algèbre et Théorie des Nombres* (Paris: Colloques Internationaux du CNRS, 1950), pp. 143–153.

[11] Cf. Halmos, P. R., and J. von Neumann, "Operator Methods in Classical Mechanics II," *Ann. Math.*, **43**, 332–350 (1942).

[12] Cf. Karlin, S., "Positive Operators," *J. Math. and Mechanics*, **8**, 907–938 (1959).

[13] Cf. Hille and Phillips, *Functional Analysis and Semigroups*, (AMS, 1957), Chap. 3. The integral developed in Dunford-Schwartz, *loc. cit.*, Chap. 3, would do just as well.

[14] Cf. Coddington and Levinson, *Theory of Ordinaru Differential Equations* (New York: Mc-Graw-Hill, 1955), pp. 186–192.

[15] Here Σ and μ are of course understood as the restrictions of Σ and μ to S_1.

[16] Cf. Dunford-Schwartz, *loc. cit.*, vol. 2.

[17] Hufford, G., "One-Parameter Semigroups of Maps," *Duke Math. J.*, **24**, 443–453 (1957).

[18] Singer, I. M., and J. Wermer, "Derivations on Commutative Normed Algebras," *Math. Ann.*, **129**, 260–264 (1955).

ANALYSE FONCTIONNELLE. — *Endomorphismes de Reynolds et théorie ergodique*. Note (*) de M. **Gian-Carlo Rota**, présentée par M. Joseph Kampé de Fériet.

Les R-endomorphismes de Reynolds ont été introduits par Mme M.-L. Dubreil-Jacotin [1] comme extension de la notion de T-endomorphisme de Reynolds, étudiée par J. Kampé de Fériet [2], et plusieurs autres auteurs après lui. Nous étudions ici, au point de vue de la théorie spectrale, ces opérateurs dans l'algèbre de Banach $L_\infty(S, \Sigma, \mu)$ sur un espace de mesure (S, Σ, μ) à mesure finie. On trouve ainsi une remarquable liaison entre la théorie de ces endomorphismes et certains problèmes classiques de la théorie ergodique.

Soit R un opérateur linéaire dans l'espace de Banach $L_\infty(S, \Sigma, \mu)$, continu dans la topologie faible induite par la dualité (L_∞, L_1) [3]. On appelle R un *opérateur de Reynolds* quand il satisfait aux conditions suivantes : (*a*) R est *positif* : $Rf \geq 0$ pour $f \geq 0$; (*b*) R satisfait à l'*axiome stochastique des moyennes* $R(fg) = Rf\,Rg + R[(f - Rf)\,(g - Rg)]$, f et g étant des « fonctions » bornées. On sait que l'espérance mathématique conditionnelle $f \to E^B(f)$, où B est un sous-corps du corps d'ensembles Σ, satisfait aux conditions (*a*) et (*b*) [4]. On obtient un exemple plus frappant d'opérateur de Reynolds à partir d'une transformation mesurable $s \to \varphi(s)$ de l'espace S dans lui-même, soumise aux conditions suivantes : l'opérateur $f(s) \to (Vf)(s) = f(\varphi(s))$ est faiblement continu dans L_∞, et son logarithme peut être défini par les méthodes usuelles du calcul opérationnel. Ces deux conditions sont satisfaites dans les cas les plus fréquents, par exemple quand la transformation φ a été obtenue en résolvant un système d'équations différentielles ordinaires. On démontre alors aisément que l'opérateur $R = z(zI - \log V)^{-1}$ est un opérateur de Reynolds pour tout z réel tel que e^z n'appartient pas au spectre de V dans L_∞. Il est intéressant de savoir s'il y a une mesure équivalente à μ et *invariante* pour φ. On démontre que ce problème est équivalent au problème suivant pour l'opérateur de Reynolds correspondant à φ : y a-t-il une mesure finie μ_1 équivalente à μ, telle que, pour la transformation adjointe R^* dans $L_1(S, \Sigma, \mu_1)$, on ait $R^*e = e$? Une telle mesure sera dite *invariante* pour R. En démontrant l'existence d'une mesure invariante pour les opérateurs de Reynolds les plus généraux, nous arriverons à une représentation concrète de ces opérateurs.

Dans la suite on considère les opérateurs de Reynolds à *spectre ponctuel*, c'est-à-dire satisfaisant aux conditions suivantes : le spectre de R dans L_{∞} est un ensemble totalement discontinu — par exemple, une succession de points avec un nombre fini de points limites — et les combinaisons linéaires des fonctions propres généralisées sont faiblement denses ([5]). Ces opérateurs généralisent les transformations à spectre discret, étudiées par Halmos et von Neumann ([6]). Notons cependant que le problème considéré ici est beaucoup plus général que le cas étudié par ces auteurs, parce qu'on démontre ici l'existence d'une mesure invariante, tandis qu'ils supposent *a priori* l'existence d'une telle mesure.

THÉORÈME 1. — *Soit* R *un opérateur de Reynolds à spectre ponctuel dans* $L_{\infty}(S, \Sigma, \mu)$. *Alors il y a une partition de l'espace* S *en deux ensembles disjoints* $S_1 + S_2$ *tels que : (a) en désignant par* R_1 *l'opérateur* $f \to \chi_{S_1} Rf$, *où* χ_{S_1} *est la fonction caractéristique de l'ensemble* S_1, *la restriction de* R_1 *à* $L_{\infty}(S_1)$ *est un opérateur de Reynolds; (b) Soit e la fonction égale à un partout dans* S_1 *et à zéro partout dans* S_2*; alors on a* $R_1 e = e$; *(c) Si* $f \geqq 0$ *et* $Rf = 0$, *le support de f est contenu dans* S_2, *et inversement, si le support de g est contenu dans* S_2, *on a* $Rg = 0$.

Dans la démonstration on définit l'ensemble S_1 comme le support de la fonction intégrable R^*e. Puis, on établit le fait que le spectre d'un tel opérateur est étalé sur le cercle $|z - 1/2| = 1/2$ du plan complexe. On démontre aussi que le sous-espace des fonctions invariantes est de la forme $L_{\infty}(S, \Sigma', \mu)$, où Σ' est un sous-corps borélien de Σ. Ceci donne aisément la conclusion (*b*).

THÉORÈME 2. — *Soit* R *un opérateur de Reynolds à spectre ponctuel, et soit* R_1 *sa restriction à l'ensemble* S_1 *du théorème* 1 (*a*). *Alors il y a sur* S_1 *une mesure* μ_1 *invariante pour* R_1 *et équivalente à* μ.

L'outil principal dans la démonstration de ce résultat est l'extension aux espaces de Banach ordonnés de la théorie classique de Frobenius des valeurs et fonctions propres des opérateurs positifs. Après avoir établi que le point $z = 1$ dans le plan complexe est un pôle du résolvant $(zI - R)^{-1}$, on démontre l'existence d'une fonction intégrable h telle que $R^*h = h$. La difficulté est de choisir une telle fonction h qui ait son support dans tout l'ensemble S_1. Cela fait, on obtient la mesure invariante μ_1 en posant

$$\mu_1(E) = \int_E h(s)\mu(ds).$$

Ce dernier théorème a pour consequence l'existence d'une mesure invariante pour les transformations φ. Plus généralement, on obtient comme corollaire l'existence d'une mesure invariante pour tout opérateur satisfaisant à l'axiome stochastique des moyennes, borné dans L_p pour une valeur fixe de p et ayant spectre ponctuel dans cet espace.

L'extension de ces résultats aux opérateurs à spectre continu ou mixte présente des problèmes tout à fait différents qui seront traités dans des publications ultérieures.

(*) Séance du 4 avril 1960.

(1) Cf. *Étude Algébrique des Transformations de Reynolds, Colloque d'Algèbre supérieure*, C. B. R. S., Louvain, 1957.

(2) J. Kampé de Fériet, *Annales de la Société Scientifique de Bruxelles*, 63, 1949, p. 156-172 et le Mémoire dans le *Congrès International de Mathématiciens*, Amsterdam, 3, 1954, p. 237 par le même auteur.

(3) Pour les résultats d'Analyse fonctionnelle utilisés ici, *voir* Dunford-Schwartz, *Linear Operators*, New-York, 1959.

(4) G.-C. Rota, *Representation of Averaging Operators, Rendiconti Padova*, 1960, p. 52-64.

(5) C'est-à-dire les fonctions bornées telles que $(zI - R)^n f = 0$ pour quelques valeurs de z et de n. Si M est un sous-espace faiblement fermé de L_∞, invariant pour R, les combinaisons linéaires de ces fonctions propres doivent être denses dans M.

(6) Halmos et von Neumann, *Annals of Math.*, 43, 1942, p. 153-166.

Extrait des *Comptes rendus des séances de l'Académie des Sciences*, t. 250, p. 2791-2793, séance du 20 avril 1960.

ANALYSE FONCTIONNELLE. — *Représentation des opérateurs de Reynolds.*

Note (*) de M. **Gian-Carlo Rota**, présentée par M. Joseph Kampé de Fériet.

Cette Note continue l'étude commencée dans la Note précédente [1]. On détermine ici tous les types d'opérateurs de Reynolds à spectre ponctuel. Nous suivons la terminologie et la notation de la Note précédente.

Rappelons qu'un opérateur de Reynolds est dit *simple* quand la seule fonction invariante est la fonction égale à un presque partout [2]. En faisant correspondre les opérateurs de Reynolds aux transformations mesurables φ comme dans la Note précédente, les opérateurs de Reynolds simples correspondent aux transformations φ *ergodiques*. Supposons que l'opérateur de Reynolds R puisse être étendu à l'espace de Hilbert $L_2(S, \Sigma, \mu)$ et qu'il soit borné dans cet espace — hypothèse qui est toujours satisfaite quand R a une mesure invariante —. En faisant usage de la théorie classique de von Neumann de la réduction des algèbres d'opérateurs, on démontre le

Théorème 3. — *Tout opérateur de Reynolds dans* $L_2(S, \Sigma, \mu)$ *est une intégrale directe d'opérateurs de Reynolds simples :*

$$R = \int^{\oplus} R_x \, d\nu(x).$$

La démonstration suit les lignes connues de ce type de théorème [3]. La seule difficulté consiste à choisir les espaces de Hilbert « différentiels » de telle manière que pour presque toute valeur de x l'opérateur irréductible R_x satisfait à l'identité de Reynolds. On achève cela en introduisant les espaces parfaits de Segal [4]. Dans le cas simple de l'espérance mathématique conditionnelle, on obtient une extension d'un résultat de Birkhoff [5], qui fit usage de la méthode de ses décompositions subdirectes.

Malheureusement, le théorème 3 ne simplifie pas l'étude des opérateurs de Reynolds. On trouve qu'une réduction préalable aux opérateurs simples ne donne pas une structure aussi fine qu'une représentation directe. Soit donc un opérateur de Reynolds R avec spectre ponctuel. Moyennant la réduction donnée par le théorème 2, on peut supposer que cet opérateur a une mesure invariante. Ceci permet de montrer que R peut être étendu à l'espace de Hilbert $L_2(S, \Sigma, \mu)$. En effet, on obtient l'identité $Rf + R^* f = 2R^* Rf$ pour toute fonction bornée f, d'où l'on conclut que $I - 2R$ est isométrique dans L_2. Comme le spectre de R est totalement discontinu, l'opérateur $I - 2R$ doit être unitaire, par suite R est *normal*. On obtient comme corollaire que l'opérateur adjoint R^* satisfait aussi à l'identité de Reynolds; cette dernière proposition est vraie seulement si R est normal.

En supposant que R est un opérateur de Reynolds normal quelconque dans L_2 — hypothèse qui est toujours satisfaite pour les opérateurs R_1 du théorème 2 — on démontre l'existence d'une partition unique $S' + S''$ de l'espace S avec les propriétés suivantes : (*a*) Les sous-espaces complémentaires $L_2(S')$ et $L_2(S'')$ sont invariants pour l'opérateur R; (*b*) La restriction de R à $L_2(S')$ est une espérance mathématique conditionnelle; (*c*) En désignant par R'' la restriction de R à $L_2(S'')$, la famille d'opérateurs $V_t = \exp(tR^{-1} - t)$ pour t réel est de la forme $V_t f(s) = f(\varphi_t(s))$, où φ_t est un groupe à un paramètre de transformations de l'espace S'' préservant la mesure. Ce résultat réduit l'étude des transformations de Reynolds à celle des transformations de l'espace S''. Notre intérêt est pourtant opposé : on fait usage des opérateurs de Reynolds pour étudier les transformations de S''.

Supposons maintenant que le spectre de R soit ponctuel. Alors on démontre que le sous-espace S'' peut être représenté d'une manière unique comme produit $S'' = I \times S'''$ de l'intervalle $I = [0, 1]$ avec la mesure de Lebesgue et d'un espace à mesure finie (S''', Σ'', μ''). Soit $f(t, s)$ une fonction à valeurs complexes dont le domaine est $I \times S'''$, avec $t \in I$ et $s \in S'''$; étendons le domaine de f de telle manière que f devienne une fonction périodique en t avec période un pour tout t réel. On a alors le

Théorème 4. — *Pour $f = f(t, s)$ ayant son support sur S'' tout opérateur de Reynolds* R *avec spectre ponctuel dans* $L_\infty(S, \Sigma, \mu)$ *est représenté par l'intégrale de Bochner* ([6])

$$(\star) \qquad Rf(t, s) = \frac{r}{e^r - 1} \int_t^{1+t} e^{r(u-t)} f(u, s)\, du, \qquad s \in S'',$$

où la constante réelle r est univoquement déterminée par R.

On arrive à la décomposition $S'' = I \times S'''$ comme il suit. Soit z_0 la valeur propre de R la plus voisine de un, différente de un; on démontre que, parmi les fonctions propres correspondant à z_0, il y en a une, appelons-la f_0, telle que $|f_0(s)| = 1$ presque partout dans S''. Soit Σ' le corps borélien engendré par f_0, et soit Σ'' le corps borélien engendré par toutes les fonctions fixes pour R. On démontre ensuite que les deux corps Σ' et Σ'' sont *indépendants* au sens du calcul des probabilités, et par suite que $\Sigma = \Sigma' \times \Sigma''$ sur S''. On définit maintenant un isomorphisme de (S, Σ', μ) sur I avec sa structure borélienne en posant $f_0 \to e^{irt}$ avec r réel convenablement choisi. Comme les polynomes en f_0 sont denses dans $L_2(S, \Sigma', \mu)$, cette application s'étend à une application de tout l'espace $L_2(S, \Sigma', \mu)$ sur $L_2(I)$; on peut démontrer que cette application est *multiplicative* sur la sous-algèbre des fonctions bornées. Par suite, en appliquant le « théorème de multiplication » de von Neumann ([7]), on trouve une application de S sur I qui préserve la mesure. Cela fait, la formule (★) s'ensuit en vérifiant que les opérateurs donnés dans les deux membres ont le même spectre et les mêmes multiplicités, d'où le théorème en appliquant le théorème spectral pour les opérateurs normaux.

L'opérateur (★) donne une généralisation de l'espérance mathématique conditionnelle, qui sera ultérieurement étudiée. Notons enfin que la théorie des opérateurs de Reynolds, en réduisant à un même ordre d'idées tant l'espérance mathématique conditionnelle que les transformations mesurables, donne un outil pour un développement unifié de la théorie de la convergence des martingales et les théorèmes des moyennes ergodiques.

(*) Séance du 4 avril 1960.

([1]) *Comptes rendus*, 250, 1960, p. 2791.

([2]) Cette notion est due à Mme Dubreil-Jacotin. Voir *loc. cit.* dans la Note précédente.

([3]) *Cf.* J. Dixmier, *Les algèbres d'opérateurs dans l'espace hilbertien*, Paris, 1956, chap. III.

([4]) *Decompositions of operator algebras*, A. M. S., 1951.

([5]) *Moyennes des fonctions bornées*, Colloque d'Algèbre et Théorie des nombres, C. N. R. S., Paris, 1950, p. 143-153.

([6]) *Voir* Hille-Phillips, *Functional Analysis and Semi-groups*, A. M. S., 1958, chap. III.

([7]) *Cf.* Halmos, *Math. Soc. Japan*, 1956, p. 45.

CALCUL DES PROBABILITÉS. — *Une généralisation de l'espérance mathématique conditionnelle qui se présente dans la théorie statistique de la turbulence.* Note (*) de M. **Gian-Carlo Rota**, transmise par M. Joseph Kampé de Fériet.

Dans l'étude du passage des équations de Navier-Stokes aux équations de Reynolds pour les vitesses $u_j(x_1, x_2, x_3, t)$ d'un fluide visqueux [1] et ses moyennes $\bar{u}_j$, on fait usage de l'identité

$$(*)\ \mathrm{R}(fg) = \mathrm{R}f\mathrm{R}g + \mathrm{R}\{(f-\mathrm{R}f)(g-\mathrm{R}g)\}$$

qui constitue l'*axiome stochastique des moyennes.* Ici R est un opérateur linéaire dans une algèbre de fonctions dont f et g sont des éléments. Pour les vitesses u_j on a $\bar{u}_j = \mathrm{R}\, u_j$. On a démontré [2] que cette identité est suffisante pour que le passage des équations de Navier-Stokes aux équations de Reynolds soit rigoureux. L'opérateur linéaire R est appelé un *endomorphisme de Reynolds.*

Dans cette Note *nous donnons toutes les formes analytiques possibles des endomorphismes de Reynolds*, en supposant que les vitesses u_j dans les équations de Navier-Stokes soient des *fonctions aléatoires*, c'est-à-dire qu'elles dépendent mesurablement d'une variable s qui parcourt un espace de probabilité $(\mathrm{S}, \Sigma, \mu)$: $u_j = u_j(x_1, x_2, x_3, t, s)$. Comme les moyennes de Reynolds doivent commuter avec les dérivations relatives aux variables x et t, nous sommes amené à considérer le cas où de tels opérateurs de Reynolds opèrent exclusivement sur la variable s. Par suite, nous sommes amené à *étudier les endomorphismes des Reynolds dans l'espace des variables aléatoires.*

Supposons donc que dans l'identité (*) f et g soient des fonctions mesurables (= variables aléatoires) essentiellement bornées définies dans $(\mathrm{S}, \Sigma, \mu)$ et prenant des valeurs réelles arbitraires, et que R est un opérateur linéaire dans l'espace vectoriel formé par ces fonctions.

Dans l'étude des moyennes de Reynolds on a fait des hypothèses supplémentaires à l'identité (*), de nature physique : telles sont la positivité ($\mathrm{R}f \geqq 0$ si $f \geqq 0$), l'identité $\int_{\mathrm{S}} \mathrm{R}f = \int_{\mathrm{S}} f$ (conservation de l'espérance mathématique) et $\mathrm{R}\, e = e$, où e est la fonction identiquement égale à 1. On peut démontrer que ces propriétés ont pour conséquence que l'opérateur R peut être étendu à toutes les variables aléatoires à carré intégrable et à valeurs complexes, et que R soit un opérateur *normal* dans cet espace.

Par suite, *nous supposons* maintenant seulement *que l'opérateur de Reynolds* R *est un opérateur linéaire borné (partout défini) de l'espace*

$L_2(S, \Sigma, \mu)$ *dans lui-même, qui satisfait à l'axiome stochastique des moyennes pour f et g bornées, et qui est un opérateur normal dans cet espace de Hilbert.*

On peut alors démontrer qu'*il n'y a que deux types de tels endomorphismes de Reynolds*, et que *tout endomorphisme de Reynolds satisfaisant à ces conditions est composé* — dans le sens que nous allons définir — *de ces deux types seuls.*

Premier type. — Si $R^2 = R$, alors (*) et la normalité de R donnent l'identité plus simple $R(fRg) = RfRg$. Les seuls opérateurs normaux satisfaisant à ces conditions sont les *espérances mathématiques conditionnelles* relatives à des sous-corps arbitraires du corps borélien Σ.

Deuxième type. — Ce type donne une *nouvelle classe d'opérateurs, qui généralisent l'espérance mathématique conditionnelle.* Soit φ_t $(t > 0)$ un semi-groupe non trivial de transformations de l'ensemble S conservant la mesure, c'est-à-dire $\mu(\varphi_t(E)) = \mu(E)$ pour tout ensemble Borélien E et pour tout $t < 0$. On a la loi de semi-groupe $\varphi_t(\varphi_s(E)) = \varphi_{t+s}(E)$ pour t, $s > 0$. Soit $U^t f(s) = f(\varphi_t(s))$. On sait que les opérateurs linéaires U^t forment un semi-groupe d'opérateurs unitaires dans l'espace de Hilbert. Soit log U le *générateur infinitésimal* de ce semi-groupe. Alors *l'opérateur linéaire défini par la formule*

$$R = (I - i \log U)^{-1},$$

où I est l'opérateur identité, *est un endomorphisme de Reynolds.*

On a le théorème de décomposition suivant :

THÉORÈME. — *Soit* R *un endomorphisme de Reynolds qui est un opérateur linéaire normal dans l'espace de Hilbert* $L_2(S, \Sigma, \mu)$ *des variables aléatoires à carré sommable. Alors il y a une décomposition unique de l'ensemble* S *dans trois sous-ensembles disjoints* $S_0 + S_1 + S_2 = S$ *telle que chacun des trois sous-espaces* $L_2(S_i)$, $(i = 0, 1, 2)$ *soit un sous-espace invariant de* R *et que :*

(*a*) *La restriction de* R *à* $L_2(S_0)$ *est l'opérateur zéro;*

(*b*) *La restriction de* R *à* $L_2(S_1)$ *est un opérateur de Reynolds du premier type, c'est-à-dire une espérance mathématique conditionnelle;*

(*c*) *La restriction* R_2 *de* R *à* $L_2(S_2)$ *est un opérateur de la forme* $R_2 = R_1 R_0$, *où* R_1 *est du deuxième type de* R_0 *est du premier type.*

Dans la démonstration, on définit S_0 comme le complément de la réunion des supports des éléments de l'image $R(L_2(S, \Sigma, \mu))$. On définit S_1 comme le plus grand sous-ensemble de $S - S_0$ tel que la restriction de R à l'intersection de $L_2(S_1)$ avec l'image de R soit l'opérateur identité. Soit enfin $S_2 = S - S_0 - S_1$. On définit R_0 comme la projection orthogonale de l'espace sur l'adhérence de l'intersection de $L_2(S_2)$ avec l'image de R. On définit R_1 comme la restriction de R à l'adhérence de l'intersection de $L_2(S_2)$ avec l'image de R.

Les opérateurs de Reynolds du deuxième type constituent une généralisation de l'espérance mathématique conditionnelle, dont la signification probabiliste reste à étudier. Son apparition naturelle dans l'étude des solutions aléatoires des équations de Navier-Stokes indique que cette généralisation n'est pas sans intérêt.

(*) Séance du 25 juillet 1960.

(1) Ce problème a été posé et étudié à plusieurs reprises par J. Kampé de Fériet, depuis 1934 : *Voir* ses Mémoires dans *Annales de la Société Scientifique de Bruxelles*, 63, 1949, p. 156 et *Congrès international des Mathématiciens*, Amsterdam, 3, 1954, p. 237, où l'on trouvera les références bibliographiques.

(2) *Voir* M.-L. Dubreil-Jacotin, *Comptes rendus*, 244, 1957, p. 2887.

(3) Pour les notations et les définitions des termes, *voir* Dunford-Schwartz, *Linear Operators*, I, Interscience, New York, 1959.

(4) *Voir* G.-C. Rota, *Comptes rendus*, 250, 1960, p. 2791 et 2831.

Comptes Rendus des Séances de l'Académie des Sciences Paris, **251** (1960), 624–626.

On Reynolds Operators in Finite-Dimensional Algebras

MARTIN BILLIK & GIAN-CARLO ROTA*

Communicated by G. BIRKHOFF

1. Definitions. Let $\mathfrak{a}$ be a commutative algebra with identity e over a field F of characteristic not equal to two. A REYNOLDS *operator* in $\mathfrak{a}$ is an everywhere defined linear operator of $\mathfrak{a}$ into itself satisfying the REYNOLDS *identity* (after the British engineer OSBORNE REYNOLDS)

$$R(fg) = (Rf)(Rg) + R[(f - Rf)(g - Rg)], \qquad f, g \in \mathfrak{a}. \tag{1}$$

The origin of this identity, as well as its numerous applications to operator theory, probability and turbulence theory, will be discussed elsewhere. The reader is referred to the works of Mme. MARIE-LOUISE DUBREIL-JACOTIN (*cf.* [6], [7] and [8]) for recent surveys of the literature related to REYNOLDS operators.

A related concept is that of a *smoothing operator*; such an operator A is an everywhere defined linear operator in the algebra $\mathfrak{a}$ which satisfies the identity (the so-called averaging identity)

$$A[f(Ag)] = (Af)(Ag), \qquad f, g \in \mathfrak{a}. \tag{2}$$

In a lattice-ordered algebra (*cf.* [4]) an order-preserving smoothing operator is called an *averaging* operator. For a survey of the theory of averaging operators, see [3]; the relationship with the notion of conditional expectation in probability theory is discussed in [12]. Order and lattice properties will not be considered in this work. Smoothing operators have been studied rather completely, and can be considered as known (cf. [3]). Smoothing operators in the algebra of continuous functions $C(S)$ over a compact Hausdorff space have been completely classified (*cf.* [2], [3] and [10]). Averaging operators in the algebra of measurable functions have also been studied (*cf.* [12]).

The following counter-example, due to MOLINARO (*cf.* [6]), shows that in general the REYNOLDS identity does not imply the averaging identity. Let $\mathfrak{a}$

* The work of one of the authors was partially supported by the Office of Naval Research Contract No. 7667.

Journal of Mathematics and Mechanics, Vol. 9, No. 6 (1960).

be the algebra of all bounded real continuous function on the real line, and let

$$(Rf)(x) = e^{-x} \int_{-\infty}^{x} f(t)e^{t}\, dt.$$

Then R is a REYNOLDS operator, but it is evidently not a smoothing operator.

The purpose of this work is to show that *in a finite-dimensional algebra* $C(S)$ of real functions on a finite set S (or, equivalently, a vector space under componentwise product) *every* REYNOLDS *operator is a smoothing operator*. This result, together with the comparatively simple characterization of smoothing operators (also derived independently below), gives a complete classification of REYNOLDS operators in a finite-dimensional algebra $C(S)$. A partial result in this direction was recently obtained by MOLINARO (*cf.* [11]), who showed that in such an algebra an order-preserving REYNOLDS operator is an averaging operator. Using similar methods J. ARBAULT (*cf.* [1]) was able to dispense with the condition of positivity. The methods used in the present work are altogether different and independent from those of MOLINARO and ARBAULT.

The exposition is self-contained. Some of the propositions below have been established independently by Mme. DUBREIL-JACOTIN and her school. Whenever possible, the proofs given below generalize or are closest to being generalizable to the case $C(S)$ of all continuous functions on a compact Hausdorff space.

2. Identities. We collect below some useful properties of general REYNOLDS operators which do not require any finiteness assumptions.

Proposition 2.1. *The following identities are equivalent to the* REYNOLDS *identity*:

$$R(fRg + gRf) = RfRg + R(RfRg) \tag{3}$$

$$2R(fRf) = (Rf)^2 + R(Rf)^2. \tag{4}$$

Proof. (3) follows from (1) upon expanding and cancelling. (4) follows from (1) by setting $f = g$. One obtains (1) from (4) by polarization, replacing f by $f + g$ and expanding.

Proposition 2.2. *Every idempotent* ($R^2 = R$) REYNOLDS *operator is a smoothing operator.*

Proof. Set $g = Rh$ in the REYNOLDS identity.

Proposition 2.3. *If f is a fixpoint of R* (i.e., $Rf = f$), *then*

$$R(fg) = fRg, \qquad g \in \mathfrak{a}. \tag{5}$$

Proof. Immediate from the REYNOLDS identity.

Proposition 2.4. *If $Rg = 0$ then*

$$R(gRf) = 0, \qquad f \in \mathfrak{a}.$$

Proof. Immediate from (3).

Proposition 2.5. *If* $Rg = e$ *(e is the identity of* $\mathfrak{a}$*) then*

$$R(gRf) = R^2f, \qquad f \in \mathfrak{a}.$$

Proof. Use identity (3) to get

$$R(fe + gRf) = Rf + R(eRf)$$

from which the conclusion is immediate.

Proposition 2.6. *Let* $\mathfrak{a}$ *be a commutative algebra with identity having no non-zero nilpotent elements. If for a given h in* $\mathfrak{a}$*, and for* $n \geqq 1$, $R^nh = 0$, *then* $R^2h = 0$.

Proof. It can be assumed that $n \geqq 3$. Setting $g = R^{n-1}h$ and $f = R^{n-2}h$ in Proposition 2.4 gives $R((R^{n-1}h)^2) = 0$. Now set $f = R^{n-2}h$ in identity (4) and obtain

$$\begin{aligned} 2R(R^{n-2}hR^{n-1}h) &= (R^{n-1}h)^2 + R(R^{n-1}h)^2 \\ &= (R^{n-1}h)^2. \end{aligned}$$

Setting $g = R^{n-1}h$ and $f = R^{n-3}h$ in Proposition 2.6 we see that the left side vanishes. Hence $(R^{n-1}h)^2 = 0$; but this implies that $R^{n-1}h = 0$ in view of the assumption.

Corollary 2.7. *Under the same assumptions on R, if f is a fixpoint of R and* $R^ng = f$, *then* $R^2g = f$.

Proof. Set $h = g - f$ in the preceding statement.

Proposition 2.8. For $n \geqq 1$, *the range* $R^n(\mathfrak{a})$ *is a subalgebra of* $\mathfrak{a}$.

Proof. Obvious from the Reynolds identity.

Proposition 2.9. *Let* $F = \{f \mid Rf = f\}$ *be the subspace of fixpoints of R. Then F is a subalgebra of* $\mathfrak{a}$.

Proof. Immediate from the Reynolds identity (1).

3. Reduction. From now on it will be always assumed that $\mathfrak{a} = C(S)$ is the algebra of all real-valued functions on a finite S.

Recall that a subalgebra $\mathfrak{a}'$ of $C(S)$ can be characterised as follows. Let $N \subset S$ be the set of all points $i \in S$ such that $f(i) = 0$ for all $f \in \mathfrak{a}'$. Then $\mathfrak{a}'$ is isomorphic to a subalgebra $\mathfrak{a}''$ of $C(S - N)$ such that for $i \in S - N$, $f(i) \neq 0$ for some $f \in \mathfrak{a}''$. It follows that one can find a unique partition $S_1, S_2, \cdots, S_p$ of the set $S - N$ with the property that $\mathfrak{a}''$ is the algebra of all functions such that $f(i) = f(j)$ if i and j belong to the same partition S_k. In particular $\mathfrak{a}''$ is isomorphic to the algebra $C(S')$, where $S' = \{S_1, S_2, \cdots, S_k\}$. The partition $N, S_1, S_2, \cdots, S_k$ is called below the partition *associated with* the sub-algebra $\mathfrak{a}'$. Note that the identity element e belongs to a subalgebra $\mathfrak{a}'$ if and only if the component N of the partition associated with $\mathfrak{a}'$ vanishes.

We denote by e_i the idempotent function such that $e_i(j) = \delta_{ij}$, where δ_{ij} is the Kronecker delta. If f belongs to $\mathfrak{a} = C(S)$, we denote by $S(f)$ the *carrier* (support) of f, that is, the set of all i in S for which $f(i) \neq 0$.

A *minimal* REYNOLDS operator is one for which the identity element e of $\mathfrak{a}$ is the only fixpoint.

Proposition 3.1. *If the* REYNOLDS *operator R is invertible, then R is the identity operator.*

Proof. Since R is one-to-one onto, we can find h such that $Rh = e$, the identity element. From Proposition 2.5 it follows that $R(hRf) = R^2f$ for all f. Since R^{-1} exists we obtain $hRf = Rf$ for all Rf. Hence $h = e$ and $Re = e$. Now in identity (3) choose g and f so that $Rg = e_i$, $Rf = e_j$ with $i \neq j$. Then $RfRg = 0$. Operating with R^{-1} on both sides of (3) we obtain $fRg + gRf = fe_i + ge_j = 0$. Multiplying both sides by e_i we get $e_ife_i = fe_i = f(i)e_i = 0$ for all $i \neq j$. Hence only the jth component of f may be different from zero, or $f = k_je_j$ with $k_j \neq 0$ since R is invertible. Similarly, $g = k_ie_i$, $k_i \neq 0$, and $Re_i = k_i^{-1}e_i$ for all i. Now $e = \sum_{i=1}^{n} k_i^{-1}e_i$ is left fixed by R; hence $k_i^{-1} = 1$ for all i, and $Re_i = e_i$ for all i, *q.e.d.*

Proposition 3.2. *Let $F = \{f \mid Rf = f\}$ be the subalgebra of fixpoints of R* (cf. *Proposition* 2.9). *Then $F = R^2(\mathfrak{a})$.*

Proof. Let $\mathfrak{a}_0 = \mathfrak{a}$, $\mathfrak{a}_{n+1} = R(\mathfrak{a}_n)$. Then $\mathfrak{a}_n$ $(n = 0, 1, 2, \cdots)$ is a descending sequence of subalgebras. It follows that for some n_0, $\mathfrak{a}_{n_0} = \mathfrak{a}_{n_0+1} = \cdots$. Clearly $F \subset \mathfrak{a}_{n_0}$; on the other hand, $\mathfrak{a}_{n_0}$ is isomorphic to $C(S')$, as remarked at the beginning of this section, and the restriction of R to $\mathfrak{a}_{n_0}$ is an onto map. It follows from the preceding Proposition 3.1 that $Rg = g$ for all g in $\mathfrak{a}_{n_0}$, and hence that $\mathfrak{a}_{n_0} = F$. Thus $R^{n_0}g$ is a fixpoint of R for all g in $\mathfrak{a}$; but this implies by Corollary 2.7 that R^2g is a fixpoint. Hence $n_0 = 2$, *q.e.d.*

Corollary 3.3. *The operator R satisfies the identity $R^2(I - R) = 0$. In particular, R^2 is an idempotent, and the only points in the spectrum of R (considered as a linear operator in the complexification of the algebra $C(S)$) are* 0 *and* 1.

Proposition 3.4. *Let f be an idempotent of $\mathfrak{a}$ (that is, $f^2 = f$) fixed under R. If $S(g) \subset S(f)$ then $S(Rg) \subset S(f)$.*

Proof. We have $fg = g$ and by Proposition 2.3, $Rg = R(fg) = fRg$, *q.e.d.*

Corollary 3.5. *If f is as above, then the subalgebra $C(S(f))$ is invariant under R.*

Theorem 1. (Reduction theorem). *Let $N, S_1, \cdots, S_k$ be the partition associated with the subalgebra F of fixpoints of the* REYNOLDS *operator R. Then each of the subalgebras $C(N)$, $C(S_1)$, $C(S_2)$, $\cdots$, $C(S_k)$ is invariant under R. Thus every* REYNOLDS *operator in $C(S)$ is the direct sum (in the algebra sense) of a nilpotent* REYNOLDS *operator and of minimal* REYNOLDS *operators.*

Proof. It follows from the classification of subalgebras given above that the idempotent f_i whose support is the set S_i is fixed under R. From the preceding Proposition 3.4 it is seen that the subalgebras $C(S_1), \cdots, C(S_k)$ are invariant under R. Evidently the idempotent f_i acts as an identity in the subalgebra $C(S_i)$; thus R is a minimal REYNOLDS operator when restricted to $C(S_i)$. It remains to be shown that $C(N)$ is invariant under R, where N is the "null set" of the fixpoint algebra F. Let f be the idempotent whose support is $S - N$. If $S(g) \subset N$ then $gf = 0$. By Proposition 2.3, $R(fg) = 0 = fR(g)$ since f is fixed under R. Therefore the support of Rg is disjoint from the support of f; in other words, $S(Rg) \subset N$, *q.e.d.*

4. Structure. The main result is now established first with minimal REYNOLDS operators and then extended to the general case. We first show that there are no nilpotent REYNOLDS operators.

Proposition 4.1. *If R is a nilpotent* REYNOLDS *operator, then $R = 0$.*

Proof. By Proposition 2.6, $R^2 = 0$. The proof is complete if it can be shown that $Re_i = 0$ for each i, where $e_i(j) = \delta_{ij}$. Since $R^2 e_i = 0$ we have by Proposition 2.4 that $R(Re_i Rf) = 0$ for all f, in particular $R((Re_i)^2) = 0$. Setting $f = e_i$ in identity (4) we obtain $2R(e_i Re_i) = (Re_i)^2$. But $e_i Re_i = ke_i$, where $k = (Re_i)(i)$ (that is, the value of the function Re_i at the point i). Hence $2kRe_i = (Re_i)^2$. Taking the values of the functions on both sides at the point i we obtain $2k^2 = k^2$, whence $k = 0$. This implies that $(Re_i)^2 = 0$ and hence that $Re_i = 0$, *q.e.d.*

Corollary 4.2. *Every* REYNOLDS *operator is a direct sum (in the algebra sense) $R = 0 \oplus R_1 \oplus \cdots \oplus R_k$, where 0 is the zero operator on $C(N)$ and R_i are minimal* REYNOLDS *operators on $C(S_1)$* (cf. Theorem 1).

Corollary 4.3. *If the carrier of $R^2 e$ is S, then $Re = e$, and conversely.*

Proof. $R^2 e$ is a fixpoint of R, by Proposition 3.2. Therefore the component N in Theorem 1 is the null set, and, as remarked in the beginning of Section 3, the identity e belongs to the algebra F of fixpoints. The converse is trivial.

Theorem 2. *Every minimal* REYNOLDS *operator is idempotent.*

Proof. It suffices to show that the range of R is spanned by the identity element e. By Proposition 3.2, $R^2 e_i = ke$; it suffices to prove that $Re_i = ke_i$. If $k = 0$ then the proof given to Proposition 4.1 can be repeated to show that $Re_i = 0$. If $k \neq 0$ then set $g = k^{-1} Re_i$ in Proposition 2.5 and obtain $k^{-1} R(Re_i Rf) = R^2 f$ for all f; in particular the choice $f = Re_i$ gives $k^{-1} R[(Re_i)^2] = R^2 e_i = ke$, or equilivalently $R[(Re_i)^2] = k_i^2 e$.

In identity (4) set $f = e_i$ and obtain $2R(e_i Re_i) = (Re_i)^2 + k^2 e$. Let $k_j = (Re_i)(j)$ (the value of Re_i at the point j). Then the last identity becomes $2k_i Re_i = (Re_i)^2 + k^2 e$; evaluating this at the point i we obtain $2k_i^2 = k_i^2 + k^2$, that is, $k_i^2 = k^2$. Evaluating the same identity at the point $j \neq i$ we obtain $2k_i k_j = k_j^2 + k^2$. This gives the quadratic equation $k_j^2 - 2k_i k_j + k_i^2 = 0$ for k_j, whence

$k_j = k_i$ for all j. Hence all the components of Re_i coincide, and $Re_i = k_i e$, *q.e.d.*
The statements of Theorems 1 and 2 together yield

Theorem 3. *Every* REYNOLDS *operator in a finite-dimensional algebra* $C(S)$ *is idempotent.*

From Corollary 4.3 we infer

Theorem 4. *The identity element* e *is fixed under* R *if and only if the carrier of* Re *is all of* S.

BIBLIOGRAPHY

[1] J. ARBAULT, Transformations de REYNOLDS sur un ensemble fini, *Corso sulla teoria della turbolenza*, Turin, (1957), Vol. I, 115–121.
[2] G. BIRKHOFF, Moyennes des fonctions bornées, *Algèbre et théorie des nombres, Colloques Internationaux du CNRS*, Paris (1950) 143–153.
[3] G. BIRKHOFF, Averaging operators, *Symposium on lattice theory*, AMS (1960) (to appear).
[4] G. BIRKHOFF & R. S. PIERCE, Lattice-ordered rings, *Anais Academia Brasileira de Ciências*, 28 (1956) 41–69.
[5] CHANDLER DAVIS, Various averaging operations onto subalgebras, *Illinois Journal of Mathematics*, **3** (1959) 538–553.
[6] M.-L. DUBREIL-JACOTIN (Mme.), Etude algébrique des transformations de REYNOLDS, *Colloque d'Algèbre supérieure*, CBRS, Louvain (1957) 9–27.
[7] M.-L. DUBREIL-JACOTIN (Mme.), Sur les axiomes des moyennes, *Corso sulla teoria della turbolenza*, Turin, (1957), Vol. I, 107–114.
[8] M.-L. DUBREIL-JACOTIN (Mme.), Sur le passage des equations de NAVIER-STOKES aux equations de REYNOLDS, *C. R. Acad. Sci., Paris*, **244** (1957) 2887–2890.
[9] P. R. HALMOS, On a theorem of DIEUDONNÉ, *Proc. Nat. Acad. Sci., U.S.A.*, **35** (1949) 38–42.
[10] J. L. KELLEY, Averaging operators on $C_\infty(X)$, *Illinois Journal of Mathematics*, **2** (1958) 214–223.
[11] I. MOLINARO, Sur les endomorphismes de REYNOLDS sur un ensemble fini, *Alger Mathématique*, **4** (1957) 87–101.
[12] SHU-TEH-CHEN-MOY (Mrs.), Characterization of conditional expectation as a transformation on function spaces, *Pacific Journal of Mathematics*, **4** (1954) 47–63.
[13] G.-C. ROTA, On the representation of averaging operators, *Rendiconti Padova*, **30** (1960) 52–64.
[14] G.-C. ROTA, Spectral theory of smoothing operators, *Proc. Nat. Acad., U.S.A.*, **46** (1960) 863–868.
[15] G.-C. ROTA, Endomorphismes de REYNOLDS et théorie ergodique, *Comptes Rendus*, **250** (1960) 2791–2793.
[16] G.-C. ROTA, Représentation des opérateurs de REYNOLDS, *ibid.* 2831–2833.

Massachusetts Institute of Technology

On the Passage from the Navier-Stokes Equations to the Reynolds Equations

Gian-Carlo Rota

In 1895, Osborne Reynolds, in his memoir on fluid flow(1), sketched a method for simplifying the study of solutions of the *Navier-Stokes equations* by taking averages of the (presumed) solutions, and for determinining the partial differential equations satisfied by these averages. Rewritten in modern notation and terminology, Reynolds' reasoning goes as follows. Let u_j be the coordinates of a velocity vector in three dimensions, varying with time; write the Navier-Stokes equations in the form

$$\frac{\partial u_j}{\partial t} + \sum_{k=1}^{3} \frac{\partial}{\partial x_k}(u_j u_k) = -\frac{1}{\rho}\frac{\partial p}{\partial x_j} + \nu\,\Delta u_j, \quad j = 1, 2, 3,$$

$$\sum_{j=1}^{3} \frac{\partial u_j}{\partial x_j} = 0, \tag{1}$$

where p is the pressure, and ν and ρ are constants. Consider the velocity vector u_j as made up out of a sum of an "average" summand $\bar{u}_j$ and a "turbulent" summand u'_j. Reynolds was led by heuristic arguments to the equations

$$\frac{\partial \bar{u}_j}{\partial t} + \sum_{k=1}^{3} \frac{\partial}{\partial x_k}(\bar{u}_j \bar{u}_k) = -\sum_{k=1}^{3} \frac{\partial}{\partial x_k}\overline{(u'_j u'_k)} - \frac{1}{\rho}\frac{\partial \bar{p}}{\partial x_j} + \nu\,\Delta \bar{u}_j, \quad j = 1, 2, 3,$$

$$\sum_{j=1}^{3} \frac{\partial \bar{u}_j}{\partial x_j} = 0, \tag{2}$$

Reprinted from R. E. Langer, ed., *Partial Differential Equations and Continuum Mechanics, Proceedings of an International Conference Conducted by the Mathematics Research Center at the University of Wisconsin, Madison, June 7-15, 1960,* University of Wisconsin Press, Madison, 1961, pp. 374–376. Some obvious errors in the original abstract are corrected by the editors.

for the average velocities and average pressure $\bar{p}$. This is again a system of Navier-Stokes equations for $\bar{u}_j$ with an external force represented by the "turbulent" part, the first term on the right side.

Various justifications have been proposed for the passage from the Navier-Stokes equations (1) to the *Reynolds equations* (2). The most straightforward one consists in considering the functions $u_j(x_1, x_2, x_3, t)$ and p as *random functions* $u_j(x_1, x_2, x_3, t, s)$ depending measurably upon a stochastic variable s ranging over a probability[(2)] space (S, Σ, μ). If we then define the operator $u_j \mapsto \bar{u}_j$, $p \mapsto \bar{p}$ as a *conditional expectation*[(3)] operator relative to a σ-subfield Σ' of the σ-field Σ, so that

$$\bar{u}_j = E^{\Sigma'} u_j \text{ and } \bar{p} = E^{\Sigma'} p,$$

then the Reynolds equations follow from the the Navier-Stokes equations upon using the identity

$$E^{\Sigma'}(uE^{\Sigma'}v) = E^{\Sigma'}u\, E^{\Sigma'}v$$

which essentially characterizes condition expectation.

It was remarked by Mme. M.-L. Dubreil-Jacotin[(4)], however, that any operator $\bar{u} = Ru$ effecting the passage from the Navier-Stokes equations to the Reynolds equations must have the following properties:

(1) it must be a linear order-preserving operator;
(2) if e is the identity function from S to the reals, then $Re = e$;
(3) it must satisfy the *Reynolds identity*

$$2R(uRu) = (Ru)^2 + R(Ru)^2,$$

where $u(s)$ is any real-valued essentially bounded measurable function on (S, Σ, μ).

It is easily verified that these conditions are satisfied by setting

$$R = E^{\Sigma'},$$

that is, for condition expectation. This special operator satisfies in addition the condition that the expectation of the random variable Ru equal the expectation of u :

$$\int_S Ru(s)\mu(ds) = \int_S u(s)\mu(ds). \tag{3}$$

The author has succeeded in determining *all* linear operators on Lebesgue spaces $L_p(S, \Sigma, \mu)$ which satisfy conditions (1) through (4), that is, all operators effecting the passage from the Navier-Stokes equations to the Reynolds equations. The result can be summarized as follows.

THEOREM. *Let R be any bounded operator on $L_p(S, \Sigma, \mu)$ (for some p, $1 \leq p < \infty$) satisfying conditions (1) through (4). Then there exists a partition of the space S into $S_1 + S_2$ such that the complementary subspaces $L_p(S_1)$ and $L_p(S_2)$ are invariant under R, and such that (a) the restriction of R to the space $L_p(S_1, \Sigma, \mu)$ is a conditional expectation; (b) the restriction to R to $L_p(S_2, \Sigma, \mu)$ is an operator of the form $I(I - \log V)^{-1}$, where $Vf(s) = f(\phi(s))$ for some measure-preserving transformation ϕ of the probability space (S, Σ, μ).*

The class of operators (b) which is determined by the preceding result is a generalization of conditional expectation whose study can be made on the basis of the ergodic theory of the measure-preserving transformation ϕ associated with R [(5)].

Notes

(1) cf. Osborne Reynolds, On the dynamical theory of incompressible viscous fluids and the determination of the criterion, *Philosophical Transactions of the Royal Society,* **186**(1895), pp. 123-164.

(2) For explanations of notations and results from operator theory, see N. Dunford and J. T. Schwartz, *Linear Operators,* Wiley-Interscience, New York, 1959.

(3) cf. M. Loève, *Probability Theory,* van Nostrand, New York, 1955, Section 24.

(4) cf. M.-L. Dubreil-Jacotin, Sur le passage des équations de Navier-Stokes aux équations de Reynolds, *Comptes Rendus de l'Académie des Sciences, Paris,* **244**(1957), pp. 2887–2890.

(5) For a further study of these operators, see the author, Spectral theory of smoothing operators, *Proceedings of the National Academy of Sciences, U.S.A.,* **46**(1960), 863–868.

REYNOLDS OPERATORS

BY

GIAN-CARLO ROTA[1]

1. **Introduction.** Let A be an algebra over a field of characteristic not equal to 2. A *Reynolds operator* (or Reynolds endomorphism) is a linear map of A into A which satisfies the following identity, for any f and g in A:

$$R(fg) = RfRg + R((f - Rf)(g - Rg)).$$

We call this identity the *Reynolds identity*. It occurs for the first time in a famous paper of Osborne Reynolds[2] on turbulence theory.

A related identity, introduced almost contemporarily by Kolmogorov and Kampé de Fériet and studied by Garrett Birkhoff, is the *averaging identity*

$$A(fAg) = (Af)(Ag).$$

A linear operator A satisfying the averaging identity is usually called an *averaging operator*. In this paper we discuss the relationship between averaging operators and Reynolds operators, and we give a representation of a Reynolds operator when A is an algebra of bounded measurable functions, provided the operator satisfies a certain mild continuity restriction.

There is an extensive literature on averaging operators, motivated largely by the fact that they characterize and generalize the operation of conditional expectation in probability theory. It is easy to see that

PROPOSITION 1. *If the algebra A has an identity* 1 *and A is an averaging operator such that $A1 = 1$, then A is a Reynolds operator.*

The proof is immediate: setting $f = 1$ in the averaging identity, we get $A^2 = A$. Expanding the Reynolds identity, we get the equivalent identity

$$R(fRg + gRf) = RfRg + R(RfRg), \tag{1}$$

from which it is clear that if A is an averaging operator such that $A^2 = A$, then $A = R$ is a Reynolds operator.

The converse of Proposition 1 is not true, and we shall see that there is an interesting variety of Reynolds operators other than conditional expectation. In fact, our first question is the following:

[1] This work was begun under contract NSF-GP-149, continued under Contract with the Office of Naval Research, and concluded while the author was a Fellow of the Sloan Foundation.

[2] Authors' names refer to the bibliography at the end.

QUESTION 1. Under what conditions is a Reynolds operator an averaging operator?

The answer (Theorem 1) is found, unexpectedly, to depend on whether or not the range of R is a closed subspace, in a suitable topology. Of course there are Reynolds operators without a closed range.

We shall be concerned with Reynolds operators on the algebra $L_\infty(S, \Sigma, m)$ of bounded measurable functions on a measure space (S, Σ, m). This choice is motivated in part by the idea of applying the results to probability and ergodic theory, in part by the fact that this algebra has a particularly simple structure. We shall not assume the measure space to be finite; it suffices to assume enough so that the Radon-Nikodym theorem is applicable; to avoid verbal detours, we shall simply assume that the measure space is σ-finite.

The choice of a topology in $L_\infty(S, \Sigma, m)$ is an old and thorny question, which we shall resolve in the easiest way. Norm-convergence in $L_\infty(S, \Sigma, m)$ bears notoriously little relation to the measure-theoretic structure. On the other hand, there is a topology which is the natural one for L_∞, namely, the topology of L_1-convergence, and *this is the topology we shall use*. Thus, all Reynolds operators are assumed continuous in this topology, as well as, of course, everywhere defined in $L_\infty(S, \Sigma, m)$. With this proviso, an answer to Question 1 is the following:

THEOREM 1. *A Reynolds operator in $L_\infty(S, \Sigma, m)$ is an averaging operator if and only if its range is closed in the L_1-topology.*

Of course the continuity restriction on R can be relaxed, but only at the cost of introducing pathological cases, even for operators acting on the sequence space l_∞, and even in the case of "conditional expectation."

It seems unexpected at first that the answer to a question which appears to be purely algebraic should depend upon a topological condition, but we hope that the proof below shows the "reason" for this phenomenon.

The second question we shall attack is the following:

QUESTION 2. Find an analytic representation for Reynolds operators.

To answer this question, we drop the assumption of continuity in the L_1-topology, but in exchange make a more stringent assumption on R, namely the following:

$$\int_S |Rf|^2\, dm \leqq \int_S |f|^2\, dm \tag{2}$$

for any f in L_∞. Such an assumption could be justified on "physical grounds" (Reynolds operators should not "increase the energy"); mathematically, it enables us to use the machinery of spectral theory in Hilbert space. At any rate, assumption (2) is more innocuous than the assumption of positivity of R ($Rf \geqq 0$ if $f \geqq 0$) that has often been made and that we shall *not* make. Strangely enough, we found that (2) implies that R is positive, but this fact is gotten only after the representation theorem. Under assumption (2), our final result (Theorem 2 below) gives a complete classification of Reynolds operators in terms of conditional expectation and one-parameter semigroups of measure-preserving transformations.

2. **Motivation and history.** Reynolds operators first arose in connection with a problem in partial differential equations which has never been fully settled. Let $N(f) = g$ be the Navier-Stokes equations, with an external acting force g (we do not write them down explicitly because their specific form is irrelevant for the present discussion). The problem is to find all operators R such that if f is a solution, then Rf is a solution of $N(Rf) = g'$, possibly with a different external acting force g'. The idea is that the function Rf should depend on "fewer degrees of freedom"; it should be in some sense an average of f. A simple computation shows that this transformation is possible by a linear operator R only if R satisfies the Reynolds identity. One thus expects the Reynolds identity to express an averaging, or smoothing, property of the operator.

Interesting though this connection with partial differential equations is, we shall not study it in the present work; our concern is largely classificatory. We hope to show in the process that the Reynolds identity is also interesting for reasons other than those that originated it.

In recent years there has been considerable interest in linear operators satisfying algebraic identities. The most spectacular developments were those obtained by Andersen-Atkinson-Baxter-Spitzer and several others, with abstract Wiener-Hopf methods. The operators considered by these authors satisfy an identity very close to the Reynolds identity; this fact alone provides a strong motivation for the study of Reynolds operators. In fact, we hope to show in a separate work (and it is easy to verify formally) that Reynolds operators are "infinitesimal analogs" of those considered by Atkinson-Baxter, etc.

To clarify the "meaning" of the Reynolds identity, we consider some examples. The simplest is conditional expectation, which in an infinite measure space must be defined with some care. Let Σ' be totally σ-finite σ-subfield of Σ (see Halmos, *Measure theory*), and let $A: f \to f'$ be the operator mapping a real function f into the Radon-Nikodym derivative of f relative to Σ'. Then A is a Reynolds operator and an averaging operator. The Radon-Nikodym derivative is the function f', unique in virtue of the Radon-Nikodym theorem, which is measurable relative to Σ' and which has the property that

$$\int_E f'\,dm = \int_E f\,dm$$

for every E of finite measure in Σ'. Of course f is assumed integrable over every set of finite measure.

Conditional expectation is the typical example of an averaging operator, and in fact, every Reynolds operator satisfying $R^2 = R$ is a generalization of conditional expectation (see the author's Padova Rendiconti paper).

Our second example is somewhat more sophisticated. Let T^t be a one-parameter semigroup of measure-preserving transformations of (S, Σ, m), and define $V^t f(S) = f(T^t S)$, as usual. Since $\|V^t\| = 1$ in L_∞, the operator

$$R = \int_0^\infty e^{-t} V^t \, dt \tag{3}$$

is well defined. To dispose of integrability questions, we may assume that V^t is L_1-continuous, and let the integral be a Bochner integral. We claim that R is a Reynolds operator. To establish this claim, we shall make use of two well-known results:

(a) V^t is a semigroup of operators "arising from" a semigroup of measure-forming transformations in the manner described above if and only if for each t, the operator V^t is a *homomorphism* of L_∞, that is, if and only if $V^t(fg) = (V^t f)(V^t g)$ for all essentially bounded f and g (see Halmos, *Ergodic theory*, for a proof of this result due to von Neumann). (b) If V^t is *any* one-parameter semigroup of bounded operators and R is defined as in (3), then $dV^t R/dt = V^t(R - I)$ (see Hille-Phillips or Dunford-Schwartz, Chapter VIII). Using these two results, we find that

$$\frac{d}{dt} V^t(RfRg) = (V^t(Rf - f))(V^t Rg) + (V^t Rf)(V^t(Rg - g)).$$

Now, integrating by parts identity (3) and using this, we find

$$\int_0^\infty e^{-t} V^t(RfRg)\, dt = [e^{-t} V^t(RfRg)]_0^\infty + \int_0^\infty e^{-t} V^t(Rg(Rf - f) + Rf(Rg - g))\, dt.$$

Suitably translated, this reads

$$R(RfRg) = -RfRg + R(Rg(Rf - f) + Rf(Rg - g)),$$

which is trivially equivalent to the Reynolds identity.

Operators of the form (3) are a generalization of conditional expectation. Roughly speaking, one "mixes" before averaging. In fact, we have pointed out elsewhere that many of the properties of conditional expectation can be extended to operators of the form (3). For example, the martingale convergence theorems (for *closed* martingales) extend to operators of the form (3), giving the Abel version of the classical ergodic theorems, and in fact giving a unified approach to martingales and ergodic theorems, with some "mixed" cases to boot.

Our second result (Theorem 2) shows that under assumption (2), all Reynolds operators "split" into a product of two commuting operators, one of which is a conditional expectation and the other is of the form (3). It would seem that this result reduces the study of Reynolds operators to ergodic theory. Actually, the converse is just as true. Several questions of ergodic theory, for example, questions of mixing and existence of invariant measure, can be conveniently formulated in terms of the Reynolds operators, where they become easier to attack. In fact, identity (3) shows that Reynolds operators are the *potentials*, in the language of probabilistic potential theory, of semigroups of measure-preserving transformations.

Finally, the Reynolds identity bears a close connection with the identity $D(fg) = fDg + gDf$ characterizing derivations. If R is a Reynolds operator and R^{-1} exists, perhaps as an unbounded operator, then $R^{-1} - I$ is a derivation. Unfortunately, R^{-1} seldom exists, as the example of conditional expectation shows, and Theorem 1 can be compared to the theorem of Singer and Wermer on the nonexistence of bounded derivations, at least for the algebra L_∞ (a version of the Singer-Wermer Theorem for L_∞ is a corollary of Theorem 1 and their entire theorem could be proved by these methods).

Algebraically, the Reynolds identity combines a feature of homomorphisms ($V(fg) = VfVg$), since the range of R is evidently a subalgebra; and a feature of derivations, namely, the "local operator" property, since if $Rf = 0$ then $Rg = 0$ for any g in the ideal generated by f in the range of R that is, any g of the form fRh, as follows easily from the Reynolds identity. These unusual properties make the Reynolds identity in many respects richer and more inclusive than either the "homomorphism" or the "derivation" identities.

A word about the history of the subject. Since Osborne Reynolds, it was Kampé de Fériet who first recognized the importance of studying the averaging and Reynolds identities, and began their study in a series of papers stretching over a period of thirty years. Curiously enough, the averaging identity was being studied at about the same time as Kolmogorov's *Foundations of probability* became known, but the connection with conditional expectation was brought out only many years later in Mrs. Moy's thesis. The Reynolds identity, on the other hand, first made its appearance because of the difficulty of producing averaging operators that commuted with differentiation in Euclidean spaces. The first algebraic study of the identity is Mme. Dubreil's. The first study of averaging operators by the methods of functional analysis is due to Garrett Birkhoff.

3. **Analytic preliminaries.** The two lemmas we shall use from analysis are: the "discontinuous" version of the Stone-Weierstrass Theorem, and the operator characterization of conditional expectation, suitably extended to infinite measure spaces. In both cases the extensions to infinite measure proceed along much the same lines as is done in the author's note in the Padova Rendiconti, and we shall only sketch the arguments here.

LEMMA 1. *Let (S, Σ, m) be a σ-finite measure space, and let A be a subalgebra of $L_\infty(S, \Sigma, m)$. Then the closure of A in the L_1-topology of L_∞ is a subalgebra of the form $L_\infty(E, \Sigma', m)$, for some measurable subset $E \subseteq S$.*

If in addition $m(S) < \infty$ and A contains the identity function, then the closure of A in $L_2(S, \Sigma, m)$ is of the form $L_2(S, \Sigma', m)$, for some σ-subfield Σ'.

SKETCH OF PROOF. Let E be a measurable set of finite measure. Consider the subalgebra $\chi_E A$, where χ_E is the characteristic function (or indicator) of the set E. Then an argument like that of Lemma 1 on page 55 of the mentioned paper shows that the L_1-closure of $\chi_E A$ is of the form $L_\infty(E, \Sigma_E, m)$, where Σ_E is a σ-subfield of the restriction of Σ to E. If F is a measurable set and $F \supseteq E$, then it follows easily that Σ_E is the restriction of Σ_F to E. Thus, we can define Σ' as the inductive limit of the Σ_E as E runs through an increasing sequence of sets of finite measure whose union is S. It is then easy to see that the L_1-closure of A is $L_\infty(E, \Sigma', m)$ for some subset $E \in \Sigma$.

Note that the σ-subfield Σ' need not be totally σ-finite, but will evidently be totally σ-finite if A has a set of generators which are square-integrable.

The following lemma is essentially due to Blackwell:

LEMMA 2. *Let A be a selfadjoint projection in $L_2(S, \Sigma, m)$ whose range is $L_2(S, \Sigma', m)$ with a totally σ-finite σ-subfield Σ'. Then A is the conditional expectation relative to Σ'.*

Again the proof results from routine extensions of the arguments given in the above-quoted paper. The reduction to the finite-measure case is accomplished as follows. Let $E, F \in \Sigma'$ be sets of finite measure, and let $g \in L_2(S, \Sigma, m)$. Then, by the selfadjointness of A,

$$\int_S \chi_E \chi_F A g \, dm = \int_S \chi_E \chi_F g \, dm = \int_S \chi_E A(\chi_F g) \, dm.$$

Since this holds for all E, and since Ag and $A(\chi_F g)$ both belong to $L_2(S, \Sigma', m)$, it follows that $\chi_F A g = A(\chi_F g)$. This reduces the proof to the case where S is of finite measure, and in fact proves that the operator A satisfies the averaging identity.

A word about notation. If E is a subset of S and Σ' a σ-subfield of Σ, we write $L_p(E, \Sigma', m)$ to denote the restriction of Σ' to E.

All Banach spaces are taken over the complex field, unless otherwise specified. If R is a Reynolds operator defined on L_∞ over the *real* field, then define $R_1 f = Ru + iRv$ for $f = u + iv$. It is easily verified that R_1 is again a Reynolds operator over the complex field. Similarly for averaging operators.

All our conventions relative to the use of functional-analytic details follow Dunford-Schwartz or Halmos.

4. **Reduction to averaging operators.**

We begin by deriving some algebraic identities which are consequences of the Reynolds identity.

PROPOSITION 1. *Let R be a Reynolds operator in any commutative algebra over a field of characteristic zero, and let n be an integer $\geqq 1$. Then, for any element g the following identity holds:*

$$(*) \qquad nR(g(Rg)^{n-1}) = (n-1)R((Rg)^n) + (Rg)^n.$$

PROOF. For $n = 1$ the identity is trivial; for $n = 2$ it follows from the Reynolds identity by setting $f = g$. For $n > 2$ we proceed by induction: assuming the identity

$$(n-1)R(g(Rg)^{n-2}) = (n-2)R((Rg)^{n-1}) + (Rg)^{n-1},$$

we establish (*). In the Reynolds identity, set $f = -(n-2)(Rg)^{n-1} + (n-1)g(Rg)^{n-2}$. It follows from the induction hypothesis that $Rf = (Rg)^{n-1}$. Therefore, substituting this choice of f in the Reynolds identity, we obtain

$$R(-(n-2)(Rg)^n + (n-1)g(Rg)^{n-1} + g(Rg)^{n-1}) = (Rg)^n + R((Rg)^n).$$

Simplifying, identity (*) follows at once.

PROPOSITION 2. *Let R be as in Proposition* 1, *let $p(x)$ be any polynomial in the variable x without constant term. Then the following identity holds for any Reynolds operator, for g in A:*

$$R(p(Rg)) = p(Rg) + R(Rgp'(Rg)) - R(gp'(Rg)),$$

where p' is the derivative of p.

PROOF. If $p(x) = x^n$ the statement reduces to that of Proposition 1. The general case follows at once by linearity.

PROPOSITION 3. *Let R be a continuous Reynolds operator in $L_\infty(S, \Sigma, m)$, and let $q(x)$ be a continuously differentiable function of a real variable defined for x in a compact interval such that $q(0) = 0$. For f in L_∞, let the range of f be contained in the domain of q. Then the function $q(f)$ is in L_∞, and*

$$R(q(Rf)) = q(Rf) + R(Rfq'(Rf)) - R(fq'(Rf)).$$

PROOF. Let D be the domain of q, a compact interval. By a variant of the Stone-Weierstrass Theorem, we can find a sequence p_n of polynomials without constant term such that $p_n \to q$ and $p_n' \to q'$ uniformly on D. In view of the assumed continuity of R, the result follows at once from the preceding proposition.

REMARK. The preceding result is not the most general of its kind, one can easily extend it to the case of several elements, to functions q which are not continuously differentiable, etc.

COROLLARY 1. *Let f be in the range of R and let q be as above. Then $q(f)$ is in the range of R.*

COROLLARY 2. *Let $f(s) \geqq \varepsilon > 0$ for s in $E \subseteq S$, and let f be in the range of R. Then there is a function g in the range of R such that $g(s) \geqq \delta > 0$ for almost all s in E, and $g(s) \geqq 0$ elsewhere.*

PROOF. Let the range of f be contained in a compact interval J containing 0. Choose a continuously differentiable function q such that $q(0) = 0$ and $q(x) > 0$ for $x \neq 0$ in J. Such a function obviously exists. Then $g = q(f)$ is the desired function by Corollary 1.

COROLLARY 3. *If for every $F \subseteq S$ with $m(F) > 0$ there is a function f in the range of R and an $\varepsilon > 0$ such that if $f(s) \geqq \varepsilon$ on F, then there is a g such that $Rg(s) \geqq \varepsilon > 0$ for almost all s in S.*

Proof. Here we use for the first time the fact that the algebra on which R acts is L_∞. It is convenient to use the compactification of L_∞, so that we can talk about the value of a function at a point. Thus, we may and will assume that R acts on $C(E)$, the space of all continuous functions on an extremally disconnected compact Hausdorff space E. For every s in E there is an open set G_s containing s and nonnegative g_s such that $Rg_s(t) \geqq \varepsilon$ for t in G_s. The family G_s as s ranges through E forms a covering of E, hence it has a finite subcovering $G_{s_2}, \cdots, G_{s_n}$. Then the function $g = g_{s_1} + \cdots + g_{s_n}$ will do the trick.

Corollary 4. *Let R be as in the preceding corollary on the space $L_\infty(E, \Sigma, m)$. Then $R\chi_E = \chi_E$, where χ_E is the indicator of E.*

Proof. Choose a function $f = Rg$ as in the preceding Corollary 3, and choose q so that $q(0) = q'(0) = 0$ and $q(x) = 1, q'(x) = 0$ for $x \neq 0$ in the range of f. Then $q(f) = \chi_E$, and from Proposition 3 it follows that

$$R\chi_E = \chi_E + R((f - g)q'(f)) = \chi_E, \qquad \text{q.e.d.}$$

Theorem 1. *Let R be a continuous Reynolds operator in $L_\infty(S, \Sigma, m)$. Then R is an averaging operator if and only if it has closed range.*

Proof. If R is an averaging operator, as well as a Reynolds operator, then $R^2 = R$, so R is a projection and hence its range is closed.

Conversely, suppose R is a Reynolds operator and its range is closed. Since the range is also closed under products, it follows from Lemma 1 that the range of R is of the form $L_\infty(E, \Sigma', m)$, where Σ' is a σ-subfield of Σ. It follows that R maps $L_\infty(E, \Sigma, m)$ into itself, and by Corollary 4 that $R\chi_E = \chi_E$. Hence $R(\chi_E f) = \chi_E Rf$, and we can and will assume that $S = E$ and $R1 = 1$. The proof results from the following steps:

(1) Suppose that Σ' is the trivial σ-subfield, containing only the null set and the whole space. Then Rf is a constant function, say α, and hence $R^2 f = \alpha R1 = \alpha = Rf$. Hence R is a projection.

(2) Suppose now that Σ' is not the trivial σ-subfield, and let A be a set of positive measure belonging to Σ'. Let χ_A be the indicator of A. By assumption, $Rf = \chi_A$ for some f. We shall prove that $R\chi_A = \chi_A$. To this end, let $g(x)$ be a continuously differentiable function of the real variable x, for $-1 \leqq x \leqq 2$, with the following properties: (a) $g(x) = 0$ for $-1 \leqq x \leqq \frac{1}{2}$; (b) $g(x) = 1$ for $\frac{3}{4} \leqq x \leqq 2$. Then $g(\chi_A) = \chi_A$, since χ_A takes only the values 0 and 1. Hence, by Proposition 3 we have

$$R(\chi_A) = \chi_A + R((Rf - f)g'(\chi_A))$$

where f is a function such that $Rf = \chi_A$. But $g'(\chi_A)$ vanishes identically, because $g'(0) = g'(1) = 0$. This concludes the proof of this step.

(3) If $R\chi_A = \chi_A$ for the indicator of every set A, it follows by linearity that $Rf = f$ for every step function f in the range of R, and by continuity that $Rg = g$ for every function g in the range of R. But this amounts to saying that $R^2 f = Rf$

for every f. Hence R is idempotent. (Since in addition we are assuming that $R1 = 1$, we obtain by setting $g = 1$ in the Reynolds identity that $R^2 = R$.) Now set $g = Rf$ in the Reynolds identity, and get $R(fRf) = (Rf)^2$. Polarizing, the averaging identity follows.

COROLLARY. *Let the algebra $L_\infty(S, \Sigma, \mu)$ be finite-dimensional. Then every Reynolds operator is an averaging operator.*

Similar results can be obtained for operators acting on the sequence space l_∞.

REMARK. The reasoning that led to Theorem 1 can be generalized without much difficulty (but with lengthy technical details) to $C(S)$, the space of all continuous functions on a compact Hausdorff space S. One then obtains a generalization of the Singer-Wermer Theorem on derivations.

5. **Representation.** In the following the continuity of R in L_∞ need not be assumed; we do, however, assume that R is everywhere defined in L_∞, and maps L_∞ into L_∞.

THEOREM 2. *Let R be a Reynolds operator in $L_\infty(S, \Sigma, m)$, $m(S) < \infty$, satisfying the condition $R1 = 1$ and*

$$(*) \qquad \int_S |Rf|^2\, dm \leqq \int_S |f|^2\, dm, \qquad f \in L_\infty(S, \Sigma, m).$$

Then there exist:

(1) *A unique σ-subfield Σ_1 of Σ;*

(2) *A unique positive conditional expectation operator A onto $L_\infty(S, \Sigma_1, m)$;*

(3) *A unique semigroup V^t of measure-preserving transformations defined for $t \geqq 0$ on $L_\infty(S, \Sigma_1, m)$, such that:*

(a) *The algebra $L_\infty(S, \Sigma_1, m)$ is invariant under R;*

(b) *The operators A and R commute; and the range of A is the closure of the range of R;*

(c) *The action of R for f in $L_\infty(S, \Sigma, m)$ is given by the formula*

$$Rf = \int_0^\infty e^{-t} V^t A f\, dt.$$

PROOF. (1) From assumption (*) it follows that the domain of R can be uniquely extended to the space $L_2(S, \Sigma, m)$. We denote again by R the resulting operator, since this can cause no confusion. In view of (*), R is an operator of norm one. We now apply a theorem of Nagy (cf. Riesz and Nagy), and infer that $R^*1 = 1$, where R^* is the adjoint operator of R, again acting on $L_2(S, \Sigma, m)$.

(2) It follows from step (1) that the operator $2R - I$ is isometric: that is, for all f in $L_2(S, \Sigma, m)$,

$$\int_S |2Rf - f|^2\, dm = \int_S |f|^2\, dm.$$

This can be established as follows. By the Reynolds identity applied to essentially bounded f and g we have

$$\int_S R(fRg + gRf)\,dm = \int_S [RfRg + R(RfRg)]\,dm.$$

Passing to adjoints, we have

$$\int_S (R^*1)(fRg + gRf)\,dm = \int_S [RfRg + (R^*1)(RfRg)]\,dm.$$

Using the conclusion of (1), this simplifies to

$$\int_S (fRg + gRf)\,dm = \int_S 2RfRg\,dm,$$

and passing to adjoints again (f and g may be assumed real without loss of generality)

$$\int_S g(R^*f + Rf)\,dm = \int_S 2gR^*Rf\,dm.$$

Since this is true for all functions g, we infer that

$$R^*f + Rf = 2R^*Rf,$$

for all f in L_∞, and hence, by continuity, also for all f in L_2. Now let $U = 1 - 2R$; then we obtain from the above identity that $U^*U = 1$. Hence $\int (Uf)^2\,dm = \int fU^*Uf\,dm = \int f^2\,dm$, which shows that U is isometric, at least on real functions. The extension to complex functions is immediate.

(3) Since the operator U is isometric in L_2, the ergodic theorem applies to it (cf. Dunford-Schwartz, Riesz-Nagy) and gives an orthogonal projection operator $I - A$ as a strong limit of the Cesàro averages $(1 + U + U^2 + \cdots + U^n)/(n+1)$. The projection $I - A$ has the following properties. (a) The range of $I - A$ consists of the fixpoints of $U = 1 - 2R$, that is, the set of all f such that $Rf = 0$; (b) the range of the orthogonal complement $A = I - (I - A)$ consists of the closed subspace spanned by functions of the form $f + Uf$, that is, the closure of the range R; in particular, since $R1 = 1$, we have $A1 = 1$, (c) A commutes with R. But the range of R, restricted to L_∞, is a subalgebra of L_∞. Hence we can apply Lemma 1 of §3, and conclude that the range of R is a subspace of the form $L_2(S, \Sigma_1, m)$, where Σ_1 is a σ-subfield of Σ. By Lemma 2 of §3, it now follows that A is the conditional expectation operator relative to Σ_1.

(4) It follows from the preceding step that R, restricted to the invariant subspace $L_2(S, \Sigma_1, m)$ is a one-to-one operator. More is actually true, namely, the range of R, restricted to $L_2(S, \Sigma_1, m)$ is dense in $L_2(S, \Sigma_1, m)$. For if it were not, then there would be a function f in $L_2(S, \Sigma_1, m)$ orthogonal to it: for such a function, $R^*f = 0$, that is, $U^*f = f$. But U^* is of norm one, hence, again by the Nagy theorem quoted above, $U^{**}f = Uf = f$. This in turn implies that $Rf = 0$, which is impossible if f lies in $L_2(S, \Sigma_1, m)$.

(5) Restrict R to the invariant subspace $L_2(S, \Sigma_1, m)$, from now on. We first verify that the unbounded closed operator $D = 1 - R^{-1}$ is antisymmetric, that is, D^* is an extension of $-D$. From the identity, already established in step (2), namely,

$$\int_S (Rf - f)Rg\, dm = -\int_S Rf(Rg - g)\, dm$$

we obtain, writing $p = Rf$ and $q = Rg$,

$$\int_S (1 - R^{-1})pq\, dm = -\int_S p(1 - R^{-1})q\, dm.$$

This holds for all p and q in the domain of the operator $D = I - R^{-1}$, and proves that $1 - R^{-1}$ is antisymmetric and the domain of $1 - R^{-1*}$ includes the domain of $I - R^{-1}$.

(6) The spectrum of D lies in the half plane $y \leqq 0$ in the complex plane $(z = x + iy)$. This follows from a variant of the spectral mapping theorem: the spectrum of U is inside the unit disk, hence the spectrum of R lies inside the circle of radius $\frac{1}{2}$ centered at $x = \frac{1}{2}$, and $D = I - R^{-1}$.

(7) From (5) and (6), together with the Hille-Yosida theorem, we conclude that D is a semigroup generator. Let V^t be the one-parameter semigroup generated by D. Since $R = (1 - D)^{-1}$, it follows from a well-known formula of semigroup theory (cf. Hille-Phillips or Dunford-Schwartz) that

$$Rf = \int_0^\infty e^{-t} V^t f\, dt, \qquad f \in L_2(S, \Sigma_1, m).$$

(8) We now prove that $V^t fg = V^t f V^t g$ for bounded f and g. Since the range of R is dense, it will suffice to show this for $f = Rp$ and $g = Rq$.

First, recall that if V^t is *any* one-parameter semigroup of bounded operators and R is defined as in (6), then $dV^t R/dt = V^t(R - I)$ (cf. Hille-Phillips), as follows easily from the relationship between R and V^t given in (6). We adapt to the present setup an argument due to Feller. Consider the function

$$h(s, t) = V^s((V^t Rf)(V^t Rg)),$$

for fixed f and g and variable s and t. An easy computation, using the Reynolds identity, the above remark, and the fact that R commutes with V^t, shows that $\partial h(s, t)/\partial s = \partial h(s, t)/\partial t$, hence $h(s, t) = q(s + t)$ for some function q. The conclusion amounts to saying that $h(t, 0) = h(0, t)$.

(9) The representation in (6) was obtained for functions of class L_2; however, it is easily seen that it can be extended to all functions of class $L_p(S, \Sigma, m)$, $1 \leqq p \leqq \infty$. This comes from the fact that V^t, being a semigroup of measure-preserving transformations, is norm-preserving in all L_p.

(10) The proof of the theorem is now almost complete, except for the passage from the σ-subfield Σ_1 to the σ-field Σ. The projection A is a conditional expectation; therefore it can be extended to any L_p so that it remains bounded. Clearly

$Rf = RAf = ARf$ for any f for which the expression is defined. Therefore on the right of the displayed formula in (7) we can replace f by Af, q.e.d.

REMARK (*Added in proof*). The restriction $m(S) < \infty$ is not essential, and can be removed with a little extra work.

BIBLIOGRAPHY

1. J. Arbault, *Nouvelles propriétés des transformations de Reynolds*, C.R. Acad. Sci. Paris, **239** (1954), 858–860.

2. ———, *Sur les transformations de Reynolds quasi régulières*, C.R. Acad. Sci. Paris, **239** (1954), 949–951.

3. ———, *Transformations de Reynolds sur les ensembles finis*, Conferenze, Centro Internazionale di Matematica Estivo, Varenna, September, 1957.

4. R. R. Bahadur, *Measurable subspaces and subalgebras*, Proc. Amer. Math. Soc. **6** (1955), 565–570.

5. Martin Billik and Gian-Carlo Rota, *On Reynolds operators in finite-dimensional algebras*, J. Math. Mech. **9** (1960), 927–932.

6. G. Birkhoff, *Moyennes des fonctions bornées*, Colloque d'Algèbre et de Théorie des Nombres, pp. 143–153, Centre National de la Recherche Scientifique, Paris, 1949.

7. Barron Brainerd, *On the structure of averaging operators*, J. Math. Anal. Appl. **5** (1962), 347–377.

8. ———, *Sur la structure des opérateurs moyennes*, C.R. Acad. Sci. Paris **252** (1960), 2058–2060.

9. J. Dieudonné, *Sur le théorème de Lebesgue-Nikodym*. III, Ann. Univ. Grenoble Sect. Sci. Math. Phys. (N.S.) **23** (1948), 25–53.

10. M. L. Dubreil-Jacotin, *Propriétés algébriques des transformations de Reynolds*, C.R. Acad. Sci. Paris **236** (1953), 1950–1951.

11. ———, *Propriétés algébriques des transformations de Reynolds*, C.R. Acad. Sci. Paris **236** (1953), 1136–1138.

12. ———, *Propriétés générales des transformations de Reynolds*, C.R. Acad. Sci. Paris **239** (1954), 856–858.

13. ———, *Étude algébrique des transformations de Reynolds*, Colloque d'Algèbre Supérieure, Brussels, December, 1956.

14. ———, *Sur le passage des équations de Navier-Stokes aux équations de Reynolds*, C.R. Acad. Sci. Paris **244** (1957), 2887–2890.

15. ———, *Sur les axiomes des moyennes*, Corsò sulla teoria della turbolenza Vol. 1, pp. 107–114, Centro Internazionale di Matematica Estivo, Libreria Editrice Universitaria Levrotto e Bella, Turin, 1957.

16. N. Dunford and J. Schwartz, *Linear operators*, Interscience, New York, 1958.

17. P. R. Halmos, *Lectures on ergodic theory*, Publications of the Mathematical Society of Japan, no. 3, The Mathematical Society of Japan, 1956.

18. ———, *Measure theory*, rev. ed., Van Nostrand, New York, 1952.

19. ———, *On a theorem of Dieudonné*, Proc. Nat. Acad. Sci. U.S.A. **35** (1949), 38–42.

20. Rudi Hirschfeld, *Sur les semi-groupes de transformations de Reynolds*, C. R. Acad. Sci. Paris **245** (1957), 1493–1495.

21. G. C. Hufford, *One-parameter semigroups of maps*, Duke Math. J. **24** (1957), 443–453.

22. J. Kampé de Fériet, *L'état actuel du problème de la turbulence*, La science Aeriennce **3** (1934), 9–34; II, ibid. **4** (1935), 12–52.

23. ———, *Les fonctions aléatoires stationnaires et la théorie statistique de la turbulence homogène*, Ann. Soc. Sci. Bruxelles Sér. I **59** (1939), 145–194.

24. ———, *La notion de moyenne dans les équations du mouvement turbulent d'un fluide*, 6ième Congrès International de Mécanique Appliquée, Paris, 1946.

25. ———, *Sur un problème d'algèbre abstraite posé par la définition de la moyenne dans la théorie de la turbulence*, Ann. Soc. Sci. Bruxelles Sér. I **63** (1949), 165–180.

26. ———, *Turbulencia*, Vol. I, Instituto Nacional Esteban Terradas de Tecnica Aeronautica, Madrid (to appear).

27. ———, *Averaging processes and Reynolds equations in atmospheric turbulence*, J. Meteorol. **8** (1951), 358–361.

28. ———, *Construction des transformations de Reynolds régulières*, C.R. Acad. Sci. Paris **239** (1954), 934–936.

29. ———, *Transformations de Reynolds opérant dans un ensemble de fonctions mesurables non negatives*, C.R. Acad. Sci. Paris **239** (1954), 787–789.

30. ———, *La notion de moyenne dans la théorie de la turbulence*, Rend. Sem. Mat. Fis. Milano **27** (1955–1956).

31. J. L. Kelley, *Averaging operators on* $C_\infty(X)$, Illinois J. Math. **2** (1958), 214–223.

32. A. Kolomogoroff, *Grundbegriffe der Wahrscheinlichkeitsrechnung*, Springer, Berlin, 1933.

33. V. J. Mizel and M. M. Rao, *Nonsymmetric projections in Hilbert space*, Dept. of Math. Tech. Rep. No. 47, Carnegie Institute of Technology, Pittsburgh, Pa., 1960.

34. Italico Molinaro, *Détermination d'une R-transformation de Reynolds*, C.R. Acad. Sci. Paris **244** (1957), 2890–2893.

35. ———, *Sur les endomorphismes de Reynolds de fonctions définies sur un ensemble fini*, Publ. Sci. Univ. Alger. Sér. A **4** (1957), 87–101.

36. Shu-Teh Chen Moy, *Characterizations of conditional expectations as a transformation on function spaces*, Pacific J. Math. **4** (1954), 47–63.

37. M. Nakamura, *A remark on the integral decomposition of a measure*, Mem. Osaka Univ. Lib. Arts Ed. Ser. B **3** (1954), 25–28.

38. C. W. Oseen, *Das Turbulenzproblem*, 3ième Congrès International de Mécanique Appliquée, Stockholm, tome 1, pp. 3–18, 1930.

39. O. Reynolds, *On the dynamical theory of incompressible viscous fluids and the determination of the criterion*, Philos. Trans. Roy. Soc. Ser. A **186** (1895), 123–164.

40. Gian-Carlo Rota, *On the representation of averaging operators*, Rend. Sem. Mat. Univ. Padova, **30** (1960), 52–64.

41. ———, *Remarks on stochastic solutions of differential equations*, Proc. Sympos. Partial Differential Equations, U.S. Army Research Center, Madison, Wis., 1960.

42. ———, *Spectral theory of smoothing operators*, Proc. Nat. Acad. Sci. U.S.A. **46** (1960), 863–868.

43. ———, *Une généralisation de l'espérance mathématique conditionnelle qui se présente dans la théorie statistique de la turbulence*, C.R. Acad. Sci. Paris **251** (1960), 624–626.

44. ———, *Une théorie unifiée des martingales et des moyennes ergodiques*, C.R. Acad. Sci. Paris **252** (1961), 2064–2066.

45. I. E. Segal, *Equivalence of measure spaces*, Amer. J. Math. **73** (1951), 275–313.

46. ———, *Abstract probability space and a theorem of Kolmogoroff*, Amer. J. Math. **76** (1954), 721–732.

47. Zbyněk Sidak, *On relations between strict sense and wide sense conditional expectations*, Teor. Verojatnost. i Primenen. **2** (1957), 283–288.

48. I. M. Singer and J. Wermer, *Derivations on commutative normed algebras*, Math. Ann. **129** (1955), 260–264.

49. J. Sopka, *On the characterization of Reynolds'-operators on the algebra of all continuous functions on a compact Hausdorff space*, Thesis, Harvard Univ., Cambridge, Mass., 1950.

50. F. B. Wright, *Generalized means*, Trans. Amer. Math. Soc. **98** (1961), 187–203.

51. E. Sparre Andersen, *On the fluctuations of random variables*, Math. Scand. **1** (1953), 263–285; II, ibid. **2** (1954), 195–223.

52. F. Spitzer, *A combinatorial lemma and its applications to probability theory*, Trans. Amer. Math. Soc. **82** (1956), 323–337.

53. G. Baxter, *An operator identity*, Pacific J. Math. **8** (1958), 649–663.

54. ———, *An analytic problem whose solution follows from a simple algebraic identity*, Pacific J. Math. **10** (1960), 731–762.

55. F. V. Atkinson, *Some aspects of Baxter's functional equation*, J. Math. Anal. Appl. 7 (1963), 1–30.
56. E. Hille and R. Phillips, *Functional analysis and semigroups*, Amer. Math. Soc. Colloq. Publ. Vol. 31, Amer. Math. Soc., Providence, R.I., 1958.
57. F. Riesz and B. Nagy, *Leçons d'analyse fonctionnelle*, 2nd ed., Akademiai Kiado, Budapest, 1953.

MASSACHUSETTS INSTITUTE OF TECHNOLOGY,
CAMBRIDGE, MASSACHUSETTS

INDIANA UNIVERSITY,
BLOOMINGTON, INDIANA

Reprinted from

Proceedings of Symposia in Applied Mathematics, Volume XVI

STOCHASTIC PROCESSES IN MATHEMATICAL PHYSICS AND ENGINEERING

Reynolds and Averaging Operators

Jean Dhombres

1. The background to Rota's work

There is little doubt that it was Garrett Birkhoff who, at the beginning of Gian-Carlo Rota's career, turned his interest towards Reynolds operators. These linear operators are named after Osborne Reynolds, who, at the end of the 19th Century, was able to distinguish between two regimes for fluid flows by the now famous Reynolds criterion involving the Reynolds number [40]. The mathematical object behind Reynolds' work is the Navier-Stokes equation, but Reynolds had no explanation of how turbulent behavior could be derived from the equation.

The stimulus for Rota did not come directly from papers of Reynolds, but a paper [8] of Birkhoff. In earlier papers, several French algebraists had investigated ways of dealing with the Navier-Stokes equation. Their idea was to obtain regularly varying solutions of the equation by smoothing or averaging solutions in an algebraic way with linear operators which commute with the nonlinear Navier-Stokes operator. (A survey of this work can be found in M.-L. Dubreil-Jacotin's paper [16].) In his paper [8], presented at a conference on algebra and number theory held in Paris in 1950, Birkhoff simplified some of the algebra. Much more importantly, he and his doctoral student John Sopka introduced into the subject ideas of Stone and Gelfand about representations of function algebras [57]. It then became clear that both algebra and analysis were needed. Reynolds operators are *algebraically* defined by a functional equation and their role is to produce "simplified" solutions of the Navier-Stokes *differential* equation. Such solutions were called *moyennes* or means by Birkhoff, who also coined the expression "averaging operator". Since averaging operators satisfy a much simpler functional equation, the aim of Birkhoff and the French algebraists had been to reduce Reynolds operators to averaging operators, or, more precisely, semi-multiplicative symmetric operators.

Reynolds operators also arose as conditional expectation operators in probability theory. The idea of considering conditional expectations as linear operators goes back to Kolmogoroff's work (see, for example, his 1933 book [31]). In her thesis, published as the paper [36] in 1954, Shu-Teh Chen Moy considered conditional expectations as linear operators on spaces of functions defined on probability spaces. She showed that a broad class of averaging operators must be conditional expectations.

Another prominent influence during Rota's formative years was Jack Schwartz, his thesis advisor at Yale. Rota contributed to the Dunford-Schwartz project, which led to the encyclopedic three-volume treatise [19]. His work on Reynolds operators may be regarded as a attempt to derive results about specific operators from this gigantic enterprise.

2. The place of Reynolds operators in Rota's work

In [41] and [42], two of Rota's earliest papers in this area, he outlined the connections of the rather peculiar Reynolds identity to seemingly completely different situations in measure theory, ergodic theory, probability theory (martingales), Banach algebras, lattice theory, and logic (the Tarski quantification axiom). He thoroughly explored these different connections in a series of papers [41, 42, 43, 44, 46, 7, 47] ending in his 1964 paper [48]. The 1964 paper is an excellent survey of the area and provides an extensive list of references. Many of these connections had been inspired by Birkhoff, who was, like Rota, so sensitive to algebraic forms.

Rota published the papers mentioned above during his early years on the faculty at the Massachusetts Institute of Technology. The diverse mathematical domains in which Reynolds operators occur deepened Rota's broad knowledge of mathematics, and these papers exemplify the kind of *Bildungsprozess* a gifted mathematician interested in analysis, algebra and probability might undertake at that time. Such a panoramic awareness was uncommon in a century when concentration of effort on one specific area was the typical path to a successful career. Rota's papers located the central ideas then current in mathematics in the context of Reynolds operators. Thus, they show what ideas were in the minds of influential mathematicians in different subjects at that time.

Rota's approach around 1961 can best be described by quoting some lines of [46] in translation:

> Therefore one sees that Reynolds endomorphisms unify the idea of a measure-preserving transformation and the idea of conditional expectation. [The latter] operators correspond to the intuitive idea of a mean, which occurs in many applications, for example, in the study of the equations of fluid mechanics and also in the methods used by electrical engineers.

The key word here is "mean". In the 1964 paper [48], although Rota commented on the fact that a mean is naturally linked to applications, he immediately added: "Interesting though this connection with partial differential equations is, ... , our concern is largely classificatory. We hope to show in the process that the Reynolds identity is also interesting for reasons other than those that originated it." The key word here has shifted to "identity". Not that the task of finding means was given up. It has changed. Means have characteristic properties, but the task was not to find a set of such properties (preferably as identities), once and for all, but to classify all such properties, so as to find the most suitable mean in a given situation.

The idea of a mean was studied in the 1930's. Hardy, Littlewood and Pólya [23]. studied various constant means for functions defined on a finite or countable set. Kolmogoroff [31] took a more imaginative line. For him, a mean for a function was not necessarily a constant; it could be another function less irregular or varying than the initial function. Finding suitable means for turbulent functions, and specifically for *fonctions pseudo-aléatoires,* was one area of research for Jean Bass. He distilled his work in [4], published in 1962. The work of Bass was directly related to work of Kampé de Fériet, except that no probability theory was involved: the idea was to define *a priori* a class of functions containing sufficiently turbulent solutions of the Navier-Stokes equation. Rota's aim was to define *a priori* operators averaging turbulent solutions. Thus, Bass was interested in Rota's work, and he reviewed some of Rota's papers [43, 44, 46] for *Mathematical Reviews.* It was from Bass, who was my thesis director, that I learned about Rota's work and Rota's work provided a starting point and a model for my thesis in 1970, entitled "Sur les opérateurs multiplicativement liés" [11].

In his 1964 paper [48], Rota explains the advantages of the Reynolds identity. It is a "rich" identity in the sense that it combines the identities for homomorphisms and derivations. This double capability was used in his attempt to unify martingales and ergodic theory [46]. Interestingly, this linking of two apparently different subjects by Reynolds operators was carried out in two different ways in [46] and [48]. However, by 1964, Rota has become interested in the Baxter identity and its striking combinatorial properties [49]. Later, his interest was drawn to derivations in homological algebra and averaging operators in logic [50, 20]. Is it so astonishing that Rota moved to combinatorial mathematics after such experiences?

In the note [43], Rota mentioned that to prove a theorem on Reynolds operators, he needed to extend to Banach spaces the Frobenius theory of eigenvalues and eigenfunctions for positive operators. This was perhaps the initial motivation for his work on positive operators (see Chapter 2). Positive operators link lattice theory with integration theory. The study of this linkage was launched by Stone in the 1930's and emphasized by Bourbaki in the 1950's. Around 1964, Rota announced a book on positive operators, to be written with Helmut Schaefer from Tübingen. However, Rota's growing interest in combinatorial theory prevented him from pursuing this project, which Schaefer later completed alone. This shift of interest also diverted Rota from taking the natural step after [48], to move on to Reynolds and averaging operators on Banach algebras. This was done by J. B. Miller and his coauthors in the series of papers [33, 22, 21].

Finally, we remark that the need to represent Reynolds operators in different concrete situations might have fostered the development of Rota's phenomenological point of view for mathematics. He first explained his views in 1973 in a paper [51] on Husserl. See also [52].

3. The (ir)relevance of applications

When I met Rota at the International Congress of Mathematicians in Nice in the summer of 1970, he was still fascinated by Reynolds operators and, particularly, their origin in the work of an engineer. Nevertheless, Rota never worked specifically on the Navier-Stokes partial differential equations. He considered that this phenomenological perspective provided enough room for works from a purely mathematical point of view. For example, his abstract [47] for a conference on partial differential equations and continuum mechanics in 1960 announced a result in pure operator theory.

The practical origin of the Reynolds identity had already been made clear by Joseph Kampé de Fériet in 1939 [27], and later, in his 1946 paper "La notion de moyenne dans les équations du mouvement turbulent d'un fluide" [28], presented at the 6th International Congress of Applied Mechanics. (In this connection, see also [17].) In 1964, Rota dedicated his capstone paper [48] on Reynolds operators to Kampé de Fériet, who was then retiring from his position at the University of Lille. Quite appropriately, for it was Kampé de Fériet who had sparked the interest of Birkhoff in the Reynolds identity by the paper [29].

In engineering problems, the smoothing properties of Reynolds operators are tacitly used, along with other "physical" properties. From this viewpoint, there are two distinct tasks for Reynolds operators. The first is to construct from a solution f of the Navier-Stokes equation another solution Pf. The second is to construct from a turbulent solution f a less turbulent function, which is in some sense an average of f. Reynolds operators perform the first task because they commute with the Navier-Stokes operator and the second task because they are generalizations of averaging operators. Thus, attempts were made to reduce Reynolds operators to averaging operators. However, approximation theory suggests that the second task can also be done by smoothing the function f. The smoothing property is a key idea in Rota's work.

The mathematical task in Rota's mind was to understand the proper role of the identity, but not in an axiomatic or formal way. He thought that studying a variety of situations where Reynolds operators played a role in the real world (or the phenomenological world, as he would have said) would bring a fuller understanding of the meaning of the Reynolds identity. Although Rota believed that a mathematician should take inspiration from physical world to do pure mathematics, he disagreed with the point of view of those mathematicians who insisted on the primacy of problems coming from technology which had to be solved, whether or not they lead to pure mathematics results.

When I explained to Rota in 1970 my doctoral work on *opérateurs multiplicativement liés*, I did not know this was a theme he had planned to investigate, as indicated his list of works in preparation in August 1964. I vividly remember how he insisted on concrete realizations of such operators on different function algebras and how he focused on the way the Reynolds and related identities could simplify known results.

There are not many papers after 1974 with the words "Reynolds operators" in their titles. Perhaps the Reynolds identity is just too rich or complicated. Indeed, in his 1964 paper [48], Rota was no longer claiming a unification in measure theory through the use of Reynolds operators. Instead, the Reynolds identity was seen as a "game": it generated computations which could be used or illustrated by examples in functional analysis. Why a game? Because other operators, such as derivations and Baxter operators, were satisfying functional equations just as rich.

For Rota, progress in mathematics could sometimes be achieved by recognizing a pattern or working by analogy. Reynolds operators, which had been a sort of exercise for the mind, remained a source of inspiration for him. This suggests that other authors, while not specifically working on Reynolds operators, might also be stimulated by Rota's methods. A crude way of measuring this is a citation search, but mathematical papers tend to cite direct technical links rather than indirect inspirational ones.

4. Reynolds operators, semi-group theory, and means

Another early influence – perhaps better called a catalyst – was the now classic book *Functional Analysis and Semi-Groups* by Hille and Phillips [24]. The mixing of algebra and analysis in it must have offered an intriguing change from the concentrated analysis of Dunford and Schwartz [19]. In [42] and [43], Rota noticed that if

P is an operator defined in terms of a composition with a measurable transformation and the complex number λ is not in the spectrum of $\log P$, then the operator

$$\frac{\lambda}{\lambda I - \log P}$$

is a Reynolds operator. The logarithm suggests the infinitesimal generator of a semi-group. For example, Rota associates a one-parameter family of Reynolds operators to a one-parameter group of measure preserving functions on a measure space in [44]. In a sense, this result reduces the structure of Reynolds operators to the known structure of measure-preserving transformations. This is *not* Rota's view. In [44], he wrote: "Our interest, however, is in the opposite direction: we use Reynolds operators in order to study measure-preserving transformations". In the same vein, when a structural result reduces Reynolds operators to the integral of a semi-group in [48], Rota added:

> It would seem that this result reduces the study of Reynolds operators to ergodic theory. Actually, the converse is just as true. Several questions of ergodic theory, for example, questions of mixing and existence of invariant measure, can be conveniently formulated in terms of the Reynolds operators, where they become easier to attack.

Another typical remark of Rota concerns the representation of Reynolds operators using the integral of an exponential: "[The representation] shows that the Reynolds operators are the potentials, in the language of probabilistic potential theory, of semi-groups of measure-preserving transformations". I do not believe that this powerful but formal analogy has attracted any researchers. I even have the impression that Rota knew or rather sensed it might not lead anywhere.

The only work on Reynolds operators and semi-groups are the representation theorems of Andreas Neeb in 1999 for positive Reynolds operators on $C(S)$ algebras [37, 38]. Neeb used the fact that if T^t is a one-parameter semigroup of measure-preserving transformations of (S, Σ, m), then, in certain cases, the operator

$$Rf(x) = \int_0^\infty e^{-t} f(T^t x)\, dt$$

is a Reynolds operator.

5. Functional equations and operator identities

János Aczél's book, *Vorlesungen über Funktionalgleichungen und ihre Anwendungen* [1] appeared in 1961. This fundamental book was a systematic survey of works on functional equations but there was no mention of functional equations having the same form as the Reynolds identity. The introduction of the Reynolds and similar identities to the subject of functional equations came in 1979, in a book [15] by Dhombres, and in 1989, in a book [2] by Aczél and Dhombres. From then on, a stream of papers appeared, many in the journal *Aequationes Mathematicae,* investigating functional equations of the Reynolds type.

There are several natural identities a linear operator $P : A \to A$ acting on a commutative algebra A could satisfy.

(R) Reynolds identity: $P(fPg + gPf) = PfPg + P(PfPg)$.

(SMS) Semi-multiplicative symmetry: $P(fPg) = PfPg$.

(MS) Multiplicative symmetry: $P(fPg) = P(gPf)$.

(M) Multiplicative property: $P(fg) = PfPg$.

(B) Baxter identity: $P(fPg + gPf) = PfPg + P(fg)$.

(I) Interpolation property: $P(fPg) = P(fg)$.

(D) Derivation property: $P(fg) = fPg + gPf$.

Here, we have used the convention of writing Pf instead of $P(f)$; for example, $P(fPg)$ stands for the more cumbersome expression $P[fP(g)]$. The terminology used here is not always consistent with Rota's usage; in any case, Rota's own terminolgy changed, as he tried to disentangle topological and regularity properties from algebraic properties.

Functional equations can easily be derived from these identities, but in general, more than one functional equation is derivable. In the Reynolds case, for example, if a function f maps the algebra A into itself, then either of the following two identities can be derived:

$$f(xf(y) + yf(x)) = f(x)f(y) + f(f(x)f(y)),$$

or

$$f(xf(y)) + f(yf(x)) = f(x)f(y) + f(f(x)f(y)).$$

In addition, the solutions depend on the structure of A.

Only one functional equation had been studied when Rota was working on Reynolds operators. This is the equation

$$f(xf(y)) = f(x)f(y),$$

corresponding to semi-multiplicative symmetry [10]. (See [14] for a study of this equation on abelian groups.)

For many operators satisfying algebraic identities with two variables, there is a one-variable version equivalent to the two-variable identity. For example, Rota noted in [42] that the Reynolds identity is equivalent to

$$2P(fP(f)) = (Pf)^2 + P((Pf)^2).$$

However, as is usual with functional equations, the one-variable identity has many more solutions than the original identity, at least when regularity conditions are not imposed upon the function f. Thus, although there is now an extensive literature for the SMS and Reynolds functional equations in two variables, little is known about the one-variable equations.

6. Reynolds operators on finite-dimensional algebras

The simplest cases occur when A is a finite-dimensional algebra. Rota studied this case in [7], written with Martin Billik, one of his early doctoral students at the

Massachusetts Institute of Technology. Billik and Rota proved that a Reynolds operator on a finite-dimensional algebra satisfies semi-multiplicative symmetry (SMS). To prove this, it suffices to show that a Reynolds operator is an idempotent operator, or, equivalently, a projection ([7], Theorem 3). This is done by a computation, using the natural basis for $C(S)$ composed of the elementary idempotent functions and considering supports of fixed points of the Reynolds operator. This approach has the advantage of looking at the finite-dimensional algebra A as an algebra $C(S)$ of functions, even when no topological properties are used. The proof also has an algebraic aspect: a Reynolds operator is shown to be a direct sum (in the algebraic sense) of a nilpotent Reynolds operator (which must necessarily be zero) and minimal Reynolds operators on subspaces $C(S_k)$, where the sets S_k are a (finite) partition of the set S. Related results had been obtained in 1957 by Marie-Louise Dubreil-Jacotin, Jean Arbault and Italico Molinaro, the two first in a *Corso sulla teoria della turbolenza* in Turin for that year [3, 18, 34, 35].

With the addition of some recurrence formulas on operators satisfying algebraic identities and generalizing the methods in [7], Dhombres in 1972 obtained structure results for a large class of operators, including Reynolds and MS operators, on complex finite-dimensional algebras [12].

The paper [7] completely settled the finite-dimensional case for Reynolds operators. The structure of smoothing operators was already known for spaces $C(S)$ of real continuous functions on a compact Hausdorff space S. This was reported by Birkhoff in [8] and had been extended to various $C(S)$ algebras by several authors: John Sopka in his 1950 doctoral thesis [57] at Harvard University, nominally under the supervision of Birkhoff but inspired by Kampé de Fériet, Kelley in [30], Brainerd in [9], and Wright in [59].

Baxter operators appear in [12] as a singular and particularly interesting case. This had already been pointed out in probability theory by G. Baxter [5] in 1960, the very year Rota began to publish on Reynolds operators. But the functional identity was used four years earlier by F. Spitzer [58], and emphatically by Rota [49], where the algebra of Baxter operators was completely determined by the theorem that the "standard" Baxter algebra is free. Yet Reynolds and SMS operators are really different, and the analogy Rota was looking for could not be pinned down.

In certain finite-dimensional cases, Reynolds operators are semi-multiplicative symmetric (SMS) operators, so they satisfy the identity characteristic of averaging operators. Rota worked on such operators in two early papers, [41] and [42]. There is some inconsistency in Rota's terminology: the "smoothing" operators of [42], although required to satisfy the Reynolds identity, are not necessarily averaging in the sense of [41]. It is interesting to note that between the two equivalent forms of the Reynolds identity, the one then preferred by Rota is

$$P(fg) = PfPg + P((f - Pf)(g - Pg)).$$

It is this form of Reynolds identity which best links two aspects of Reynolds operators: the aspect of a multiplicative homomorphism,

$$P(fg) = PfPg,$$

and the aspect of a mean in probability theory, where the operator P acting on the product of two deviations, the deviation $f - Pf$ of Pf from f and the deviation of

Pg from g. The product $P((f-Pf)(g-Pg))$ then serves as a correction term. [From this point of view, the interesting form for the Reynolds functional equation is

$$f(xy) = f(x)f(y) + f((x - f(x))(y - f(y))).$$

This functional equation has not been studied as far as I know.] Note that if one interprets the "average" of a function as *the* statistical mean, that is, a constant, then one obtains the SMS identity

$$P(fPg) = PfPg$$

simply because Pg is a constant and P is a linear operator.

7. Reynolds operators on L_p and L_∞ spaces

In [41], Rota worked with operators on L_p spaces. Because such spaces are not algebras, additional technical conditions are required. The simplest requirement is that the averaging identity holds for all functions f and g belonging to a dense subalgebra of L_p, such as L_∞ when the measure space is a probability space. However, this would have meant also requiring that L_∞ be invariant under P. Rather than doing this, Rota gave up the symmetry of the SMS functional equation. His hypothesis was that if g is in L_p and f is in L_∞, then $PfPg$ is in L_p and

$$P(fPg) = PfPg.$$

Rota defined an *averaging operator* to be an operator P satisfying his hypothesis, the "contraction" condition that the norm of P as a linear operator on L_p be at most one, and $P(1) = 1$. The contraction condition was a natural property for a mean to have. Rota then proved that averaging operators map (essentially) bounded functions into bounded functions, so that L_∞ was invariant under P. What was to have been a hypothesis was then proved.

The proof used a basic method for L_p spaces, namely, playing with the duality theorem (with L_q, where $1/p + 1/q = 1$) and multiplying p by an integer n and passing from the norm in L_{np} to the norm in L_∞ by letting n go to infinity. In this way, P was proved to be a contraction operator on L_∞.

In [41], Rota noticed that P is more than just a bounded operator on L_∞. It is also a positive operator, in the sense that $Pf \geq 0$ almost everywhere whenever $f \geq 0$ almost everywhere. This could be proved directly because P is a contraction on L_∞ and $P(1) = 1$. However, Rota chose another way, using Birkhoff's idea in [8] of representing L_∞ algebras isomorphically as $C(S)$ algebras. Under such a representation, an averaging operator P on L_∞ becomes an operator Q on $C(S)$ of uniform norm one satisfying $Q(1) = 1$ and the SMS functional equation. Then Rota used easy topological arguments to show that Q is a positive operator. Since the chosen isomorphism between L_∞ and $C(S)$ is order-preserving, the proof is complete.

At this stage, we see a hesitation in Rota's thinking. If he used the technique of function algebras and their representations, he would be able to remove the restriction that the measure space be a probability space by using classical methods to obtain a σ-finite measure space from finite measure spaces. From a Banach algebra point of view, the difference between L_∞ on a σ-finite measure space and L_∞ on a finite one is the absence of an unit, and so, this approach would lead to a topological compactification question for $C(S)$ algebras. Rota chose to stay with probability spaces with their built-in σ-fields.

The main result of [41], Theorem 1, provides a definitive structure theorem for averaging operators in Rota's sense on an $L_p(S, \Sigma, m)$ space with $m(S) = 1$. For such an operator P, there exists a σ-subfield Σ' of Σ so that to every f in $L_p(S, \Sigma, m)$, Pf equals its Radon-Nikodým derivative f'. The result shows how the algebraic identity leads to the notion of a mean, and that there cannot be other means.

The converse is not stated in the theorem, but the paper was organized so that at the beginning, using results of Kolmogoroff, Birkhoff and Moy, any Radon-Nikodým derivative was observed to be an averaging operator. This was presented as a generalization of means in the finite-dimensional case, thus showing how to go from an algebraic result to a result in probability.

We have many instances in [41] of typical Rota procedures. Starting with an example expressed phenomenologically (as a mean for the finite dimensional case and then generalized to Radon-Nikodým derivatives), it is then interpreted within a specific theory (as a conditional expectation in probability theory). The proof is then completed by showing that the interpretation happened to be the general case (that of averaging operators). It is important to see that the interpretation as conditional expectation was not applied to the algebra L_∞, but to the space L_2.

The price to pay in [41] for obtaining averaging operators was to combine algebraic and topological properties. Rota's phenomenological perspective for operators satisfying algebraic identities was confirmed when reasonable contractive projections P in $C(S)$ spaces, and hence on $L_\infty(S, \Sigma, m)$ spaces as well, were proved to be MS operators [13]. Indeed, the identity

$$P(fPg) = P(PfPg),$$

a "term" in the Reynolds identity, is satisfied as well.

Once the result in [41] was obtained, Rota noticed that an averaging operator (in his sense) on $L_p(S, \Sigma, m)$ is determined by its range. However, he also noticed the range no longer determines an averaging operator even when it is a contraction operator on $L_\infty(S, \Sigma, m)$. Thus, even if the structure theorem was inspired by results from representations of function algebras, Rota felt justified in taking the measure-theoretic approach; as he put it on page 60 in [41], "The following discussion is therefore limited to those results which do not follow from such a representation technique". The representation technique used Radon measures and topological properties while the measure-theoretic technique involved working with positive finitely additive measures on the σ-field Σ, vanishing on m-null sets.

This decision not to rely on Banach function algebra techniques was not changed in [48]. Rota worked directly with the space $L_\infty(S, \Sigma, m)$, where (S, Σ, m) is a measure space for which the Radon-Nikodým theorem is applicable, for example, a σ-finite measure space. He also used a very specific topology on $L_\infty(S, \Sigma, m)$, the weak topology induced by $L_1(S, \Sigma, m)$. The main result in [48] is the following theorem: *A Reynolds operator on $L_\infty(S, \Sigma, m)$ is an averaging operator if and only if its range is closed in the $L_1(S, \Sigma, m)$-topology.* Rota was astonished that an algebraic result – the reduction of the Reynolds identity to the SMS identity – might depend on a topological condition – the range being closed in some topology. He knew, by his results in [41], that if continuity conditions are relaxed, "pathological cases" may appear, and so, he wrote that he hoped his proof shows the "reason" for this surprising dependence of algebra upon analysis. This remark indicates that he felt he had found the "right" proof of this theorem, which generalized the result in [7] on finite-dimensional algebras. The proof depends essentially on the action of Reynolds operators on the algebra's idempotents, which are the characteristic functions in a function algebra. This idea would be used again in Rota's work on valuation rings and logic.

The paper [48] could not have been Rota's final thoughts on Reynolds or averaging operators. In it, he claimed to have obtained for Reynolds operators a result similar to the famous result of I. M. Singer and J. Wermer for derivation operators [55]. This result is usually stated in the form: Every bounded derivation operator on a Banach algebra has range in the radical of the algebra. Rota wrote that he could deduce the Singer-Wermer result from his study of Reynolds operators. Indeed, this was done later by Martin Billik, who showed how to associate a family of Reynolds operators to a derivation operator [6]. Billik required the algebra to be commutative. Specifically, he showed that every bounded derivation operator on a commutative semi-simple Banach* algebra is the zero operator. The spectrum of a Reynolds operator played a role in the proof and this led to several papers on the spectral properties of Reynolds operators. A useful reference for this line of research is the paper of R. A. S. Fox and J. B. Miller, "Averaging and Reynolds operators in Banach Algebras, III" [21]. Later, Egon Scheffold studied the spectral radius of Reynolds operators on semi-simple Banach lattice algebras, and proved that the property of being a Reynolds operator or an averaging operator is preserved under taking biadjoints [54]. But there was no serious attempt to develop with Reynolds operators the rich ideas in the Singer-Wermer result in Banach algebras which led to B. E. Johnson's paper [25], in which a continuity assumption on the derivation operator is removed, his memoir [26] on cohomology in Banach algebras, and S. Sakai's survey [53] of the nonabelian Banach algebra case, in which he proved that any derivation in a W^*-algebra is inner.

8. Averaging operators in logic

In 1973, Rota exploited the idea of averaging operators in the radically different context of valuation rings of a distributive lattice in [50]. This paper is fascinating, because as Rota wrote, "the method of presentation is deliberately informal and discursive". Much of the reasoning is by analogy, but not in the structuralist or Bourbakian sense, and many of the results are only heuristic. The aim is grandiose, nothing less than the "linearization of logic".

Rota was a master at perceiving connections among disparate subjects and in this instance, he took a combinatorial point of view. In particular, he used canonical idempotents instead of characteristic functions, borrowing the idea from Louis Solomon's work on Burnside algebras [56]. But something else was occurring: Rota was no longer analysing the structure of averaging operators, but constructing averaging operators to fit the application. For the study of logic, he constructed specific "averaging" operators on valuation rings which simulated the properties of existential quantifiers. In this way, "Problems of first order logic, such as the decision problem, can be shown to be equivalent to algebraic problems for valuation rings with averaging operators". Unfortunately, it proved quite difficult to carry through this program rigorously, as one can see in the paper [20], written with David Ellerman.

9. EXAVES

Rota did not seem to have been aware of the 1968 paper [39] by A. Pełczyński. In this paper, Pełczyński systematically studies averaging operators on spaces of continuous functions on compact spaces. His ultimate objective is to classify $C(S)$ spaces and this paper proved to be an important milestone towards the geometry of Banach spaces.

Pełczyński defined *linear exaves,* which are operators generalizing *ex*tension operators and *ave*raging operators on totally disconnected spaces, such as hyperstonian spaces. The algebra of exaves was explicated by Dhombres in [13]: linear exaves correspond exactly to multiplicatively symmetric operators. Moreover, like Reynolds operators, linear exaves combine two properties, semi-multiplicative symmetry and the interpolation property. To explain this combination of properties, it is necessary to discover structural properties for exaves. Let $C(S)$ be the algebra of all complex-valued functions over a compact Hausdorff space S, and S' be a closed subspace of S. We shall only consider linear operators continuous in the uniform norm on $C(S)$. An extension operator E relative to S' is an operator $E : C(S') \to C(S)$ such that $E(f)(x) = f(x)$ for every point x in S'. This operator is an inverse of the restriction operator R which associates to every function f in $C(S)$ the function f restricted to S'. An operator $P : C(S) \to C(S)$ is interpolating if there exists a closed subspace S', the interpolating subspace, of S such that for any point x in S and any function f in $C(S)$, the value of $f(x)$ only depends on the values of f on S'. It can be proved that there exists an extension operator relative to S' if and only if there exists an interpolating operator P having S' as an interpolating space. For an SMS operator P, Birkhoff has shown that the notion of an interpolating space is equivalent to that of an averaging equivalence relation RP on S, defined by $x\,RP\,y$ if and only if $Pf(x) = Pf(y)$ for every f in $C(S)$. From this viewpoint, $P : C(S) \to C(S)$ is an SMS operator if and only if for any function f in $C(S)$, the value of $Pf(x)$ at a point x only depends upon the values of f on the averaging equivalence class $[x]$ under RP. Whenever a closed equivalence relation, such as an averaging relation, is defined on S', the space $C(S'')$, where S'' is the topological quotient of S' by the closed equivalence relation, can be regarded as a subspace of $C(S')$. In this way, it is possible to define an extension operator relative to S''. The following structure theorem, resembling the representation theorem in [48] for Reynolds operators, can now be proved: An operator $P : C(S) \to C(S)$ of norm one such that $P(1) = 1$ is multiplicatively symmetric if and only if P is the composition

$$Q \circ A \circ R,$$

where R is the restriction operator for a certain closed subspace S' of S, A is an SMS operator on $C(S')$ with exactly the same range as the restriction of P to S', and Q is an extension operator for S'. Useful as it is, this structure theorem only becomes a constructive one if one uses, as Rota did, the $L_\infty(S, \Sigma, m)$ equivalence. This was done in Chapter 5 of my thesis [11] using as tools extreme points in convex sets and the Choquet boundary for function subalgebras of $C(S)$.

10. Concluding reflections

If I have not tried to give a detailed analytical account of Rota's papers on Reynolds and averaging operators, it is because his own explanations are so clear they just would have to be repeated. In any case, these papers are conveniently gathered in the present volume. Rather, I have tried to describe Rota's own history in his early years as a mathematician, and how he achieved a mathematical style. To conclude this introduction, I would remark on his remarkable lucidity as a mathematician. He was always thinking about what his insights led him to create, and why, and he polished his proofs so that they are as clear and obvious as possible. In this way, Rota's style imparts a sense of inevitability to his methods and results.

Acknowledgement.

I thank Joseph Kung, Norton Starr, and Roberta T. Tugger for their help in the preparation of this article.

References

[1] J. Aczél, *Vorlesungen über Funktionalgleichungen und ihre Anwendungen.* Birkhäuser, Basel, 1961. English translation: *Lectures on functional equations and their applications.* Academic Press, New York and London, 1966.

[2] J. Aczél and J. G. Dhombres, *Functional equations in several variables.* Encyclopedia of Mathematics and its Applications, 31, Cambridge Univ. Press, Cambridge, 1989.

[3] J. Arbault, Nouvelles propriétés des transformations de Reynolds. *C. R. Acad. Sci. Paris* **239**(1954), 858–860.

[4] J. Bass, *Les Fonctions Pseudo-Aléatoires.* Mémor. Sci. Math., Fasc. CLIII, Gauthier-Villars, Paris, 1962.

[5] G. Baxter, An analytic problem whose solution follows from a simple algebraic identity. *Pacific J. Math.* **10**(1960), 731–742.

[6] M. Billik, Idempotent Reynolds Operators. *J. Math. Anal. Appl.* **18**(1967), 135–144.

[7] M. Billik and G.-C. Rota, On Reynolds operators in finite-dimensional algebras. *J. Math. Mech.* **9**(1960), 927–932. Reprinted in this chapter.

[8] G. Birkhoff, Moyennes des fonctions bornées. In *Algèbre et Théorie des Nombres, Colloques Internationaux du CNRS, no. 24*, CNRS, Paris, 1950, pp. 143–153.

[9] B. Brainerd, On the structure of averaging operators. *J. Math. Anal. Appl.* **5**(1962), 347–377.

[10] Z. Daróczy, Über die Funktionalgleichung $\phi(\phi(x)y) = \phi(x)\phi(y)$. *Acta Univ. Debrecen. Fiz. Chem.* **8**(1962), 125–132.

[11] J. G. Dhombres, Sur les opérateurs multiplicativement liés. Mém. No. 27. Supplément au *Bull. Soc. Math. France,* **99**, Société Mathématique de France, Paris, 1971.

[12] J. G. Dhombres, Sur les opérateurs multiplicativement liés dans les algèbres de dimension finie. *Ann. Inst. H. Poincaré Sect. B (N.S.)* **8**(1972), 333–363.

[13] J. G. Dhombres, A functional characterization of Markovian linear exaves. *Bull. Amer. Math. Soc.* **81**(1975), 703–706.

[14] J. G. Dhombres, Solution générale sur un groupe abélien de l'équation fonctionnelle $f(x*f(y)) = f(f(x)*y)$. *Aequationes Math.* **15**(1977), 173–193.

[15] J. G. Dhombres, *Some Aspects of Functional Equations.* Lecture Notes. Chulalongkorn University, Department of Mathematics, Bangkok, 1979.

[16] M.-L. Dubreil-Jacotin, Étude algébrique des transformations de Reynolds. In *Colloque d'algèbre supérieure, 19 au 22 Décembre 1956, CRBS, Louvain,* Gauthier-Villars, Paris, 1956, pp. 9–27.

[17] M.-L. Dubreil-Jacotin, Sur le passage des équations de Navier-Stokes aux équations de Reynolds. *C. R. Acad. Sci. Paris* **244**(1957), 2887–2890.

[18] M.-L. Dubreil-Jacotin, Sur les axiomes des moyennes. In *Corso sulla teoria della turbolenza,* Vol. 1, Centro Internazionale di Matematica Estivo, Libreria Editrice Universitaria Levrotto e Bella, Turin, 1957, pp. 107–114.

[19] N. Dunford and J. T. Schwartz, *Linear Operators. Part I. General Theory. Part II. Spectral Theory, Self Adjoint Operators in Hilbert Space. Part III. Spectral Operators.* Wiley, New York, 1958, 1963, 1971.

[20] D. P. Ellerman and G.-C. Rota, A measure theoretic approach to logical quantification. *Rend. Sem. Mat. Univ. Padova* **59**(1978), 227–246. Reprinted in Chapter 7.

[21] R. A. S. Fox and J. B. Miller, Averaging and Reynolds operators in Banach algebras. III. Spectral properties of Reynolds operators. *J. Math. Anal. Appl.* **24**(1968), 225–238.

[22] J. L. Gamlen and J. B. Miller, Averaging and Reynolds operators on Banach algebras. II. Spectral properties of averaging operators. *J. Math. Anal. Appl.* **23**(1968), 183–197.

[23] G. H. Hardy, J. E. Littlewood and G. Pólya, *Inequalities.* 1st edition, Cambridge Univ. Press, Cambridge, 1934.

[24] E. Hille and R. S. Phillips, *Functional Analysis and Semi-Groups.* Revised ed., Amer. Math. Soc., Providence, RI, 1974.

[25] B. E. Johnson, Continuity of derivations on commutative algebras. *Amer. J. Math.* **91**(1969), 1–10.

[26] B. E. Johnson, Introduction to cohomology in Banach algebras. In *Algebras in Analysis (Proc. Instructional Conf. and NATO Advanced Study Inst., Birmingham, 1973),* Academic Press, London, 1975, pp. 84–100.

[27] J. Kampé de Fériet, Les fonctions aléatoires stationnaires et la théorie statistique de la turbulence homogène. *Ann. Soc. Sci. Bruxelles Sér. I.* **59**(1939), 145–210.

[28] J. Kampé de Fériet, La notion de moyenne dans les équations du mouvement turbulent d'un fluide. Presented at 6ième Congrès International de Mécanique Appliquée, Paris, 1946.

[29] J. Kampé de Fériet, Sur un problème d'algèbre abstraite posé par la définition de la moyenne dans la théorie de la turbulence. *Ann. Soc. Sci. Bruxelles, Sér. I.* **63**(1949), 165–180.

[30] J. L. Kelley, Averaging operators on $C_\infty(X)$. *Illinois J. Math.* **2**(1958), 214–223.

[31] A. N. Kolmogorov, *Grundbegriffe der Wahrscheinlichkeitsrechnung.* Springer, Berlin, 1933.

[32] P. E. Kopp, D. Strauss, and F. J. Yeadon, Positive Reynolds operators on Lebesgue spaces. *J. Math. Anal. Appl.* **44**(1973), 350–365.

[33] J. B. Miller, Averaging and Reynolds operators on Banach algebras. I. Representation by derivations and antiderivations. *J. Math. Anal. Appl.* **14**(1966), 527–548.

[34] I. Molinaro, Détermination d'une R-transformation de Reynolds. *C. R. Acad. Sci. Paris* **244**(1957), 2890–2893.

[35] I. Molinaro, Sur les endomorphismes de Reynolds de fonctions définies sur un ensemble fini. *Publ. Sci. Univ. Alger. Sér. A* **4**(1957), 87–101.

[36] S.-T. C. Moy, Characterizations of conditional expectation as a transformation on function spaces. *Pacific J. Math.* **4**(1954), 47–63.

[37] A. Neeb, Positive Reynolds operators and generating derivations. *Math. Nachr.* **203**(1999), 131–146.

[38] A. Neeb, Positive Reynolds operators and moment theory. *Semigroup Forum* **59**(1999), 233–243.

[39] A. Pełczyński, Linear extensions, linear averagings, and their applications to linear topological classification of spaces of continuous functions. *Dissertationes Math. Rozprawy Mat.*, **58,** 1968.

[40] O. Reynolds, On the dynamical theory of incompressible viscous fluids and the determination of the criterion. *Phil. Trans. Roy. Soc. London. A.* **186**(1895), 123–164.

[41] G.-C. Rota, On the representation of averaging operators. *Rend. Sem. Mat. Univ. Padova* **30**(1960), 52–64. Reprinted in this chapter.

[42] G.-C. Rota, Spectral theory of smoothing operators. *Proc. Nat. Acad. Sci. U. S. A.* **46**(1960), 863–868. Reprinted in this chapter.

[43] G.-C. Rota, Endomorphismes de Reynolds et théorie ergodique. *C. R. Acad. Sci. Paris* **250**(1960), 2791–2793. Reprinted in this chapter.
[44] G.-C. Rota, Représentation des opérateurs de Reynolds. *C. R. Acad. Sci. Paris* **250**(1960), 2831–2833. Reprinted in this chapter.
[45] G.-C. Rota,Une généralisation de l'espérance mathématique conditionnelle qui se présente dans la théorie statistique de la turbulence. *C. R. Acad. Sci. Paris* **251**(1960), 624–626. Reprinted in this chapter.
[46] G.-C. Rota, Une théorie unifiée des martingales et des moyennes ergodiques. *C. R. Acad. Sci. Paris* **252**(1961), 2064–2066. Reprinted in Chapter 4.
[47] G.-C. Rota, On the passage from the Navier-Stokes equations to the Reynolds equations. In R. Langer, ed., *Partial Differential Equations and Continuum Mechanics, Proc. Int. Conf., 1960,* Univ. of Wisconsin Press, Madison, WI, 1961, pp. 374–376. Reprinted in this chapter.
[48] G.-C. Rota, Reynolds operators. In *Proc. Sympos. Appl. Math.,* Vol. XVI, Amer. Math. Soc., Providence RI, 1964, pp. 70–83. Reprinted in this chapter.
[49] G.-C. Rota, Baxter algebras and combinatorial identities. I, II. *Bull. Amer. Math. Soc.* **75**(1969), 325–329; *ibid.* **75**(1969), 330–334. Reprinted in J. P. S. Kung, ed., *Gian-Carlo Rota on Combinatorics,* Birkhäuser, Boston and Basel, 1995, pp. 471–480.
[50] G.-C. Rota, The valuation ring of a distributive lattice. In *Proceedings of the University of Houston Lattice Theory Conference (Houston, Tex., 1973),* Dept. Math., Univ. Houston, Houston, TX, 1973, pp. 574–628.
[51] G.-C. Rota, Edmund Husserl and the reform of logic. In D. Carr and E. S. Casey, eds., *Explorations in Phenomenology,* Nijhoff, The Hague, 1973, pp. 299-305.
[52] G.-C. Rota, *Indiscrete Thoughts.* F. Palombi, ed., Birkhäuser, Boston and Basel, 1997.
[53] S. Sakai, *C^*-Algebras and W^*-Algebras.* Springer, New York and Heidelberg, 1971.
[54] E. Scheffold, Über Reynoldsoperatoren und "Mittelwert bildende" Operatoren auf halbeinfachen F-Banachverbandsalgebren. *Math. Nachr.* **162**(1993), 329–337.
[55] I. M. Singer and J. Wermer, Derivations on commutative normed algebras. *Math. Ann.* **129**(1955), 260–264.
[56] L. Solomon, The Burnside algebra of a finite group. *J. Combinatorial Theory* **2**(1967), 603–615.
[57] J. J. Sopka, Jr., *On the Characterization of Reynolds Operators on the Normed Algebra of All Continuous Real-Valued Functions Defined on a Compact Hausdorff Space.* Thesis, Harvard University, Cambridge, MA, 1950.
[58] F. Spitzer, A combinatorial lemma and its application to probability theory. *Trans. Amer. Math. Soc.* **82**(1956), 323–339.
[59] F. B. Wright, Generalized means. *Trans. Amer. Math. Soc.* **98**(1961), 187–203.

CHAPTER 4

Ergodic Theory

CALCUL DES PROBABILITÉS. — *Une théorie unifiée des martingales et des moyennes ergodiques.* Note (*) de M. **Gian-Carlo Rota**, transmise par M. Joseph Kampé de Fériet.

L'analogie entre les théorèmes aux limites concernant les transformations qui conservent la mesure (théorèmes ergodiques, où, en termes probabilistes, lois des grands nombres) et les lois de convergence des martingales (1) a été notée depuis longtemps. Il y a plus de dix ans, Doob et Kakutani (2) ont posé le problème de déduire d'un seul principe général ces deux types de théorèmes. C'est ce principe que nous donnerons ici.

On arrive à un concept unificateur en *généralisant la notion d'espérance mathématique conditionnelle.* On a vu dans la Note précédente (3) qu'une telle généralisation est donnée par les endomorphismes de Reynolds (4). En voici l'idée fondamentale. Soit $L_\infty(S, \Sigma, \mu)$ l'algèbre des variables aléatoires bornées à valeurs réelles où complexes dans un espace de probabilité (S, Σ, μ), et soit D une *dérivation* dans cette algèbre $[D(uv) = u\,Dv + v\,Du]$, définie dans un sous-espace (faiblement) dense. On suppose que D est un opérateur linéaire fermé dans l'espace de Banach L_∞, dont le spectre ne contient pas tout l'axe réel. En multipliant D par une constante, on peut supposer que l'inverse de $I - D$, où I est l'opérateur identité, existe. Dans ces conditions, on démontre que l'opérateur borné $R = (I - D)^{-1}$ satisfait à l'identité (5) $R(uv) = Ru\,Rv + R[(u - Ru)\,(v - Rv)]$ pour u, v dans L_∞, qui caractérise les endomorphismes de Reynolds. En particulier, soit $U^t f = f \circ \varphi_t$ pour f dans L_∞ un semi-groupe non trivial d'opérateurs dans L_∞ engendré par un semi-groupe de transformations $\varphi_t . \varphi_s = \varphi_{t+s}$ conservant la mesure. On vérifie que le générateur infinitésimal $D = \log U$ de U^t est une dérivation; par suite, l'opérateur borné $(I - \log U)^{-1}$ est un endomorphisme de Reynolds. On peut donc associer un endomorphisme de Reynolds à chaque semi-groupe de transformations conservant la mesure, et la classification de ces semi-groupes est donc réduite à celle des endomorphismes.

Mais la réciproque n'est pas vraie : il y a des endomorphismes de Reynolds qui ne « proviennent » pas d'une transformation conservant la mesure. Ce deuxième type d'endomorphismes de Reynolds coïncide exactement avec les *espérances mathématiques conditionnelles.*

En effet, on a démontré (6) qu'une espérance mathématique conditionnelle est un opérateur linéaire A qui satisfait à l'identité $A(uAv) = Au\,Av$ pour u, v dans L_∞. Cette identité, avec la condition triviale $Ae = e$, où e est la variable aléatoire identiquement égale à l'unité, implique l'identité de Reynolds. On démontre que sous des conditions peu restrictives et « naturelles » pour R, à savoir, que R est positif et contractif $\left(\int Ru \leq \int u\right)$

— ce que nous supposerons désormais — ces deux types sont essentiellement les seuls possibles d'endomorphismes de Reynolds.

On voit donc que les endomorphismes de Reynolds unifient l'idée d'une transformation conservant la mesure et l'idée d'espérance mathématique conditionnelle. Ces opérateurs correspondent à une idée intuitive de moyenne qui se présente dans plusieurs applications, par exemple, dans les études des équations de la mécanique des fluides et dans les moyennes utilisées par les électriciens (7).

Nous arrivons à l'unification des théorèmes ergodiques et des martingales en donnant l'analogue du théorème ergodique pour les endomorphismes de Reynolds. Ceci nous amène au nouveau processus stochastique suivant. Appelons *martingale généralisée* une famille d'endomorphismes de Reynolds $R_t (t > 0)$ dont les termes sont deux à deux commutatifs, et qui satisfont à l'identité suivante :

$$(\#) \qquad (sR_t - tR_s)R_s u = (s-t)R_t R_s^2 u, \qquad s > t, \qquad u \in L_\infty.$$

On a le théorème ergodique fondamental suivant :

Théorème. — *Pour toute martingale généralisée, la limite* $\lim_{t \to 0} R_t u$ *existe presque partout pour u dans* $L_p(S, \Sigma, \mu)$, $1 < p < \infty$.

Voyons comment ce résultat contient les deux théorèmes classiques. Une martingale est une famille d'espérances mathématiques conditionnelles $A_t (t > 0)$ telles que $A_t A_s = A_s$ pour $s \geq t$. En posant $R_t = A_t$ dans (#), on voit que toute martingale est aussi une martingale généralisée; par suite, on obtient la convergence des martingales.

Soit maintenant $U^x (x > 0)$ un semi-groupe de transformations conservant la mesure. Posons $R_t u = t(tI - \log U)^{-1}$ pour $t > 0$. Pour l'opérateur $T_t = (tI - \log U)^{-1}$ on a l'identité de Hilbert $T_t - T_s = (s-t)T_t T_s$. Si l'on pose $T_t = t^{-1} R_t$ dans cette identité, ensuite on multiplie par R_s, on obtient (#). On voit donc que R_t est une martingale généralisée. Comme $T_t u = \int_0^\infty e^{-tx} U^x u \, dx$ pour u dans L_p, notre théorème donne la convergence presque partout des moyennes d'Abel :

$$\lim_{t \to 0} t \int_0^\infty e^{-tx} U^x u \, dx.$$

L'existence de cette limite implique la convergence des moyennes de Cesàro, en vertu d'un théorème Taubérien classique. Nous voici donc arrivés au théorème ergodique.

D'autres applications du théorème précédent donnent des théorèmes ergodiques « mixtes ». Il y a aussi un théorème analogue (et beaucoup plus facile à démontrer) qui correspond aux martingales inverses et à l'ergodicité d'un semi-groupe U^x pour x tendant vers zéro.

Le théorème peut aussi être généralisé « au sens large », en obtenant convergence en moyenne dans un espace de Banach réflexif quelconque, pour une suite d'opérateurs deux à deux commutatifs et à normes uniformément bornées, satisfaisant à $(\neq)$.

La démonstration du théorème dépend de la théorie générale des endomorphismes de Reynolds ([8]).

(*) Séance du 20 mars 1961.

([1]) Pour la théorie des martingales, *voir* DOOB, *Stochastic Processes*, New York, 1953. Pour la théorie ergodique, *voir* HALMOS, *Ergodic Theory*, Tokyo, 1956. Pour les résultats de la théorie des opérateurs linéaires, *voir* DUNFORD-SCHWARTZ, *Linear Operators*, New York, 1958, vol. I.

([2]) S. KAKUTANI, *Ergodic Theory, International Congress of Mathematicians, Conference in Analysis*, Cambridge, 1950, p. 128.

([3]) *Comptes rendus*, 251, 1960, p. 624.

([4]) Pour les références aux travaux de J. Kampé de Fériet et Mme M.-L. Dubreil-Jacotin, voir *Comptes rendus*, 250, 1960, p. 27, 91 et 1831.

([5]) Découverte par J. Kampé de Fériet, et nommée l'axiome stochastique des moyennes par Mme Dubreil.

([6]) S. T. C. MOY, *Pac. J. Math.*, 4, 1954, p. 47; G.-C. ROTA, *Rendiconti Padova*, 30, 1960, p. 52.

([7]) *Voir* par exemple : KAMPÉ DE FÉRIET et BETCHOV, *Nederl. Akad. van Wetenschappen, Proceedings*, Série B, 54, 1951, p. 389; R. M. FANO, *Journal Acoustical Soc.*, 22, septembre 1950.

([8]) Cf. *Proc. Nat. Acad. Sc. U. S. A.*, 46, 1960, p. 863.

Comptes Rendus des Séances de l'Academie des Sciences
t. 252 (1961), p. 2064–2066

AN "ALTERNIERENDE VERFAHREN" FOR GENERAL POSITIVE OPERATORS[1]

BY GIAN-CARLO ROTA[2]

Communicated by Felix Browder, October 25, 1961

1. **Introduction.** Two types of limiting processes involving linear operators are of frequent occurrence. One type is associated with ergodic limits of Cesaro averages $n^{-1}(1+T+\cdots+T^{n-1})f$ of a linear operator T, for f in suitable function spaces, the so-called "ergodic theorems." The strongest such results establish almost everywhere convergence for f in a suitable Lebesgue space, under suitable hypotheses on T. This type of theorem is, nowadays, fairly well understood.[3]

A second type of limiting theorem of frequent occurrence concerns the limit of *products* $T_1T_2 \cdots T_nf$ of a sequence of operators T_n. The two noteworthy special cases are (a) Limiting theorems of the type $\lim_{t\to\infty} P^tf$, where P^t is a semigroup of selfadjoint operators. Results of this type express the "tendency to equilibrium" in certain physical processes (typically, in diffusion theory). (b) One considers two or more noncommuting projections (most commonly *conditional expectations* in a probability space) say F_1 and F_2, and asks for $\lim_{n\to\infty} (F_1F_2)^nf$. This is an abstract rendering of the so-called "*alternierende Verfahren*," (see [**8**]) which originated in function theory and has recently merged with various probabilistic and other considerations.

While results relating to the *mean convergence* of such iterations are easily obtained (for (a) and (b) they were obtained independently by von Neumann and Wiener [**10**; **15**] in the 1930's), the corresponding statements relating to *almost everywhere* convergence have been missing. A result in this direction has recently been obtained by Burkholder and Chow [**1**], but their results, though more general in some respects, are limited to L_2.[4]

---

[1] I wish to thank Professors J. Feldman, P. R. Halmos and C. C. Moore for listening to expositions of the contents of this paper, and offering much helpful advice.

[2] Work supported in part by Grant No. NSF G-19684 with the National Science Foundation and in part by the Office of Naval Research. The contents of this paper were delivered in an invited address at the Symposium in Ergodic Theory held at Tulane University in November 1961.

[3] For the most refined results see [**2**; **7**].

[4] Another important set of results has been recently obtained by E. M. Stein (personal communication, whose methods are of an altogether different character from the present ones; to him are due the statement and a first proof Theorem 2b below.

It is our present purpose to establish a single general procedure which yields *almost everywhere* convergence in L_p of certain products of operators. The main result stated below (Theorem 1) is general enough to include the case of $\lim_{n\to\infty} P^n f$, as well as the limit of iteration of several noncommuting conditional expectations.

2. **Doubly stochastic operators.** Given a *positive measure space* (S, Σ, μ), we consider a linear operator P whose domain and range are certain subspaces of the space $TM(S, \Sigma, \mu)$ of all real-valued totally measurable functions on (S, Σ, μ), as specified below. We make the following assumption on the operator P.

(1) P is defined on all of $L_1(S, \Sigma, \mu)$ and on all of $L_\infty(S, \Sigma, \mu)$, maps each of those spaces into itself. It is a *positive*[5] operator (that is, $Pf \geqq 0$ if $f \geqq 0$), and it is a contraction (that is, is of norm at most one) in each of these spaces. In other words,

$$\sup_{s\in S} | Pg(s) | \leqq \sup_{s\in S} | g(s) | ,$$

and

$$\int_S | Pf(s) | \mu(ds) \leqq \int_S | f(s) | \mu(ds)$$

for bounded and integrable g and f, respectively. (It follows by the Riesz convexity theorem that P will be a contraction in $L_p(S, \Sigma, \mu)$, $1 \leqq p \leqq \infty$.)

(2) The adjoint operator P^*, bounded by the identity $\int_S fPg d\mu = \int_S P^* f g d\mu$, is also defined in $L_1(S, \Sigma, \mu)$ and $L_\infty(S, \Sigma, \mu)$.

It follows that P^* is also a contraction in each of these spaces.

Evidently the various conditions specified in (1) and (2) are far from independent from one another, and could in fact be reduced in many ways to smaller sets of conditions. The redundance of these specifications is more than made up for by the ease with which they can be verified.

(3) Let 1 denote the function defined on S and with real values, which takes identically the value 1. We assume that $P1 = 1$ and that $P^*1 = 1$.

It follows, by an easy argument, which we omit, that in a finite measure space $P1 = 1$ implies $P^*1 = 1$ for an operator satisfying (1) and (2).

We shall call an operator satisfying conditions (1), (2), (3) a *doubly stochastic operator.*

[5] *Positive* operators should not be confused with *positive-definite* operators in Hilbert space, which are also sometimes called positive operators.

Consider now a sequence P_n $(n=1, 2, 3, \cdots)$ of doubly stochastic operators in a probability space (S, Σ, μ). The notion of the *path space* associated with this sequence[6] is an easy generalization of the classical probabilistic construction of the sample space of a Markov process (which in turn is a special case of the present notion), as found for example in Doob [**4**] or Loève [**9**]. It is defined as follows.

Consider the *measurable space* (S', Σ'') consisting of the one-sided infinite product of replicas of the measurable space (S, Σ); that is, $S' = \prod_{n=0}^{\infty} S_n$, and $S_n = S$ for all n. The σ-field Σ'' is the ordinary product σ-field.

A particular class of measurable real-valued functions on (S', Σ'') consists of functions $F(s_0, s_1, s_2 \cdots) = f_0(s_0)f_1(s_1)f_2(s_2) \cdots$, where each of the functions f_n is a measurable real-valued function defined on S, and where $f_n = 1$ for almost all n (that is, all integers n except for a finite number). We shall denote by $\boldsymbol{A}$ the algebra generated by these functions, when all the f_n are assumed to be essentially bounded functions.

The algebra $\boldsymbol{A}$ inherits a natural (and obvious) *lattice structure* from $L_\infty(S, \Sigma, \mu)$, which makes it into a lattice-ordered algebra.

With the aid of the given sequence P_n of doubly stochastic operators, we can define an L-space structure[7] on $\boldsymbol{A}$, considered as a vector space. A positive linear functional L on functions of the form $F(s_0, s_1, s_2, \cdots) = f_0(s_0)f_1(s_1)f_2(s_2) \cdots$ belonging to the algebra $\boldsymbol{A}$ is defined by the formula

$$(*) \qquad L(F) = \int_S f_0 P_1[f_1 P_2[f_2 P_3[f_3 \cdots]] \cdots]d\mu,$$

and then extended to all of $\boldsymbol{A}$ by linearity. The linear functional L is *well-defined* on $\boldsymbol{A}$. First, note that the integral in (*) involves only a *finite* number of terms, since $f_n = 1$ for almost all n and $P_n 1 = 1$ for all n; hence the integral makes sense. Secondly, the extension to $\boldsymbol{A}$ is well-defined, because $\boldsymbol{A}$ is essentially a tensor product of algebras isometric to $L_\infty(S, \Sigma, \mu)$, and L in (*) is compatible with the distributive laws defining the tensor product. Thirdly, from the definition of the lattice structure of $\boldsymbol{A}$ it is clear that $|F| = |f_0| \cdot |f_1| \cdot |f_2| \cdot \cdots,$

[6] The notion of a path space, and much of the theory to follow, can be studied for an arbitrary measure space, which need not even be σ-finite. In the present account we have refrained from such generalizations, for the sake of clarity of exposition. In many cases the reader himself may easily supply the necessary steps to achieve the needed generalization.

[7] In the sense of Kakutani, cf. Day [**3**, p. 107 ff.]. Day calls these (AL)-spaces.

where $|f|$ denotes the function equal to $|f(s)|$ at the point s. Furthermore, the formula

$$\|F\| = P(|F|) = \int_S |f_0| P_1[|f_1| P_2[|f_2| P_3[|f_3| \cdots]] \cdots] d\mu$$

defines a seminorm, and this seminorm is of "L-type" in the sense of Kakutani, (in other words, $\|F+G\| = \|F\| + \|G\|$ whenever both F and G are positive elements of $\boldsymbol{A}$). Factoring off the subspace of elements F such that $\|F\| = 0$, and completing, we obtain an abstract L-space X. By Kakutani's theorem,[8] the space X is isometric to (and will thereafter be identified with) $X = L_1(S'', \Sigma', \nu)$ for some (structurally unique) measure space (S'', Σ', ν).

The measure space (S'', Σ', ν) is the *path space* of the sequence of operators P_n. The definition of the path space does not use the full strength of the assumption that the P_n are doubly stochastic operators. All that is needed, for the definition to make sense, is that $P_n 1 = 1$ for all n, and that P_n is a positive operator.

It is clear that we can identify the set S'' with S', and the σ-field Σ with a σ-subfield of Σ''. We shall regard this identification (whose purpose is purely heuristic) as made, and "picture" the space (S', Σ', ν) as a space of sequences of points of S, with a certain measure. From $L(1) = 1$ it follows that (S', Σ', ν) is a probability space.

We next note that there is a natural imbedding of (S, Σ, ν) obtained by identifying $f \in L_1(S, \Sigma, \mu)$ with the function $F(s_0, s_1, s_2, \cdots) = f(s_0)$ of $\boldsymbol{A}$; the definition of L shows that this identification is faithful. Accordingly, we shall consider this imbedding as accomplished, and set $S = S'$, and $\Sigma \subset \Sigma'$, writing sometimes the path space as (S, Σ', ν). We shall also set $d\mu = d\nu$, when convenient. Thus, the path space is obtained essentially by *refining* the given σ-field Σ to a large enough σ-field.

3. **Main theorem.** Our main result is the following:

THEOREM 1. *Let P_n $(n = 1, 2, 3, \cdots)$ be a sequence of doubly stochastic operators in a probability space (S, Σ, μ). Then for every $p > 1$ and for every $f \in L_p(S, \Sigma, \mu)$, the sequence $P_1 P_2 \cdots P_n P_n^* P_{n-1}^* \cdots P_1^* f$ converges almost everywhere as $n \to \infty$.*

PROOF. We consider the subspace L_n of $L_1(S', \Sigma', \nu)$ generated by all functions $F(s_0, s_1, s_2, \cdots, s_n, \cdots) = f_0(s_0) f_1(s_1) \cdots f_n(s_n) \cdots$, where all f_n are essentially bounded, and where $f_0 = f_1 = \cdots = f_{n-1}$

[8] Cf. Day, loc. cit.

$=1$. Evidently the L_1-closure of L_n is also an L-space; it is therefore representable in the form $L_1(S', \Sigma_n, \nu)$, where Σ_n is a σ-subfield of Σ', uniquely determined by L_n. If $n>m$, then $\Sigma_n \subset \Sigma_m$. Let E_n be the *conditional expectation*[9] operator of $L_1(S', \Sigma', \nu)$ onto $L_1(S', \Sigma_n, \nu)$, and let E be the conditional expectation projecting onto $L_1(S, \Sigma, \mu)$. If f is any function of class $L_p(S', \Sigma', \nu)$, then the sequence of functions $f_n = E_n f$ forms a martingale reversed sequence, (cf. Loève loc. cit. p. 306 ff.) which is uniformly integrable. By the martingale convergence theorem, we infer that $f_\infty = \lim E_n f$ exists *almost everywhere* for $p>1$. Furthermore, by the maximal ergodic theorem for martingales there is a function f^* in $L_p(S', \Sigma', \nu)$ such that $\sup_{n\geq 0} |f_n(s)| \leq f^*(s)$. In other words, the sequence f_n converges boundedly in L_p to f_∞. Since conditional expectation preserves bounded almost everywhere convergence, we conclude that the sequence $EE_n f = Ef_n$ converges almost everywhere to Ef_∞.

We have therefore proved that *for every function f in $L_p(S, \Sigma, \mu)$, $p>1$, the sequence $EE_n f$ converges almost everywhere to some function g in $L_p(S, \Sigma, \mu)$.*

The proof of the theorem will be complete, if we can show that $EE_nE = P_1 \cdots P_{n-1}P_nP_n^*P_{n-1}^* \cdots P_1^*$. This can actually be done by a simple direct computation.

To calculate $E_n f$, for f in $L_p(S, \Sigma, \mu)$, we make use of the fact that $E_n f$ is the *unique* function such that

$$(**) \qquad \int_{S'} gE_n f \, d\nu = \int_{S'} gf \, d\nu, \qquad g \in L_q(S', \Sigma_n, \nu),$$

where q is the index conjugate to p.

To verify that (**) holds, it suffices to show that it holds when g ranges over a dense subset of $L_q(S', \Sigma_n, \nu)$. Such a dense subset will be taken to be the subspace L_n defined above. We therefore need only verify (**) when $g=F$, where F is in L_n, and F is of the "product" form

$$F(s_0, s_1, s_2, \cdots) = f_0(s_0)f_1(s_1)f_2(s_2) \cdots .$$

We shall in fact find that $E_n f = G$, where

$$G(s_0, s_1, s_2, \cdots) = P_n^* P_{n-1}^* \cdots P_1^* f(s_n) = h(s_n).$$

Now,

---

[9] The reader unfamiliar with this notion may consult Loève [9, Chapter VII], or, for an "operatorial" treatment, [11]. The particular properties that single out conditional expectations among all projections are essential in what follows.

$$\int_{S'} FG d\nu = \int_S f_0 P_1[f_1 P_2[\cdots P_n[(hf_{n+1})P_{n+1}[f_{n+2} \cdots]]\cdot] d\mu.$$

Since $f_0 = f_1 = \cdots = f_{n-1} = 1$, the right side simplifies to

$$\int_S P_1 P_2 \cdots P_n[hf_n P_{n+1}[\cdots]] d\mu.$$

Using the assumption that $P_k^* 1 = 1$ for all k the last expression in turn simplifies to

$$\int_{S'} FG d\nu = \int_S hf_n P_{n+1}[f_{n+2} P_{n+2}[\cdots]] d\mu$$

$$= \int_S (P_n^* P_{n-1}^* \cdots P_1^* f)(f_n P_{n+1}[f_{n+1} P_{n+2}[\cdots]]) d\mu.$$

Passing from the operator P_k^* to their adjoints P_k, this last integral becomes identical with

$$\int_S f(P_1 P_2 \cdots P_n[f_n P_{n+1}[f_{n+1} P_{n+2} \cdots]]) \, d\mu = \int_{S'} fF \, d\nu.$$

We have therefore shown that

$$\int_{S'} GF \, d\nu = \int_{S'} fF \, d\nu$$

for all F of the type indicated. But, as mentioned above, this amounts to saying that $G = E_n f$.

Next, we shall compute the conditional expectation EH, where $H \in L_p(S', \Sigma_n, \nu)$ is of the form $H(s_0, s_1, s_2, \cdots) = h(s_n)$ for some $h \in L_p(S, \Sigma, \mu)$. We shall in fact find that $EH = P_1 P_2 \cdots P_n h = g$. Again, this amounts to verifying that

$$\int_S gf \, d\mu = \int_{S'} fH \, d\nu$$

for all $f \in L_p(S, \Sigma, \mu)$. Now, by definition of $d\nu$, we have

$$\int_{S'} fH d\nu = \int_S fP_1 P_2 \cdots P_n h \, d\mu = \int_S gf \, d\mu,$$

which is exactly what we want.

Combining the results of the two computations obtained in the two above paragraphs, we obtain for $f \in L_p(S, \Sigma, \mu)$, that $EE_nf = P_1P_2 \cdots P_nP_n^*P_{n-1}^* \cdots P_2^*P_1^*f$, and the proof of convergence is therefore complete.

4. **Applications.** The two main applications of the above theorem are (a) the "alternierende Verfahren" for sequences of noncommuting conditional expectations, and (b) the "tendency to equilibrium" theorems for *powers* of a positive doubly stochastic operator. We begin with the second result.

THEOREM 2. *Let P be a doubly stochastic operator which is selfadjoint in $L_2(S, \Sigma, \mu)$. Then:*

(a) *There is a dilation of the sequence of operators P^{2n} into a martingale E_n.*

(b) *For f in $L_p(S, \Sigma, \mu)$, $p>1$, $\lim_{n\to\infty} P^{2n}f$ exists almost everywhere.*

PROOF. It suffices to choose all the $P_n = P$ in Theorem 1.

The "alternierende Verfahren" is actually an application of Theorem 2 when only two conditional expectations, say F_1 and F_2 are involved. Setting $P = F_1F_2F_1$ one obtains a doubly stochastic selfadjoint operator, and hence the convergence of $(F_1F_2F_1)^nf$ follows under the same conditions as in Theorem 2.

More generally, an infinite sequence F_n $(n = 1, 2, \cdots)$ of noncommuting conditional expectation operators is a special case of a sequence of operators satisfying the conditions of the Main Theorem; we therefore obtain that $\lim_{n\to\infty} F_1F_2 \cdots F_{n-1}F_nF_{n-1} \cdots F_1f$ exists almost everywhere. Actually, in this result the F_n can be arbitrary doubly stochastic selfadjoint operators.

If only a *finite* sequence $F_1, F_2, \cdots, F_m$ of noncommuting conditional expectations (or again, more generally, arbitrary doubly stochastic selfadjoint operators) are involved, we may obtain from the Main Theorem various convergent sequences of operators yielding almost everywhere convergence. For example, we may take $(F_1 \cdots F_m)^n(F_mF_{m-1} \cdots F_1)^n$, or else $(F_1F_2F_1F_3F_2 \cdots F_m)^n \cdot (F_m \cdots F_2F_3F_1F_2F_1)^n$, etc., etc.

5. **Generalizations and extensions.** The method outlined above is capable of numerous generalizations, among which we mention: a "weak type L_1" theorem, a convergence theorem with reversed indices (that is, for $\lim_{n\to\infty} P_nP_{n-1} \cdots P_1P_1^* \cdots P_n^*f$), more general semigroups, (for example the continuous P^t, t real). Such developments will be given in a forthcoming publication, which will exploit a gen-

eral theory of "dilation" in measure spaces by conditional expectation.[10]

Bibliography

1. Burkholder and Chow, *Iterates of conditional expectation operators*, Proc. Amer. Math. Soc. **12** (1961), 490–495.

2. Chacon and Ornstein, *A general ergodic theorem*, Illinois J. Math. **4** (1960), 153–160.

3. Day, *Normed linear spaces*, Springer, Berlin, 1958.

4. Doob, *Stochastic processes*, Wiley, New York, 1953.

5. Dunford-Schwarz, *Linear operators*, Vol. I, Interscience, New York, 1958.

6. Halmos, *Lectures on ergodic theory*, Mathematical Society of Japan, Tokyo, 1956.

7. E. Hopf, *On the ergodic theorem for positive linear operators*, Crelle's J. **205** (1960), 101–106.

8. Hurwitz-Courant, *Funktionen Theorie*, Springer, Berlin, 1929.

9. Loève, *Probability theory*, 2nd ed. Van Nostrand, Princeton, 1960.

10. Von Neumann, *Functional operators*, 2 vols., Princeton Univ. Press, Princeton, N. J., 1950.

11. Rota, *On the representation of averaging operators*, Rend. Sem. Mat. Univ. Padova **30** (1960), 52–64.

12. Wiener, *On the factorization of matrices*, Comment. Math. Helv. **29** (1955), 97–110.

Massachusetts Institute of Technology

---

[10] A proof of the *Ergodic* theorem (for almost everywhere convergence) for a doubly stochastic operators along the lines of the present paper is easily obtained by using the well-known sample space construction of a Markov process, as has been known for some time (cf. [**4**]).

Bulletin of the American Mathematical Society, **68** (1962), 95–102.

ON THE CLASSIFICATION OF PERIODIC FLOWS

GIAN-CARLO ROTA[1]

1. Introduction. Let (S, Σ, μ) be a probability space,[2] and let ϕ and ϕ' be measure-preserving transformations of (S, Σ, μ), that is, functions from Σ to Σ such that $\phi(E) \in \Sigma$ if $E \in \Sigma$ and $\mu(\phi(E)) = \mu(E)$. One of the central questions of ergodic theory is: when are ϕ and ϕ' conjugate? We recall that two measure-preserving transformations ϕ and ϕ' are said to be (algebraically) conjugate (cf. Halmos [2, p. 44]) when $\phi' = \alpha \circ \phi \circ \alpha^{-1}$ for some transformation α such that both α and its inverse α^{-1} are measure-preserving transformations. We are considering here, for greater generality, *set* transformations of the Boolean σ-algebra Σ, which may or may not be induced by *point* transformations of the set S (cf. Halmos [2, pp. 42–45]).

To answer this question, several conjugacy invariants of measure-preserving transformations have been introduced, such as the spectrum of the induced unitary operator, entropy, etc. However, these invariants are at present very far from being a complete set, which would allow an immediate decision on whether or not two measure-preserving transformations are conjugate.

Another approach to the question consists in relating a measure-preserving transformation to some one among a set of "canonical forms," selected for their simplicity. The result of Halmos and von Neumann (cf. Halmos [2, p. 48]) goes far in this direction: it is shown that an ergodic measure-preserving transformation with discrete spectrum is conjugate to a rotation on a compact abelian group.

The result of the present note lies in this second direction: we shall prove that *every* periodic group of measure-preserving transformations, or "flow," ergodic or not, is conjugate to a "restriction" of the simplest such flow, namely, the flow obtained on the lateral surface of unit area of a cylinder by rotating the cylinder around its axis.

2. Definitions. A *periodic flow* (or simply *flow*) is a semigroup ϕ_t, $0 \leq t \leq \infty$, of transformations ϕ_t of Σ into Σ, such that $\mu(\phi_t(E)) = \mu(E)$ for every E in Σ and for every t, $0 \leq t \leq \infty$. We shall assume the semigroup property $\phi_t(\phi_s(E)) = \phi_{t+s}(E)$, and the periodicity condition $\phi_0 = \phi_1$. Clearly, some sort of smoothness condition relative to t of

Received by the editors June 7, 1961.

[1] Work done under contract with the Office of Naval Research.

[2] We follow the notation in Dunford-Schwartz [1]. Thus S is a set, Σ is a σ-field (or σ-algebra) of "measurable" subsets of Σ, and μ is a measure defined on sets of Σ such that $\mu(S) = 1$.

the flow must be assumed. We shall make the weakest such assumption, as follows. Consider the measure space (C, Σ_1, ν) obtained by taking the product of the given measure space (S, Σ, μ) with the interval $[0, 1]$, endowed with ordinary Lebesgue measure. A point of C can be written in the form (t, s), where $0 \leq t \leq 1$ and $s \in S$. For $E \in \Sigma$, let E_1 be the set $\{(t, s) \mid s \in \phi_t(E),\ 0 \leq t \leq 1\}$. Our smoothness assumption is that E_1 shall be a measurable set, namely, that $E_1 \in \Sigma_1$, for every $E \in \Sigma$. In other words, as t runs from 0 to 1, $\phi_t(E)$ describes a measurable set in the product.

The above description of a periodic flow can be summarized as follows: a periodic flow is a measurable *action* of the circle group on a probability space. Any continuous action of the circle group on any topological space which is also endowed with a finite regular measure will satisfy our requirements; it is thus easy to produce "many" examples of periodic flows, of the most disparate varieties.

It follows from the definition that ϕ_t^{-1} is well-defined, and since $\phi_t \circ \phi_{1-t} = I$, that $\phi_t^{-1} = \phi_{1-t}$. Defining $\phi_t = \phi_{1-t}$ for $t < 0$, we can obtain a group defined for $-\infty < t < \infty$.

Two flows ϕ_t and ϕ_t', defined on *different* measure spaces (S, Σ, μ) and (S', Σ', μ'), are (algebraically) *conjugate* when $\phi_t = \alpha \circ \phi_t' \circ \alpha^{-1}$, where α is a measure-preserving transformation of Σ onto Σ' with measure-preserving inverse α^{-1}. This implies in particular that the measure-algebras (S, Σ, μ) and (S', Σ', μ') are isomorphic.

A σ-subalgebra Σ'' of Σ is said to be *invariant* under a flow ϕ_t on (S, Σ, μ), whenever $\phi_t(E) \in \Sigma''$ for every $E \in \Sigma''$ and for every t. If Σ'' is an invariant σ-subalgebra of ϕ_t, then clearly one obtains from ϕ_t a new flow ϕ_t'' on the probability space (S, Σ'', μ) by *restricting* ϕ_t to the σ-subalgebra Σ''. The question stated at the end of §1 can now be reformulated as follows: what other flows can one obtain (to within conjugacy) from a given flow ϕ_t, by restricting ϕ_t to each one of its proper invariant σ-subalgebras? If the restriction ϕ_t'' of a flow ϕ_t to some one of its invariant σ-subalgebras Σ'' is conjugate to another flow ϕ_t', we shall say that ϕ_t is a *model* for ϕ_t'.

3. Our main result is then the following

THEOREM. *Let (S, Σ, μ) be a separable nonatomic probability space. Then there exists a universal periodic flow Φ_t in (S, Σ, μ), namely, a flow which is a model for every other periodic flow.*

PROOF. Recall the definition of a nonatomic separable measure-algebra, as given in Halmos [3, pp. 168–173]. From the discussion in Halmos, the following facts are easily derived: (a) the measure algebras of any two nonatomic separable probability spaces are isomorphic; (b) the product of two (and, in fact, of a countably infinite number of) nonatomic separable probability spaces is again a nonatomic separable probability space. (c) In particular, if (S, Σ, μ) is as in the statement of the Theorem, then the measure algebra of the "cylinder" space (C, Σ_1, ν) is isomorphic to that of (S, Σ, μ). We shall write a point of C as the pair (t, s), where $s \in S$ and $-\infty < t < \infty$, agreeing to identify two points whose t-coordinates differ by an integer. We now define Φ_t' on (C, Σ_1, ν) as follows: set $\Phi_t'(E_1) = \{(u, s) \mid (u+t, s) \in E_1\}$. It is obvious that Φ_t' is a flow. Let Φ_t be a flow conjugate to Φ_t' and defined on (S, Σ, μ). Such a flow exists by (c). We claim that Φ_t is the desired universal flow.

To show this, it suffices to show that Φ_t' is a universal flow, or, more simply, by (a), that Φ_t' is a *model* for every flow ϕ_t on (S, Σ, μ).

Given ϕ_t, consider the family Σ'' of sets $E'' = \{(t, s) \mid s \in \phi_t(E), E \in \Sigma\}$. It is clear that Σ'' is a σ-subalgebra of Σ_1, and since Σ is nonatomic, so is Σ''. Furthermore, $\nu(E'') = \mu(E)$. Indeed, $\nu(E'')$ can be calculated by Fubini's theorem, using the fact that the measure ν is the product of Lebesgue measure dt with the measure μ. This gives $\nu(E'') = \int_0^1 \int_S \phi_t(E) d\mu dt$. The inner integral gives $\mu(\phi_t(E)) = \mu(E)$, by the measure-preserving property of ϕ_t. Thus the double integral simplifies to $\nu(E'') = \int_0^1 \mu(\phi_t(E)) dt = \int_0^1 \mu(E) dt = \mu(E)$.

Now define $\alpha: \Sigma \to \Sigma''$ as $\alpha(E) = E''$. Then α is an isomorphism of the measure-algebra of (S, Σ, μ) with that of (C, Σ'', ν). Let Φ_t'' be the restriction of Φ_t' to Σ''. To complete the proof of the theorem, we must establish that Σ'' is an invariant σ-subalgebra of Φ_t' and that the diagram

$$\begin{array}{ccc} (C, \Sigma'', \nu) & \xrightarrow[\Phi_t'']{} & (C, \Sigma'', \nu) \\ \alpha \updownarrow \alpha^{-1} & & \alpha^{-1} \updownarrow \alpha \\ (S, \Sigma, \mu) & \xrightarrow[\Phi_t]{} & (S, \Sigma, \mu) \end{array}$$

is commutative. But, for $E \in \Sigma$, we have by definition

$$\begin{aligned} \Phi_t''(\alpha(E)) &= \Phi_t''(E'') = \{(u, s) \mid (u + t, s) \in E''\} \\ &= \{(u, s) \mid s \in \phi_{u+t}(E)\} = \{(u, s) \mid s \in \phi_u(\phi_t(E))\}. \end{aligned}$$

This shows that $\Phi_t''(\alpha(E)) = \alpha(\phi_t(E))$, or $\alpha^{-1} \circ \Phi_t'' \circ \alpha = \phi_t$, and completes the proof.

4. Comments. The preceding result can be considered as a "strict sense" analog of a result obtained in the author's note [6]. It is clear that the result can be generalized to any nonseparable but homogeneous measure algebra. However, the statement is significant within the context of separable measure spaces, because it amounts to a restriction on the conjugacy classes of periodic flows.

We conclude with a number of questions suggested by the present situation, which would contribute to the solution of some current problems in ergodic theory.

(1) Suppose ϕ_t is a model for ϕ_t' and ϕ_t' is a model for ϕ_t. What can be said about ϕ_t and ϕ_t'? Are they conjugate? If not, as is probable, then the equivalence relation thus obtained should be studied.

(2) What is the structure of the invariant σ-subalgebras of a measure-preserving transformation ϕ_t?

(3) The usual "functorial" considerations suggest that there should be a dual result for a discrete group, that is, for a single measure-preserving transformation and its powers. In this direction, a method used by Oxtoby [5, p. 128], taking infinite direct products of probability spaces, is suggestive.

Bibliography

1. Dunford-Schwartz, *Linear operators*, Vol. I, Interscience, New York, 1951.
2. Halmos, *Lectures on ergodic theory*, The Mathematical Society of Japan, Tokyo, 1956.
3. Halmos, *Measure theory*, Van Nostrand, New York, 1950.
4. Jacobs, *Neuere Methoden und Ergebnisse der Ergodentheorie*, Springer, Berlin, 1960.
5. Oxtoby, *Ergodic sets*, Bull. Amer. Math. Soc. 58 (1952), 116–136.
6. Rota, *On models for linear operators*, Comm. Pure Appl. Math. 13 (1960), 469–472.

Massachusetts Institute of Technology

Proceedings of the American Mathematical Society, **13** (1962), 659–662

ON THE MAXIMAL ERGODIC THEOREM FOR ABEL-LIMITS[1]

GIAN-CARLO ROTA

The powerful and ingenious method of Chacón and Ornstein [1], later simplified by E. Hopf [6], has led to a proof of the ergodic theorem for general operators which is in many ways definitive. The purpose of this note is to show that if Abel convergence is used instead of Cesàro convergence, then a proof of the corresponding limit theorem is obtained which is of extreme simplicity. The argument below shows that Abel convergence, rather than Cesàro convergence, is the natural limiting process to be associated with the method of Chacón-Ornstein-Hopf.

Our simplification of Hopf's proof can be given in a few lines; we follow some of the steps in Hopf's basic Lemma 1. The proof given below can however be read without knowledge of Hopf's paper.

Let T be a positive operator ($Tf \geqq 0$ if $f \geqq 0$) in $L_1(S, \Sigma, \mu)$ and $L_\infty(S, \Sigma, \mu)$, where (S, Σ, μ) is an arbitrary measure space. Assume that T is a contraction ($\|T\| \leqq 1$) in the first space, and that it is bounded in the second. The Abel-ergodic theorem states that for every f in $L_p(S, \Sigma, \mu)$, $1 < p < \infty$, the limit

$$\lim_{\lambda \to 1} (1 - \lambda) \sum_{n=0}^{\infty} T^n f(s) \lambda^n = \lim_{\lambda \to 1} (1 - \lambda) R_\lambda f(s)$$

exists for almost all s in S. The associated Abel-maximal theorem states that the function

$$f^*(s) = \sup_{0 < \lambda < 1} \sum_{n=0}^{\infty} T^n f(s) \lambda^n$$

is of class L_p whenever the function f is of class L_p, $\infty > p > 1$. The maximal theorem can be used to prove the Abel-ergodic theorem by using standard techniques (cf. Dunford-Schwartz [3, Chapter VIII], and Hille-Phillips [5, Chapter 18]). Equally standard techniques (cf., for example, Dunford-Schwartz, loc. cit., Theorem VIII. 6.8, which is easily adapted to Abel limits) show that the Abel-maximal theorem depends on the following weak-type L_1 estimate:

Basic Lemma. *Let T be a positive contraction operator in $L_1(S, \Sigma, \mu)$, and let f be an integrable real function on the measure space (S, Σ, μ). Let $A = \{s \mid \sup_{0<\lambda<1} R_\lambda f(s) > 0\}$. Then $\int_A f(s)\mu(ds) \geqq 0$.*

Received by the editors July 30, 1962.

[1] By contract with grant NSF-GP-149 and the Office of Naval Research.

It is in proving this lemma (or variants thereof) that the difficulty of the ergodic theorem lies. The following is a short proof. We omit the argument s in functions ($f(s)=f$), unless it is indispensable. Following E. Hopf, we define the functions h_n by the recursion

$$h_{n+1} = Th_n^+ - h_n^-, \qquad h_0 = f,$$

where $f^+=\max(f, 0)$ and $-f^-=\min(f, 0)$. It is obvious that $h_{n+1}^- \leq h_n^-$. Letting $h^\pm(\lambda) = \sum_{n=0}^\infty h_n^\pm \lambda^n$, we obtain the equation

$$(*) \qquad (I - \lambda T)h^+(\lambda) = f + (1 - \lambda)h^-(\lambda),$$

that is, shuffling terms,

$$(1 - \lambda)^{-1}f = (1 - \lambda T)(1 - \lambda)^{-1}h^+(\lambda) - h^-(\lambda).$$

Taking coefficients of the power series we see that $f = \sum_0^n h_k^+ - T\sum_0^{n-1} h_k^+ - h_n^-$ whence $\int f \geq -\int h_n^-$ (using the fact that $\int Tg \leq \int g$ for all positive g). But $\int_A f \geq \int f$, since f is negative on the complement of A, hence the proof will be complete if we can show that h_n^- decreases to zero, or, equivalently (since $h^+h^-=0$), that for every point s in A, there is an n for which $h_n^+(s)>0$. But this is a trivial consequence of the main hypothesis! Indeed, from (*) we get, dropping a positive term, $(I-\lambda T)h^+(\lambda) \geq f$, whence, inverting (the inverse exists and preserves order for $0<\lambda<1$ since T is a positive contraction) $h^+(\lambda) \geq \sum_{n=0}^\infty T^n f \lambda^n$. The hypothesis now gives $\sup_{0<\lambda<1} h^+(\lambda)>0$ on A, so that for every s, some coefficients $h_n^+(s)$ must be positive, q.e.d.

It would be interesting to use the same method to give a proof of the Abel analog of Chacón's ergodic theorem [2].

REFERENCES

1. R. Chacón and D. Ornstein, *A general ergodic theorem*, Illinois J. Math. 4 (1960), 153–160.

2. R. Chacón, *An ergodic theorem for operators satisfying norm conditions*, J. Math. Mech. 11 (1962), 165–172.

3. N. Dunford and J. Schwartz, *Linear operators*. I, Interscience, New York, 1958.

4. P. R. Halmos, *Lectures on ergodic theory*, The Mathematical Society of Japan, 1956 (reprinted by Chelsea, New York, 1960).

5. E. Hille and R. S. Phillips, *Functional analysis and semigroups*, Amer. Math. Soc. Colloq. Publ. Vol. 31, rev. ed., Amer. Math. Soc., Providence, R. I., 1957.

6. E. Hopf, *On the ergodic theorem for positive linear operators*, J. Reine Angew. Math. 205 (1960), 101–106.

7. K. Jacobs, *Neuere Methoden und Ergebnisse der Ergodentheorie*, Springer, Heidelberg, 1960.

MASSACHUSETTS INSTITUTE OF TECHNOLOGY

Proceedings of the American Mathematical Society, 14 (1963), 722–723

A Limit Theorem for the Time-Dependent Evolution Equation

GIAN-CARLO ROTA

1. INTRODUCTION

The topic of the present lecture is slightly at variance with those treated in the lecture series of this course. The main theme has been the abstract differential equation

$$\frac{du}{dt} = A(t)u, \quad u \in X, \quad t \geq 0 \tag{1}$$

where X is a Banach space, and where $A(t)$ is a family of unbounded linear operators.

The main problem has been the *solution* of (1) under very weak conditions on $A(t)$. Under suitable conditions, which we shall specify below, the solution $u(t)$ of (1) which for $t = 0$ satisfies the initial condition $u(0) = f$, where $f \in X$, can be expressed in the form

$$u(t) = P(t, 0)f, \tag{2}$$

where $P(t, s)$ is a family of *bounded* linear operators satisfying the *evolution equation*

$$P(t, s)P(s, r) = P(t, r), \ P(t, t) = 1, \quad t, s, r \geq 0. \tag{3}$$

Thus, we shall understand the expression "solving the initial value problem for the equation (1)" as meaning that a family of bounded linear operators P can be determined which satisfies (3) and which gives a solution of (1) by formula (2).

Our present point of view will be to take the existence question of (1) for granted, and to investigate instead another problem related to the evolution equation, which for many investigations is as important, if not actually more, as the existence-uniqueness question. This is the *limiting behavior* of solutions $u(t)$ as $t \to \infty$. To get an idea of the kind of question we have in mind, let us consider first briefly the case where $A(t) \equiv A$ is independent of t. This case has been thoroughly studied: it is the theory of one-parameter semigroups of Hille and Phillips. The evolution operators (3) become in this case

$$P(t, s) = P^{t-s},$$

Seminar Notes, Equazioni Differenziali Astratte, Varenna, May 31 – June 8, 1963, Centro Internazionale Matematico Estivo.

where the right side is given by a one-parameter semigroup P^t whose infinitesimal generator is the operator A. In this case it is well-known that the solution $u(t)$ of

$$\frac{du}{dt} = Au, \quad u(0) = f, \quad t \geq 0, \tag{4}$$

is given by $u(t) = P^t f$.

The behavior at infinity of $u(t)$ is described by the classical *ergodic theorem.* Under suitable assumptions on the semigroup P^t - which can be equivalently stated as conditions on the infinitesimal generator A - the limit

$$\lim_{T\to\infty} \frac{1}{T}\int_0^T P^t f\,dt = \lim_{T\to\infty} \frac{1}{T}\int_0^T u(t)dt \tag{5}$$

exist in the norm of the given Banach space X. (Cf. Dunford-Schwartz, *Linear Operators,* Vol. I, Ch. VIII).

The limiting relation (5) is of fundamental importance in many investigations, and expresses a property of the semigroup which is analogous to a "boundary behavior" of sorts. In fact, to quote Norbert Wiener, the ergodic theorem is a kind of fundamental theorem of calculus at infinity.

We shall be concerned below with the following two questions:

(a) When can the Cesàro limit in (5) be replaced by an ordinary limit?

(b) Is there a "natural" analog of the ergodic theorem for the general evolution equation, corresponding to the time-dependent differential equation (1)?

These questions of course have innumerable answers, which depend largely on the kind of convergence that is required (norm convergence, weak convergence, etc.). The type of convergence we shall require below is *dominated convergence* in $L_p(S,\Sigma,\mu) = X$, $p > 1$. Specifically, we take $u(t)$ and f to belong to some Banach space $L_p(S,\Sigma,\mu)$ where $\mu(S) = 1$; the operators $A(t)$ and $P(t,s)$ will operate on these spaces only and will be subjected to the further conditions specified below.

We say that a sequence f_n in $L_p(S,\Sigma,\mu)$ converges *dominatedly* to f (in symbols, $f_n \underset{d}{\to} f$) when:

(I) $f_n(s) \to f(s)$ for almost every s in S, that is, f_n converges to f pointwise; and

(II) the function $f^*(s) = \sup_n |f_n(s)|$ belongs to $L_p(S,\Sigma,\mu)$.

We hasten to add that from this point on all equalities will be understood in the "almost everywhere" sense. Thus, a function is defined almost everywhere, etc.

Our main concern will be to give an answer to questions (a) and (b) when convergence is understood in the dominated sense. Questions related to dominated convergence appear in many contexts in both analysis and probability theory, and are usually considerably more difficult than the analogous questions for norm convergence; we need only recall, beside the G. D. Birkhoff ergodic theorem, the Riesz-Calderón-Zygmund theory of Hilbert transforms and singular integrals, the martingale theorem (cf. below), the results relating to the almost-everywhere convergence of Fourier series and of more general orthogonal expansions (such as the Rademacher-Menchoff theorem), the various strong laws of large numbers, etc.

Dominated convergence is the most satisfactory type of convergence that can be hoped for in problems of both analysis and probability; it is the nearest to actual pointwise convergence. It is very desirable to have a unified theory relating to this kind of convergence, in much the same way as the theory of Banach spaces gives a unified approach to questions of convergence *in the mean*; unfortunately, no such theory exists at present, and we have to rely on various methods of varying degrees of complication.

We can now proceed to state our main result. First, we consider the *discrete* analog of equation (1). This is the difference equation (for integer n)

$$\Delta U_n = (P_n - I)U_n, \quad n \geq 1. \tag{6}$$

The solution of this difference equation which satisfies the initial condition $u_1 = f$ is easily seen to exist uniquely when the P_n form a sequence of bounded operators in the Banach space X. Indeed, since $\Delta u_n = u_{n+1} - u_n$ we easily get

$$U_{n+1} = P_n U_n = P_n P_{n-1} U_{n-1} = \ldots = P_n P_{n-1} \ldots P_1 f. \tag{7}$$

The discrete analog of the evolution operators $P(t,s)$ of (3) are the operators

$$P(n,k) = P_n P_{n-1} \cdots P_k, \quad P(n,n) = 1, \quad n > k. \tag{8}$$

We now consider an *adjoint equation* to (6) (for fixed n), that is, the equation

$$\Delta V_k = \left(P^*_{n-k} - I\right) V_k, \quad k \geq 1, \tag{9}$$

where P^*_k is the adjoint operator of P_k. The solution of this equation that at $k = 1$ takes a given value g is given by the following expression

$$V_k = P^*_{n-k} P^*_{n-k+1} \cdots P^*_n g. \tag{10}$$

We shall call (9) the adjoint equation of (6). The continuous analog of equation (9), related to (1), is the equation (for fixed t)

$$\frac{dw}{dr} = A^*\,(t-r)\,w, \qquad r \leq t \tag{11}$$

where $A^*\,(r)$ is the adjoint of the operator $A(r)$. Luckily, here we can make the change of variable $t - r = s$, thereby getting

$$\frac{dv}{ds} = -A^*\,(s)\,v, \quad s \geq 0. \tag{12}$$

We shall call (12) the *adjoint equation* to (1). The family of evolution operators which give the solution of (12), call them $P_{ad}(s,q)$, are easily obtained in terms of (3), as follows.

Consider the family of operators $P^*(t,s)$, $t > s$. Because $(AB)^* = B^*A^*$, the operators $P^*(t,s)$ do not satisfy the evolution equation (3), but

$$(13) \qquad P^*(s,r)P^*(t,s) = P^*(t,r), \;\; P^*(s,s) = I, \quad t \geq s \geq r \geq 0.$$

We now set $P_{ad}(s,q) = P^*(q,s)$. It is then easy to verify that P_{ad}, thus defined, is the family of evolution operators which solves (12).

We can now state the "natural" solutions of (12). The expression to be considered is the following:

$$(14) \qquad \lim_{t\to\infty} P_{ad}(t,0)P(t,0)f.$$

We shall see that, under suitable general conditions, this limit will exist. We shall derive the existence of the limit in the sense of dominated convergence, but other similar theorems for other types of convergence are easily stated and proved; in a sense, the theorem we prove below is best possible (cf. Burkholder, "Successive conditional expectations of an integrable function", *Ann. Math. Statist. 33* (1962), 887–893).

2. ANALYTIC PRELIMINARIES

Before we state our main theorem concerning the convergence of (14) and that of its discrete analog, we must review some of the analytic tools which will be essential steps in the proof. These are the following:

(α) *Doubly stochastic operators.* Let (S,Σ,μ) be a probability space, namely, a measure space of measure one. A doubly stochastic operator P is one that has the following properties: (1) $P1 = 1$, where 1 is the function identically equal to one; (2) $P^*1 = 1$, where P^* is the adjoint of P; (3) if $f \geq 0$ in $L_p(S,\Sigma,\mu)$, then $Pf \geq 0$, that is, P is a *positive operator.* Doubly stochastic operators are bounded operators of norm one in all $L_p(S,\Sigma,\mu), 1 \leq p \leq \infty$.

The theory of doubly stochastic operators is not well-developed except when (S,Σ,μ) is an atomic measure space with a finite number of atoms having equal measure. In this case we obtain the classical case of *doubly stochastic matrices,* which are being thoroughly studied. More generally, if the atoms do not necessarily have equal measure, the notion of doubly stochastic operators and their study is related to the extensive theory of *matrices with the unimodular property.* It would be interesting to extend this theory to the case of a non-atomic measure space, in view of the difficulty of extending the Birkhoff-von Neumann theorem to this case.

Probabilistically, doubly stochastic operators are simply Markov processes with discrete parameters and with an invariant measure, for which the adjoint process also has an invariant measure. Such processes occur in the most disparate connexions.

(β) *Conditional expectation.* A projection E in $L_2(S, \Sigma, \mu)$ which is doubly stochastic is called a conditional expectation. There is an extensive theory and representation of such operators (cf. the texts of Doob and Loéve, or the author's paper in *Padova Rendiconti,* 1959); however, we shall only need the following property: conditional expectation preserves dominated convergence. In other words if $f_n \underset{d}{\rightarrow} f$, then $Ef_n \underset{d}{\rightarrow} Ef$.

Roughly speaking, a conditional expectation has much the same properties as an integral, for example, the Lebesgue bounded convergence theorem holds.

(γ) *Martingales.* An increasing (decreasing) martingale E_n is a sequence of conditional expectations such that $E_n E_m = E_n$ for $n \geq m$, $(n \leq m)$. There is an analogous notion of martingales with a continuous parameter. The theory of martingales has been developed largely by Doob. The main result is the *martingale theorem*, which states that if E_n is a martingale, than $E_n f \underset{d}{\rightarrow} f$ for f in L_p, $p > 1$. (Actually the general theorem is more far-reaching.)

The notion of a martingale is one of the most fundamental in analysis; there is no telling how vast their applications will be in future years, once analysts begin to realize their power. So far the notion has been used largely by probabilists.

(δ) *Abstract L-spaces.* An abstract L-space is a Banach lattice X such that if $f, g \geq 0$ in X, then $\|f + g\| = \|f\| + \|g\|$. For the theory of these spaces, see Day, *Normed Linear Spaces.*

The main result on abstract L-spaces we shall need is the following. Let X be a Banach lattice such that $\|1\| = 1$, where 1 is a lattice identity. (Cf. Birkhoff, *Lattice Theory.*) Then there exists a structurally unique probability space (S, Σ, μ) such that X is latticially isomorphic to $L_1(S, \Sigma, \mu)$. This theorem is due to Kakutani.

(ϵ) *Infinitesimal doubly stochastic operators.* In keeping with the spirit of this Symposium, we consider the conditions to be satisfied by the operator function $A(t)$ in (1) in order that the evolution operators (3) be doubly stochastic. These conditions were essentially determined by Phillips ("Semi-groups of positive contraction operators, *Czechoslovak Math. J.* *12* (1962), 294–313), and we shall limit ourselves to stating them without proof, since we shall have no occasion of using them. They are: (1) $A(t)1 = 0$; (2) $A^*(t)1 = 0$; (3) for f in the domain of $A(t)$, and f in $L_p \cap L_q$ $(p^{-1} + q^{-1} = 1)$, we have $(A(t)f, f^+) \geq 0$, where $f^+ = \max(f, 0)$.

3. MAIN THEOREM

After all these preliminaries we can now proceed to state our main result, in both discrete and continuous forms.

Theorem. (1) *Let $P_1, P_2, \ldots, P_n, \ldots$ be an infinite sequence of doubly stochastic operators in $L_p(S, \Sigma, \mu)$, where $p > 1$ and where (S, Σ, μ) is a probability space. Then for f in $L_p(S, \Sigma, \mu)$,*

$$\lim_{n \to \infty} P_1 P_2 \cdots P_n P_n^* P_{n-1}^* \cdots P_1^* f \tag{15}$$

exists in the sense of dominated almost everywhere convergence.

(2) *Let $P(t,s)$ be a family of doubly stochastic evolution operators in $L_p(S,\Sigma,\mu)$, where again (S,Σ,μ) is a probability space (that is, operators satisfy (3)). Then, for f in $L_p(S,\Sigma,\mu)$ the limit*

$$\lim_{t\to\infty} P_{ad}(t,0)P(0,t)f = \lim_{t\to\infty} P^*(0,t)P(0,t)f \tag{16}$$

exists in the sense of dominated almost everywhere convergence.

Proof. The main device in the proof consists in reducing the convergence result to an application of the martingale theorem. For simplicity we shall only prove the discrete case. The continuous case can be proved either by reducing it to the discrete case, or else by a similar construction to the one we shall use below for the discrete case. At any rate, the difficulties involved in passing from the discrete to the continuous case are purely technical and involve no new ideas.

We begin by constructing what we shall call the *path space* of the sequence $P_1, P_2, \ldots$. To this end, we consider an infinite product of replicas (S_n, Σ_n, μ_n) of (S,Σ,μ) for $n = 0,1,2,\ldots$, namely

$$(S',\Sigma',\mu') = \prod_{n=0}^{\infty}(S_n,\Sigma_n,\mu_n).$$

It is well-known (cf. Halmos, *Measure Theory*) that this infinite product space is a well-defined probability space. We shall now consider an algebra $\mathcal{A}$ of real valued functions on (S',Σ',μ') defined as follows. A function F on (S',Σ',μ') is a function of infinitely many variables $F(s_0,s_1,s_2,\ldots)$, $s_i \in S_i$. We say that F belongs to $\mathcal{A}$ if F is a sum of finite products of the form

$$f_0(s_0)f_1(s_1)\cdots f_h(s_h), \tag{17}$$

where $f_i \in L_\infty(S_i,\Sigma_i,\mu_i)$.

Clearly $\mathcal{A}$ is an *algebra.* We define a linear functional L on $\mathcal{A}$ as follows. We first define L on functions F of the form (17) by the formula

$$L(F) = \int_S f_0 P_1 \left[f_1 P_2 \left[f_2 \cdots \left[P_h f_h\right]\right]\cdots\right] d\mu$$

and then extend L by linearity. It is easy to see, using the fact that $P_n 1 = 1$ for all n, that L is *well-defined.*

We shall now verify that L has a very important property; it is a *positive* linear functional. In other words, we shall prove that if F is in $\mathcal{A}$ and takes only non-negative values, then $L(F) \geq 0$. This appears at first sight not to be a trivial statement, because a function F in $\mathcal{A}$ is a linear combination of functions of the form (17), and each of the summands may take negative values, even though the sum is always non-negative. To prove this statement, we notice that if

$$F(s_0,s_1,s_2,\ldots) = \sum_i f_0^{(i)}(s_0)f_1^{(i)}(s_1)\cdots f_k^{(i)}(s_k) \geq 0$$

for all $s_0, s_1, s_2, \dots, s_k$, then, changing the variable s_k to s and remembering that P_k applies only to functions of the variable s, we get

$$(18) \qquad P_k\left(\sum_i f_0^{(i)}(s_0) f_1^{(i)}(s_1)\cdots f_{k-1}^{(i)}(s_{k-1}) f_k^{(i)}(s)\right)$$

$$= \sum_i f_0^{(i)}(s_0)\, f_1^{(i)}(s_1)\cdots f_{k-1}^{(i)}(s_{k-1})\, P_k f_k^{(i)}(s).$$

This expression is non-negative for every value of the variables $s_0, s_1, s_2, \dots, s_{k-1}, s$ because P is a positive operator. Next, we change the variable s_{k-1} to s, and then apply the operator P_{k-1}, that is

$$(19) \qquad P_{k-1}\left(\sum_i f_0^{(i)}(s_0) f_1^{(i)}(s_1)\cdots f_{k-2}^{(i)}(s_{k-2}) f_{k-1}^{(i)}(s) P_k f_k^{(i)}(s)\right)$$

$$= \sum_i f_0^{(i)}(s_0)\, f_1^{(i)}(s_1)\cdots f_{k-2}^{(i)}(s_{k-2})\, P_{k-1}\left[f_{k-1}^{(i)} P f_k^{(i)}\right](s),$$

and again this expression is positive, because P_{k-1} is a positive operator. Proceeding in this way down to $k = 1$, we finally obtain that

$$\sum_i f_0^{(i)} P_1\left[f_1^{(i)} P_2\left[\cdots\left[P_k f_k^{(i)}\right]\right]\cdots\right](s) \geq 0$$

for all s in S. Hence, integrating, we let $L(F) \geq 0$ as we wanted to show. We now remark that the algebra $\mathcal{A}$ has a very important property. If F belongs to $\mathcal{A}$, and $F^+ = \max(F, 0)$, then F^+ belongs to $\mathcal{A}$. This is clear if F is of the form (17) for then $F^+(s_0, s_1, s_2, \dots)$ is a linear combination of terms of the form

$$f_0^{\pm}(s_0) f_1^{\pm}(s_1)\cdots f_k^{\pm}(s_k),$$

where there are an even number of negative signs in each term. Therefore, to establish the assertion, it suffices to consider the case when both F and G depend upon two coordinates only: for then an easy induction will yeld the general statement. Let $F^- = -\min(F, 0)$. Then $F = F^+ - F^-$, and furthermore the functions F^+ and F^- have disjoint supports. From this, it is easy to see that if $F = P - Q$, where P and Q are non-negative functions with disjoint supports, then $P = F^+$ and $Q = F^-$. Using this fact, the statement we are to prove reduces to proving that $F + G$ can be written as the difference of two non-negative functions with disjoint supports. Let

$$\begin{aligned}
F &= \sum_i f_0^{(i)}(s_0) f_1^{(i)}(s_1) \\
&= \sum_i [f_0^{(i)+}(s_0) - f_0^{(i)-}(s_0)][f_1^{(i)+}(s_1) - f_1^{(i)-}(s_1)] \\
&= \sum_i [f_0^{(i)+}(s_0) f_1^{(i)+}(s_1) + f_0^{(i)-}(s_0) f_1^{(i)-}(s_1)] - \\
&\qquad \sum_i [f_0^{(i)-}(s_0) f_1^{(i)+}(s_1) + f_0^{(i)+}(s_0) f_1^{(i)-}(s_1)] \\
&= P - Q,
\end{aligned}$$

say. The functions P and Q are non-negative and it is an easy verification that their supports are disjoint. Thus, $P = F^+$ and $Q = F^-$ as we wanted to show.

At this point we can define a *seminorm* on $\mathcal{A}$ as follows:

(20) $$\|F\| = L\left(F^+ + F^-\right), \quad L(1) = 1.$$

Since L is a positive linear functional, the positivity and triangle inequality follow at once. Clearly, those G in $\mathcal{A}$ whose norm in 0 form a linear subspace N; taking the quotient vector space $\mathcal{A}/N = B$ we obtain canonically a norm on B. We now complete B, thereby obtaining a Banach space C.

We now claim that C is an abstract L-space. This is intuitively clear, but for the sake of completeness let us verify the statement. First we claim that C is a Banach lattice. Now, $\mathcal{A}$ is certainly lattice-ordered; thus, to show that C is a Banach lattice, it suffices to show that B is a lattice, since the completion of a lattice-ordered vector space is again lattice-ordered, and is in fact a Banach lattice. Thus, it all boils down to showing that if $F \geq 0$ and $0 \leq G \leq F$, with $L(F) = 0$, then $L(G) = 0$. But this is an obvious consequence of the positivity of L, which gives $L(F) \geq L(G) \geq 0$.

The fact that $\|x + y\| = \|x\| + \|y\|$ for x and $y \geq 0$ in C is also an evident consequence of (20). Thus, C is an L-space.

We now apply Kakutani's theorem and represent C as $L_1\left(S_\infty, \Sigma_\infty, \mu_\infty\right)$, where $(S_\infty, \Sigma_\infty, \mu_\infty)$ is a probability space. This is the space in which we shall now define a martingale. We shall need the following well-known lemmas (Cf. e.g. Rota, *Padova Rendiconti,* 1959).

Lemma 1. *Let $(S_\infty, \Sigma_\infty, \mu_\infty)$ be a measure space, and let:*

(a) Y be a closed subspace of $L_1\left(S_\infty, \Sigma_\infty, \mu_\infty\right)$ which is also closed under the lattice operations (that is, a closed Banach sublattice of $L_1(S_\infty, \Sigma_\infty, \mu_\infty)$).

(b) Z be a subspace of $L_\infty\left(S_\infty, \Sigma_\infty, \mu_\infty\right)$.

Then:

(a) There exists a σ-subfield Σ of the σ-field Σ_∞ such that $Y = L_1(S_\infty, \Sigma, \mu_\infty)$.

(b) There exists a σ-subfield Σ' of Σ_∞ such that the closure of Z in $L_1\left(S_\infty, \Sigma_\infty, \mu_\infty\right)$ is $L_1\left(S_\infty, \Sigma', \mu_\infty\right)$.

Lemma 2. *Let T be a linear operator from $L_1\left(S, \Sigma, \mu\right)$ into $L_1\left(S_\infty, \Sigma_\infty, \mu_\infty\right)$ (both* *over probability spaces) such that* $T1 = 1$ *and* $\| Tf \| = \| f \|$ *for all* $f \geq 0$. *Then there is a* σ*-subfield* Σ_0 *of* Σ_∞ *such that* $L_1(S, \Sigma, \mu)$ *is isomorphic to* $L_1(S_\infty, \Sigma_0, \mu_\infty)$.

Lemma 3. *Let* Σ *be a* σ*-subfield of* Σ_∞*, and let* E *be the orthogonal projection of* $L_2(S_\infty, \Sigma_\infty, \mu_\infty)$ *onto* $L_2(S_\infty, \Sigma, \mu_\infty)$. *Then* E *is a conditional expectation. In particular,* E *can be extended to a bounded operator in* $L_p(S_\infty, \Sigma_\infty, \mu_\infty)$, $1 \leq p \leq \infty$.

We shall make use of these lemmas presently. First, notice that there is a natural mapping of $L_1(S, \Sigma, \mu)$ into $L_1(S_\infty, \Sigma_\infty, \mu_\infty)$ given by $f(s) \to f(s_0) \to x$, where $x \in C$. This mapping preserves the norms of all positive functions. Hence, by Lemmas 2 and 1(a), there exists a σ-subfield of Σ_∞ isomorphic to Σ: we shall identify such a σ-subfield with Σ altogether, and identify $L_1(S, \Sigma, \mu)$ with $L_1(S_\infty, \Sigma, \mu_\infty)$.

We now come to the crucial point. Consider the algebra $\mathcal{A}_n$ of all functions of $\mathcal{A}$ which are "independent" of the coordinates $s_0, s_1, s_2, \ldots s_n$. Under the canonical map into B and C successively the closure of $\mathcal{A}_n$ in $L_1(S_\infty, \Sigma_\infty, \mu_\infty)$ is mapped into a subspace of the form $L_1(S_\infty, \Sigma_n, \mu_\infty)$, $n = 0, 1, 2, \ldots$, where $\Sigma_0 = \Sigma$. Let E_n be the orthogonal projection of $L_2(S_\infty, \Sigma_\infty, \mu_\infty)$ into $L_2(S_\infty, \Sigma_n, \mu_\infty)$.

The crucial part of the proof consists in verifying that for f in $L_p(S_\infty, \Sigma, \mu_\infty)$, $E_0 E_n f = P_1 P_2 \cdots P_n P_n^* P_{n-1}^* \cdots P_1^* f$. Once this is verified, the proof is completed by (a) remarking that E_n is a martingale; (b) applying the martingale convergence theorem; (c) remarking that the conditional expectation operator E_0 preserves dominated convergence in L_p.

Let us first compute $E_n f$, remembering that f is in $L_p(S_\infty, \Sigma, \mu_\infty)$. Actually, it can be assumed that $f \in L_p \cap L_2$, since this subspace is dense. Let

$$P_n^* P_{n-1}^* \cdots P_1^* f(s_n) = g(s_n). \tag{21}$$

We shall prove that $E_n f(s_0, s_1, s_2, \ldots) = g(s_n)$. We have to show that g is the unique function such that

$$\int_{S_\infty} hf \, d\mu_\infty = \int_{S_\infty} hg \, d\mu_\infty, \tag{22}$$

for all $h \in L_\infty(S_\infty, \Sigma_n, \mu_\infty)$. But once (22) has been verified for h ranging over a dense subset of $L_\infty(S_\infty, \Sigma_n, \mu_\infty)$ it will be true for all h. We shall take this dense subset to be $\mathcal{A}_n$ (or rather its image in L_∞, but for simplicity we are identifying the two). Thus, it suffices to choose $h(s_0, s_1, s_2, \ldots) = f_n(s_n) f_{n+1}(s_{n+1}) \ldots f_{n+k}(s_{n+k})$. Then the right side of (22) equals, for $g = P_n^* P_{n-1}^* \cdots P_1^* f(s_n)$,

$$\int_S P_1 P_2 \cdots P_n [(P_n^* P_{n-1}^* \cdots P_1^* f) f_n P_{n+1} [f_{n+1} \ldots]] d\mu.$$

Now, for a doubly stochastic operator, $\int Pq \, d\mu = \int (P^* 1) q \, d\mu = \int q \, d\mu$. Hence this last expression simplifies to:

$$\begin{aligned}
&\int_S (P_n^* P_{n-1}^* \cdots P_1^* f) f_n P_{n+1}[f_{n+1} \cdots] d\mu \\
= &\int_S f P_1 P_2 \cdots P_n [f_n P_{n+1}[f_{n+1} \cdots]] d\mu \\
= &\int_{S_\infty} h f d\mu_\infty,
\end{aligned}$$

as we wanted to show. This shows that $E_n f = g$.

Next - and last - we prove that if $q = q(s_n)$, then $E_0 q = P_1 P_2 \cdots P_n q(s_0)$. This is very easy. It suffices to verify that for $f = f(s_0)$ we have

$$\int_{S_\infty} f P_1 P_2 \cdots P_n q d\mu = \int_{S_\infty} f q d\mu.$$

But the integral on the right is (by definition!)

$$\int_S f P_1 P_2 \cdots P_n q d\mu,$$

hence the proof is complete.

4. APPLICATIONS

It is easy to get now a solution of our problem (a). We have that $P^{2n} f$ converges dominatedly if P is selfadjoint. This last fact was discovered independently by E. M. Stein ("On the maximal ergodic theorem", *Proc. Nat. Acad. Sci. U. S. A. 47* (1961), 1894–1897).

In the continuous case, we infer the dominated convergence of $P^t f$ as $t \to \infty$ if P^t, in addition to being doubly stochastic, is also selfadjoint.

The reader may perhaps wonder how the result of the main theorem (originally published in "An 'alternierende Verfahren' for general positive operators", *Bull. Amer. Math. Soc., 68* (1962), 95–102) was arrived at. The answer is that doubly stochastic operators are related to conditional expectation operators in much the same way as contractions operators in Hilbert space are related to orthogonal projections. It is this analogy that has suggested the present result, although of course the proof has to be based upon entirely different principles. As an interesting unsolved problem we shall mention the convergence of

$$(P_1 P_2 \cdots P_k)^n f,$$

where P_i are doubly stochastic operators or even conditional expectations.

BULLETIN OF THE
AMERICAN MATHEMATICAL SOCIETY
Volume 77, Number 5, September 1971

FROBENIUS RECIPROCITY IN ERGODIC THEORY

BY K. LANGE, A. RAMSAY AND GIAN-CARLO ROTA

Communicated December 3, 1970

1. Introduction. The analogy between group actions and group representations, which has guided algebraists' intuition since the turn of the century, has been slow in influencing analysis. To be sure, the analogy between "strict sense" and "wide sense" notions, first noticed in probability, can be traced to be a natural analog of the preceding analogy, and in fact has led to the discovery of a host of new results (for example by Nelson[1] and Rota where the strict sense analogs of well-known Hilbert-space theorems of Nagy and Naĭmark are worked out). More recently, a conjecture of Rota (1962) regarding a strict-sense analog of Schreiber's Theorem has been settled in the affirmative by McCabe and Shields, using Ornstein's deep results relating entropy to conjugacy. Despite this and much other work, however, a systematic "spectral theory in the strict sense" for ergodic trans-

AMS 1970 *subject classifications.* Primary 18A40, 22D40, 28A65, 46M15; Secondary 20L10, 22D30, 54H20.

Key words and phrases. Induced ergodic action, locally compact group, group action, ergodic, "strict sense", groupoid, virtual group, adjoint functor, Mackey functor, induced representation.

[1] Authors' names refer to the bibliography at the end. The results of the present announcement were first presented by Rota at the Symposium on Functional Analysis held at the U. S. Naval Postgraduate School in Monterey, California, in October 1969. We are grateful to Professor Carroll Wilde for permission to reproduce parts of Professor Rota's lecture outside the Proceedings of the Symposium. Professor Rota's work was carried out under O.N.R. Contract N00014-67-A-0204-0016 and Professor Ramsay's under NSF Grant GP 11622.

formations of measure spaces remains to be developed. The far-reaching ideas recently introduced by Mackey, and arrived at by analogies between transitive actions of finite groups and ergodic actions of locally compact groups, offer renewed promise for this program. The notion of virtual group, introduced by Mackey with the purpose of developing an "action" analog of the notion of induced representation, has already found a multitude of applications (Lange and Ramsay) and, we surmise, will substantially influence and deflect the development of ergodic theory.

Our present purpose is to strengthen and develop one of Mackey's constructions by placing it in the context of categorical algebra. It has been known for some time (see e.g. Lang) that the induced representation functor is the left adjoint of the restriction (to a subgroup) functor. Guided by this analogy, we are led to surmise that the construction of an induced action should be, in a suitable context, an adjoint functor. This turns out to be the case (Theorem 2). To establish it, a detailed detour into the treacherous waters of pointwise measure theory seems necessary. Some of the basic spadework has been done by Mackey, but a few notions need to be retouched. For this and other reasons, we have found it prudent to give all definitions, some of which—in particular that of virtual group—slightly differ from Mackey's.

The present work points to several further developments, notably: a "strict sense" analog of the imprimitivity theorem; a corresponding "right adjoint" to the functor R; and a development of the functor R and its left adjoint M, the "Mackey functor", along the lines of the theory of triples. We hope to consider them in forthcoming publications.

2. **Definitions.** Recall that a group is a small category with one object, where every morphism has an inverse. A *groupoid* is a small category where every morphism has an inverse.

We refer to Moore for the definitions of *Borel space*, *Borel map*, *countably generated Borel space*, *analytic Borel space*, *equivalent σ-finite measures*, *measure class* (denoted by $\mathbf{C}$). Let T be a Borel map; if $T^{-1}(A)$ is a null set whenever A is a null set then the measure class is *invariant* under T. An *analytic Borel groupoid* is a groupoid, together with an analytic Borel space structure on the set of morphisms, and a measure class $\mathbf{C}$ (this induces a similar structure on the set U of objects of V by identifying an object with the corresponding identity morphism), such that:

(a) The domain D of composition of two morphisms is a Borel subset of $V\times V$, under the product Borel structure.

(b) Composition $(f, h)\to fh$ and inversion $f\to f^{-1}$ are Borel maps of $D(V)$ to V, and inversion leaves invariant the measure class $\mathbf{C}$.

Let d and r be the (Borel) maps of a morphism to its "domain" and "range"; then d induces a measure class $\bar{\mathbf{C}}$ on U, setting $\tilde{\mu}(A) = \mu(d^{-1}(A))$ for μ in $\mathbf{C}$. Any μ in C satisfies $\mu = \int \mu_s \tilde{\mu}(ds)$, where μ_s is a measure on V living on $d^{-1}(s)$, and, for fixed Borel set A in V, the function $s\mapsto\mu_s(A)$ is Borel on U. The fibering is unique: if $\mu = \int \nu_s \tilde{\mu}(ds)$, then $\nu_s = \mu_s$ for $\bar{C}$—almost all s; furthermore, changing μ to an equivalent measure does not change the measure class of μ_s for $\bar{\mathbf{C}}$—almost all s; this leads to measure classes C_s on almost all fibers $d^{-1}(s)$.

(c) For $s\in U$ and $r(f) = s$, the map $h\mapsto hf$ carries $d^{-1}(s)$ biuniquely to $d^{-1}(d(f))$ and C_s to $C_s f$.

We require that $C_{r(f)}f = C_{d(f)}$ for all f with $r(f)$ and $d(f)$ in some *co-null* (complement of a null) Borel set U_0 of U. Say that $(V, \mathbf{C})$ is *ergodic* whenever every real Borel function ϕ on U such that $\phi(d(f)) = \phi(r(f))$ for almost all f in V is $\bar{\mathbf{C}}$ a.e. constant. A *virtual group* is an *ergodic analytic Borel groupoid*. In a virtual group $(V, \mathbf{C})$, let U_0 be a co-null Borel set of U. Taking all $f\in V$ s.t. both $d(f)$ and $r(f)$ are in U_0, we obtain another virtual group, the *inessential contraction* $V|U_0$. Again, say that two objects u and v are *equivalent* when $d(f) = u$ and $r(f) = v$ for some morphism f; if $A\subseteq U$, write $[A]$ for the *saturation* of A under this equivalence relation; note that $[A] = r(d^{-1}(A))$ and that $[A]$ is naturally an analytic Borel space if A is a Borel set.

3. **Categories and functors.** A *strict homomorphism* ψ between virtual groups $(V_1, \mathbf{C}_1)$ and $(V_2, \mathbf{C}_2)$ is a functor from V_1 to V_2 which is also a Borel map, and s.t. if $\bar{\psi}$ is the associated map of the objects U_1 of V_1 to the U_2 of V_2, then $\bar{\psi}^{-1}(A)$ is a $\bar{\mathbf{C}}_1$-null set for every saturated analytic $\bar{\mathbf{C}}_1$-null set A. A *homomorphism* of $(V_1, \mathbf{C}_1)$ to $(V_2, \mathbf{C}_2)$ is a functor which is a Borel map, and whose restriction to some inessential contraction of $(V_1, \mathbf{C}_1)$ is a strict homomorphism. Two homomorphisms ψ_1 and $\psi_2: (V_1, \mathbf{C}_1)\to(V_2, \mathbf{C}_2)$ are *strictly similar* if $\theta(r(f))\psi_1(f) = \psi_2(f)\theta(d(f))$ for all $f\in V_1$ and for some Borel map $\theta: U_1 \to V_2$ for which both sides are defined; ψ_1 and ψ_2 are *similar* if there is an inessential contraction of V_1 on which they are strictly similar. (*Note.* The composition of two homomorphisms is not necessarily a homomorphism!) Similarity is an equivalence relation, and similarity classes $[\phi]$ of homomorphisms ϕ are preserved under composition. Taking virtual groups as objects and similarity classes of homomorphisms as morphisms, one obtains a category. A locally compact separable group G is a virtual group when endowed with its Haar measure class. A virtual group V, together with a homomorphism $\pi: V\to G$, briefly V_π, will be called a *virtual group over* G. The *Mackey*

category $\mathbf{M}(G)$ of G has the virtual groups over G as objects, and as morphisms the similarity classes of homomorphisms which make the obvious triangle over G commute, namely, which commute with the action of $[\pi]$. The category $\mathbf{R}(G)$ *of ergodic actions* of G has as objects the transformation spaces T of G, namely:

(a) T is an analytic Borel space;

(b) the map $(t, x)\to tx$ of $T\times G\to T$ is Borel;

(c) T has a measure class $\mathbf{C}$ which is invariant under the set of Borel automorphisms $t\to tx$;

(d) $(T, \mathbf{C})$ is ergodic: the only invariant Borel sets are null or co-null.

The morphisms of $\mathbf{R}(G)$ are equivalence classes of maps, as follows: T_1 and T_2 being objects, consider all Borel maps $\phi: T_1\to T_2$ s.t.

(a) There is a co-null invariant analytic subset of T_1 on which ϕ is G-equivariant, that is, $\phi(tx)=\phi(t)x$.

(b) If N is a null set in T_2, so is $\phi^{-1}(N)$.

Two such maps ϕ and ψ are *equivalent* if there is a Borel map $\alpha: T_1\to G$ such that $\phi(s)\alpha(s)=\psi(s)$ and $\alpha(sx)=x^{-1}\alpha(s)x$ for all s in some co-null invariant analytic subset of T_1. This equivalence is preserved under composition of maps; the equivalence classes are the morphisms of $\mathbf{R}(G)$. For a given ergodic action of G on T, give $T\times G$ the product Borel structure and product measure class and define $\pi: T\times G\to G$ as the projection. Defining $(s, x)(t, y)=(s, xy)$ whenever $sx=t$ gives $T\times G$ a groupoid structure whose objects are naturally identical with the set T. This construction maps the objects of $\mathbf{R}(G)$ into the objects of the Mackey category $\mathbf{M}(G)$.

THEOREM 1. *There is a functor $R: \mathbf{R}(G)\to\mathbf{M}(G)$ extending the above construction, which is faithful on objects and morphisms and whose image is a full subcategory of the Mackey category.*

The easy proof, following Mackey's techniques, is omitted.

4. Main result.

THEOREM 2. *The functor R has a left adjoint M.*

We call M the *Mackey functor*; a sketch of the proof follows, leaving fine measure-theoretic points to a detailed publication to follow. All homomorphisms and similarities are tacitly taken as *strict*. Given V_π, we first construct a transformation space T by generalizing a construction of Mackey, which is, in turn, patterned after a classical construction of induced representations. Take $U\times G$, where U is the object of V, and define a G-action by $(u, y)x=(u, x^{-1}y)$. Set (u, x)

$\sim(v, y)$ iff $(r(f), x) = (u, x)$ and $(d(f), x\pi(f)) = (v, y)$ for some $f \in V$. The G-action commutes with the equivalence $\sim$, so that the quotient $U \times G/\sim$ is a G-space (but unfortunately, not an analytic Borel G-space under the quotient Borel structure). It is however possible (using some results of Mackey) to replace $U \times G/\sim$ by a Borel space $M(V_\pi)$ having the same measure algebra, and the action of G is easily seen to be ergodic. This defines the functor M on the objects; next, we define it on morphisms. Given $[\phi]: V_1, \pi_1 \to V_2, \pi_2$ in $\mathbf{M}(G)$, the map $(u, x) \to (\phi(u), x\theta(u))$ of $U_1 \times G \to U_2 \times G$, where $\theta(r(f))\pi_2 \circ \phi(f) = \pi_1(f)\theta(d(f))$, factors through $\sim$ to give a map $U_1 \times G/\sim \to U_2 \times G/\sim$. If θ is changed to another θ_1 implementing the similarity, then $(\phi(u), x\theta(u))x\theta(u)\theta_1(u)^{-1}x^{-1} = (\phi(u), x\theta_1(u))$, and the map $(u, x) \to x\theta(u)\theta_1(u)^{-1}x^{-1}$ factors to give a map $\alpha: U_1 \times G/\sim \to G$ satisfying $\alpha(tx) = x^{-1}\alpha(t)x$. Thus, $(u, x) \to (\phi(u), x\theta(u))$ and $(u, x) \to (\phi(u), x\theta_1(u))$ factor to give G-equivariant maps $U_1 \times G/\sim \to U_2 \times G/\sim$ which belong to the same morphism class in $\mathbf{R}(G)$. Again, if $\gamma(r(f))\phi(f) = \phi_1(f)\gamma(d(f))$, then $u \to \theta(u)\pi_2 \circ \gamma(u)^{-1}$ implements the similarity between $\pi_2 \circ \phi_1$ and π_1 (recall $V_i: \pi_i \to G$). The maps $(u, x) \to (\phi_1(u), x\theta(u)\pi_2 \circ \gamma(u)^{-1})$ and $(u, x) \to (\phi(u), x\theta(u))$ yield the same map $U_1 \times G/\sim \to U_2 \times G/\sim$. Thus, the map of morphisms in $\mathbf{M}(G)$ to morphisms in $\mathbf{R}(G)$ is welldefined. It can be verified that the definition preserves composition of morphisms. Next, given $[\phi]: V_\pi \to R(T)$, we must associate a morphism $M(V_\pi) \to T$. Let $\bar{\phi}$ be the projection of ϕ onto T and β be the projection of $T \times G \to G$. For $\theta(r(f))\beta \circ \phi(f) = \pi(f)\theta(d(f))$, the map $(u, x) \to \bar{\phi}(u)\theta(u)^{-1}x^{-1}$ is constant on $\sim$-equivalence classes and G-equivariant, so that it induces a map $U \times G/\sim \to T$; it is again checked as above that changing θ or ϕ does not change the equivalence class of maps. Conversely, to a morphism $[\phi]: M(V_\pi) \to T$ we must associate a morphism $V_\pi \to R(T)$. Let ρ be the natural projection of $U \times G \to U \times G/\sim$, and set $\hat{\phi}(f) = (\phi \circ \rho(r(f), e), \pi(f))$, when e is the identity of G; one can show that $\hat{\phi}(fh) = \hat{\phi}(f)\hat{\phi}(h)$, that the required diagram commutes, and that changing to an equivalent map changes $\hat{\phi}$ only up to similarity. Lastly, one must verify that the two constructions are inverses and naturality of the whole construction. All this can be done with little difficulty.

Bibliography

1. L. Auslander, and C. Moore, *Unitary representations of solvable Lie groups*, Mem. Amer. Math. Soc. vol. 62, Amer. Math. Soc., Providence, R. I., 1966. MR **34** #7723.

2. J. L. Doob, *Stochastic processes*, Wiley, New York, 1953. MR **15**, 445.

3. B. Eckmann (editor), *Seminar on triples and categorical homology theory*, Springer, Berlin, 1969. MR **39** #1511.

4. S. Lang, *Algebra*, Addison-Wesley, Reading, Mass., 1965. MR **33** #5416.

5. K. Lange, Doctoral Dissertation, M.I.T., Cambridge, Mass., 1971.

6. G. W. Mackey, *Ergodic theory, group theory, and differential geometry*, Proc. Nat. Acad. Sci. U.S.A. **50** (1963), 1184–1191. MR **29** #2325.

7. ———, *Ergodic theory and virtual groups*, Math. Ann. **166** (1966), 187–207. MR **34** #1444.

8. ———, *Virtual groups*, Sympos. Topological Dynamics (Colorado State Univ., Ft. Collins, Colo., 1967), Benjamin, New York, 1968, pp. 335–364. MR **39** #2907.

9. ———, *Induced representations of groups and quantum mechanics*, Benjamin, New York, 1968.

10. Robert McCabe, Doctoral Dissertation, Boston University, Boston, Mass., 1971.

11. B. Mitchell, *Theory of categories*, Academic Press, New York, 1965. MR **34** #2647.

12. E. Nelson, *The adjoint Markoff process*, Duke Math. J. **25** (1958), 671–690. MR **21** #365.

13. D. Ornstein, *Bernoulli shifts with the same entropy are isomorphic*, Advances in Math. **4** (1970), 337–352.

14. A. Ramsay, *Virtual groups and group actions*, Advances in Math. **6** (1971), 253–321.

15. G.-C. Rota, *An Alterniende Verfahren for general positive operators*, Bull. Amer. Math. Soc. **68** (1962), 95–102. MR **24** #A3671.

16. J. Westman, *Cohomology for ergodic groupoids*, Trans. Amer. Math. Soc. **146** (1969), 465–471.

17. R. McCabe and P. Shields, *A class of Markov shifts that are Bernoulli shifts*, Advances in Math. **6** (1971), 323–328.

18. F. W. Lawvere, *Equality in hyperdoctrines and comprehension schema as an adjoint functor*, Proc. Sympos. Pure Math. vol. 17, Amer. Math. Soc., Providence, R. I. 1970, pp. 1–14.

UNIVERSITY OF COLORADO, BOULDER, COLORADO 80302

Frobenius Reciprocity

Arlan Ramsay

The paper "Frobenius reciprocity in ergodic theory" had its origins in G. W. Mackey's introduction of virtual groups as objects that might facilitate the handling of various mathematical situations that involved ergodicity. In his treatment of unitary representations of group extensions, the case in which the action of the larger group on the dual of the normal subgroup allowed non-transitive ergodic measures could not be handled completely by use of stabilizers of points, the 'little groups'.

Gian-Carlo Rota soon began to consider alternative conceptualizations of that theory. Virtual groups were by definition equivalence classes of ergodic groupoids, and studying groupoids for themselves made some sense. Mackey had discussed the induction of the action of a group by the action of a subgroup, reminiscent of the induction of a representation. At Rota's suggestion, his student Kenneth Lange undertook to state and prove a functorial version of the theorem about restricting an induced representation for the analogous induced (non-linear) actions, known as the Frobenius Reciprocity Theorem, although the functorial point of view was not in the formulation of Frobenius.

This paper, published in the *Bulletin of the American Mathematical Society,* is an announcement of the result.

A Unified Theory for Ergodic and Martingale Theorems

M. M. Rao

It has long been observed that the Birkhoff ergodic theorem and the strong law of large numbers for strictly stationary integrable sequences (or processes) are very similar both in their statements and proofs. If the process is in addition one of independent random variables, then the latter statement is a consequence of a standard (decreasing indexed) martingale theorem (Doob [3], p. 341). The question of a martingale proof of the preceding (stationary) case was raised by Doob on p. 342 of the same reference. The striking similarity of proofs of ergodic and martingale theorems was already observed by Kakutani [11] who raised the possibility of *unifying* both these theories. Of the many attempts at solving the latter problem, Rota's approach has a particular novelty, and to appreciate this, some of the other methods, together with their promise and limitations, will be briefly discussed first. Then Rota's solution of the problem announced in the paper "Une théorie unifiée des martingales et des moyennes ergodiques" and its extensions will be considered.

The preceding question was sharpened by M. Jerison [9], who expressed the classical ergodic averages as a decreasing martingale on an infinite measure space constructed as follows. Let $L^1(\Sigma)$ be the Lebesgue space on a probability triple (Ω, Σ, P) and $T : L^1(\Sigma) \to L^1(\Sigma)$ be a bounded linear operator. If $f \in L^1(\Sigma)$ and $(S, \mathcal{S}, \mu) = (\mathbb{N} \times \Omega, \mathcal{N} \otimes \Sigma, \zeta \otimes P)$ is the product space where $(\mathbb{N}, \mathcal{N}, \zeta)$ is the measure space of non-negative integers with ζ as counting measure, then let $h : S = \mathbb{N} \times \Omega \to \mathbb{R}$ be defined as $h(k, \omega) = (T^k f)(\omega)$, and consider the sequence $h_n : S \to \mathbb{R}$ given by:

$$h_n(k,\omega) = \begin{cases} \dfrac{1}{n}\displaystyle\sum_{j=0}^{n-1}(T^j f)(\omega), & \text{if } (k,\omega) \in \{0,1,\cdots,n-1\} \times \Omega, \\ h(k,\omega), & \text{if } (k,\omega) \in \{k : k \geq n\} \times \Omega. \end{cases}$$

It is verified that

$$\{h_n = E^{\mathcal{B}_n}(h)\, a.e.\, [\mu], n \geq 1\}$$

is a martingale, where $\mathcal{B}_n = \mathcal{F}_n \otimes \Sigma$, with $\mathcal{F}_n$ as the σ-algebra generated by $\{0, 1, \cdots, n-1\}, \{k : k \geq n\}$ so that $\mu|\mathcal{B}_n$ is σ-finite (thus the conditional expectation $E^{\mathcal{B}_n}$ exists), and hence $\{h_n, n \geq 1\}$ converges a.e. $[\mu]$ if and only if the ergodic averages

$$\frac{1}{n}\sum_{j=0}^{n-1} T^j f$$

converge. However, there is no known martingale theorem on infinite measure spaces that assures the convergence of the above martingale. This circumstance prompted Jerison to conclude that it seems most unlikely that a proof of an individual ergodic theorem might be based on a martingale convergence of the desired kind. On the other hand, in the converse direction, one may obtain some new type of martingale theorems. This aspect has been explored recently by Kachurovskii [10] to find a unified formulation with the following procedure.

Let $\{\mathcal{C}_n\}_1^\infty$ be a monotone sequence of σ-algebras from (Ω, Σ, P), and $\tau : \Omega \to \Omega$ an automorphism, i.e., a bijective measurable and measure preserving transformation (in both directions). If $T^k\colon f \mapsto f \circ \tau^k$, $k = 0, 1, \cdots, (\tau^0 = id)$, consider the ergodic averages

$$A_n f = \frac{1}{n} \sum_{k=0}^{n-1} T^k f$$

for $f \in L^1(\Sigma)$. By the Birkhoff ergodic theorem $A_n f \to f^*$ *a.e.* Let

$$f^*_\infty = \lim_{n \to \infty^+} E^{\mathcal{C}_n}(f^*),$$

a limit which exists. Then a version of the Kakutani problem obtained recently by Kachurovskii on a probability space (Ω, Σ, P) may be stated as follows:

Theorem 1. *If $f \in L \log L$ (the Orlicz-Zygmund space so that one has $L^1(\Sigma) \supset L \log L \supset L^p(\Sigma)$, $p > 1$), then $E^{\mathcal{C}_n}(A_n f) \to f^*_\infty$ a. e. and in mean. The strong and weak inequalities for $A_n f$ are:*

$$\|\sup_n E^{\mathcal{C}_n}(A_n f)\|_p \leq C_1 \|f\|_p, \ p > 1, \tag{1}$$

and

$$P[\sup_n |E^{\mathcal{C}_n}(A_n f)| \geq a] \leq C_2 \frac{\|f\|_p^p}{a^p} \ a > 0, p \geq 1. \tag{2}$$

Moreover, if T is ergodic, then f^*_∞ is a constant $(= E(f))$ a.e. for any $f \in L \log L$.

This statement unifies the ergodic and martingale theorems, the former if all $\mathcal{C}_n$ are identical, and the latter if $\tau = id$. [Kachurovskii has recently informed me that one can have

$$C_1 = \left(\frac{p}{p-1}\right)^2$$

and

$$C_2 = \left(\frac{p}{p-1}\right)^p,$$

with $p > 1$.] However, the proof of the theorem as well as the identification of the limit both depend on the ergodic theorem and the martingale maximal inequality. The point of this result is that the statement itself is independent of the latter results, so that a simple unified formulation is obtained. The last fact brings one to look for a generalized maximal theory that includes both the ergodic and martingale results. Such a general theorem in the form of a class of operators is available and it is described next for the solution from a different perspective.

Let $\mathcal{D}$ be the class of Dunford operators (i.e., linear contractions on $L^1(\mathcal{S})$ and $L^\infty(\mathcal{S})$, hence defined on all $L^p(\mathcal{S})$, p ≥ 1) on a σ-finite space $(S, \mathcal{S}, \mu)$, which map the bounded functions of $L^p(\mathcal{S})$ into themselves. Then a basic maximal inequality that admits specialization to both ergodic and martingale theories is given by the following abstract theorem due to Tulcea and Tulcea [8]:

Theorem 2. *Let $T_i \in D$, $0 \leq i \leq r$ be such that $T_0 = I$ $(= id)$, $T_j T_{j-1} = T_j$ with $T_j^0 = I$, $1 \leq j \leq r$. Define for each $a > 0$ and $f \in L^p(S)$, $p \geq 1$, the sets*

$$G_f^*(a) = \left\{ \omega : \sup_{n \geq 1,\, j \leq r} \left| \frac{1}{n} \sum_{k=0}^{n-1} (T_j^k f)(\omega) \right| > a \right\}; \tag{3}$$

$$G_f(a) = \{\omega : |f(\omega)| > a\}.$$

Then for any measurable set F such that $G_f(a) \subset F \subset G_f^*(a)$ a.e., one has the basic inequality:

$$\mu(F) \leq \frac{1}{a} \int_F |f(\omega)|\, d\mu(\omega) < \infty. \tag{4}$$

From this result, taking $T_0 = I, T_1 = T \in \mathcal{D}$, and $r = 1$, one has the maximal inequality of ergodic theory, namely, if one defines the set

$$M_a = \{ \sup_{n \geq 1} \left| \frac{1}{n} \sum_{j=0}^{n-1} T^j f \right| (\omega) > a > 0 \},$$

then

$$\mu(M_a) \leq \frac{1}{a} \int_{M_a} |f|(\omega)\, d\mu(\omega),$$

and if $T_i \in \mathcal{D}$, $i = 1, \cdots, r$, $T_i{}^2 = T_i \in \mathcal{D}$ and
$N_a = \{\omega : \sup_{1 \leq i \leq r} |T_i f\ |(\omega) > a > 0\}$, the martingale result

$$\mu(N_a) \leq \frac{1}{a} \int_{N_a} |f|(\omega)\, d\mu(\omega)$$

follows.

These and related inequalities may be employed to obtain limit theorems of ergodic theory and martingale results in the usual manner. Other specializations were also discussed in the paper.

An approach entirely different from the above ones can be considered if one can find an operator family that contains both the conditional expectation and ergodic type semi-group of contractions. This will unify both the theories under discussion.

It is remarkable that such an operator class exists, and it was located by Rota. Its main implication in this context was presented in the paper under discussion. Some motivation for it may be useful to see why it is the correct class.

The classical Cesàro $(C,1)$ and Abel A-convergence methods are related as the former implies the latter, and which in turn implies the former provided the arithmetical averages of the sequence are, moreover, bounded. (Cf., e.g., Hardy [6], p. 43.) Applying it to the $\{(T^n f)(\omega), n \geq 0\}$ sequence, the $(C,1)$-sequence becomes an ergodic average, and the A-convergence becomes

$$\lim_{x\nearrow 1}(1-x)\sum_{k=0}^{\infty} x^k (T^k f)(\omega) = \lim_{x\nearrow 1}(1-x)[(I-xT)^{-1}f](\omega). \tag{5}$$

Replacing x by y^{-1} this may be identified as $\lim_{y\searrow 1}(y-1)(R_y f)(\omega)$, when the ergodic averages are bounded. Here R_y denotes the resolvent operator of T. Replacing x by $e^{-\lambda}$ and setting $U(t) = T^n$ for $n \leq t < n+1$, $n \geq 0$, (5) may be expressed simply as:

$$\lim_{\lambda\downarrow 0} \lambda R_\lambda f = \lim_{\lambda\downarrow 0} \lambda \int_0^\infty e^{-t\lambda} U(t) f\, dt, \ f \in L^p(\Sigma). \tag{6}$$

So the $(C,1)$-average holds for the sequence $\{T^n f, n \geq 1\}$ if $\{\lambda R_\lambda f, \lambda > 0\}$ has the Abel limit and a suitable boundedness condition holds. If T arises from a measurable and measure preserving transformation $\tau\colon \Omega \to \Omega$, then for the associated operator family $\{U(n), n \geq 0\}$, each member becomes a homomorphism additionally. Thus when one defines, using the operational calculus, $U(t)$ for all $t \geq 0$, it forms a strongly continuous semi-group of contraction operators on $L^p(\Sigma)$ and $U(t)(fg) = (U(t)f)(U(t)g)$ for f or g bounded and in $L^p(\Sigma)$, with the integral in (6) understood in the abstract Riemann sense. An important property of the R_λ-family so obtained is that the operator $V_\lambda = \lambda R_\lambda$ satisfies an operator identity, called the Reynolds identity. Although Reynolds operators appeared in studies of turbulence and viscous flows among others since the early 1900's, their use in ergodic or martingale theories was not noticed. This is perhaps due to the fact that their structure, as operators acting on $L^p(\Sigma)$-spaces, was not known until Rota's analysis, which he later published in the context of $L^2(\Sigma)$ in 1964. This will be discussed in Theorem 3 below. A special class of Reynolds operators is the class of averaging operators and it was found to be intimately related to conditional expectations. The general family of Reynolds operators, when their structure is determined, is seen to be another, and in fact a direct, way of unifying both the ergodic and martingale parts even for the continuous parameter problems (i.e., flows) whereas the previous results are confined to the discrete case. This may be explained more precisely as follows.

A bounded linear mapping R on L^p is a *Reynolds operator* if it satisfies the algebraic identity:

$$R(fg) = (Rf)(Rg) + R[(f-Rf)(g-Rg)], \quad f,g \in L^p \cap L^\infty(\Sigma). \tag{7}$$

Then using some standard, but non-trivial, results on semi-groups of operators and related differential calculus on them, one shows that $V_\lambda = \lambda R_\lambda$ of (6) (without the limits) satisfies (7). The following structure theorem for R_λ, $\lambda > 0$, on $L^p(\Sigma)$ is basic for this work, and it also shows that V_λ is actually a *composition operator* made up of a certain strongly continuous semi-group of contractions and conditional expectations:

Theorem 3. *Let (Ω, Σ, μ) be a σ-finite space and V_λ be a linear contraction operator on $L^p(\Sigma)$, $p \geq 1$ satisfying (7), so that it is a Reynolds operator for each $\lambda > 0$ (and is also weakly compact if $p = 1$). Suppose there are $A_n \in \Sigma$, $0 < \mu(A_n) < \infty$, $\Omega = \cup_{n=1}^\infty A_n$, such that*

$$V_\lambda \chi_{A_n} = \chi_{A_n}, \; n \geq 1$$

and

$$V_\lambda(L^p \cap L^\infty(\Sigma)) \subset L^p \cap L^\infty(\Sigma).$$

Then there exist (a) a σ-algebra $\mathcal{B}_\lambda$ ($\subset \Sigma$) containing $\{A_n, n \geq 1\}$, (b) a conditional expectation $E^{\mathcal{B}_\lambda}$ on $L^p(\Sigma)$, and (c) a strongly continuous semi-group $\{U(t), t \geq 0\}$ of operators induced by a measure preserving transformation $\tau : \Omega \to \Omega$ in terms of which one has

$$V_\lambda f = \lambda \int_0^\infty e^{-t\lambda} U(t) E^{\mathcal{B}_\lambda}(f)\, dt, \quad f \in L^p(\Sigma), \tag{8}$$

the integral being in Bochner's sense. Moreover, V_λ is invariant on $L^p(B_\lambda)$ with a dense range, and $V_\lambda E^{\mathcal{B}_\lambda} = E^{\mathcal{B}_\lambda} V_\lambda = V_\lambda$ there.

It should be noted that this representation uses a special case of von Neumann's mean ergodic theorem, namely, the result for linear isometries in Hilbert space for which a direct and interesting proof due to Sz.-Nagy is available. In fact Rota [18] establishes the above theorem for the important $L^2(\Sigma)$-case with $\mu(\Omega) < \infty$, and notes at the end that the proof can be extended to σ-finite measures "with a little extra work". This extension for the L^p spaces ($p \geq 1$) was given in Rao [14] (Theorem 8) together with the Abel mean convergence of $\lambda R_\lambda f$ (for f in certain Orlicz spaces) with $\mu(\Omega) < \infty$, and later in Rao [16] for general measures on L^p-spaces. (A detailed account of the representation as well as the convergence analysis is included in my book [17].) The convergence statement of Rota's theorem under discussion, in a general form that unifies both the ergodic and martingale theorems, may be presented as follows. Now let us say that a pairwise commuting family of Reynolds operators $\{R_\lambda, \lambda > 0\}$ defines a *generalized martingale* on an L^p-space if the following algebraic identity holds:

$$(\lambda_2 R_{\lambda_1} - \lambda_1 R_{\lambda_2}) R_{\lambda_2}\, f = (\lambda_2 - \lambda_1) R_{\lambda_1} R_{\lambda_2}^2\, f, \; 0 < \lambda_1 < \lambda_2, \; f \in L^1 \cap L^\infty.$$

Theorem 4. *Let*

$$\{V_\lambda f : \lambda > 0, f \in L^p(\Sigma), p \geq 1\}$$

be a generalized martingale, with each V_λ satisfying the hypothesis of Theorem 3 above relative to a fixed family $\{A_n, n \geq 1\}$. Then

$$\lim_{\lambda \to 0+} V_\lambda f$$

exists both pointwise a.e.[μ] and in L^p-mean. [An analogous statement holds if $\lambda \to \infty$.] This in particular implies the statements:

(i) Martingale case. The limit

$$\lim_{\lambda \to 0+} E^{\mathcal{B}_\lambda}(f)$$

exists a.e. and in L^p-mean with $A_n \in \mathcal{B}$, $n \geq 1$ for each $\lambda > 0$, and

(ii) Birkhoff's ergodic case. The limit

$$\lim_{t \to \infty} \frac{1}{t} \int_0^t U(s) f \, ds$$

exists a.e. and in L^p-mean, where $\{U(s), s > 0\}$ is the semi-group determined by the V_λ.

The pointwise limits for the Abel averages, which are needed in the proof of the above result, depend on a maximal theorem. However, it is based on a classical theorem due to Banach (cf. Dunford-Schwartz [4], VIII.9.3 and IV.11.2) and Hille [7], which is of a different type than that considered in Theorem 2 above. Theorem 4, which is the culmination of all the preceding work, was first presented by Rota in the paper under discussion if $1 < p < \infty$ and $\mu(\Omega) < \infty$, the crucial case. The maximal theorem for Abel averages, needed here, was independently and at about the same time considered by Rao [15] and Edwards [5]. Thus Rota's result unifies and includes both the Birkhoff ergodic theorem as well as the martingale convergence theorem. He also presented some other material related to ergodic and martingale theorems in several publications in the early 1960s.

With this achievement, Rota considered other operator identities on certain algebras, especially on a class which he termed 'Baxter algebras' (cf. Rota [19]), after which he was more or less completely attracted to combinatorial theory, spending most of his time, energy, and interests making deep contributions to that area. Noting the publication of my paper (Rao [14]), which contains details of mean convergence for his theorem in the context of certain Orlicz spaces including the L^p spaces, Rota asked me to send him several reprints of it. He indicated to me that instead of writing up the details of his paper, he would distribute the reprints along with his other papers on the subject. Further extensions of the result when the L^p and Orlicz spaces are replaced by more general Banach function spaces [L^ρ-spaces, Riesz spaces, or Köthe spaces] were made later (cf. Rao [15]). For a sharper analysis, however, one has to specialize the subject, and the independent study of ergodic problems is an important branch of analysis (cf., e.g., Cornfeld, Fomin, and Sinai [1], or Krengel [12]), and similarly martingale theory develops into another distinct branch (cf., e.g., Dellacherie and Meyer [2] or Liptser and Shiryayev [13]). Thus the generalized and unified theory has esthetic interest, but specializations result in individual and (often) difficult problems as well as sharpen techniques in both subjects which highlight current research.

References

[1] I. P. Cornfeld, S. V. Fomin and Ya. G. Sinai, *Ergodic Theory*. Springer-Verlag, Berlin and New York, 1982.

[2] C. Dellacherie and P.-A. Meyer, *Probabilités et potentiel, Partie B: Théorie des martingales*. Hermann, Paris, 1980.

[3] J. L. Doob, *Stochastic Processes*. Wiley, New York, 1953.

[4] N. Dunford and J. T. Schwartz, *Linear Operators, Part I: General Theory*. Wiley-Interscience, New York, 1958.

[5] D. A. Edwards, A maximal ergodic theorem for Abel means. *J. Functional Anal.* **7** (1971), 61–70.

[6] G. H. Hardy, *Divergent Series*. Clarendon Press, Oxford, 1949.

[7] E. Hille, Remarks on ergodic theorems. *Trans. Amer. Math. Soc.* **57** (1945), 246–269.

[8] A. Ionescu Tulcea and C. Ionescu Tulcea, Abstract ergodic theorems. *Trans. Amer. Math. Soc.* **107** (1963), 107–124.

[9] M. Jerison, Martingale formulation of ergodic theorems. *Proc. Amer. Math. Soc.* **10** (1959), 531–539.

[10] A. G. Kachurovskii, A martingale ergodic theorem. *Mat. Zemetki* **64** (1998), 311–314; English translation in *Math. Notes* **64** (1998), 266–269.

[11] S. Kakutani, Ergodic theory. In *Proceedings of the International Congress of Mathematicians, Cambridge, Mass. 1950, vol. 2*, Amer. Math. Soc., Providence RI, 1952, pp. 128–142.

[12] U. Krengel, *Ergodic Theorems*. Walter de Gruyter, Berlin, 1985.

[13] R. Sh. Liptser and A. N. Shiryayev, *Theory of Martingales*. Kluwer, Dordrecht, 1989.

[14] M. M. Rao, Interpolation, ergodicity, and martingales. *J. Math. Mech.* **16** (1966), 543–568.

[15] M. M. Rao, Generalized martingales. In *Contributions to Ergodic Theory and Probability*. Lecture Notes in Math. 160, Springer-Verlag, Berlin and New York, 1970, pp. 241–261.

[16] M. M. Rao, Abstract martingales and ergodic theory. In *Multivariate Analysis, III*, Academic Press, New York, 1973, pp. 45–60.

[17] M. M. Rao, *Foundations of Stochastic Analysis*. Academic Press, New York, 1981.

[18] G.-C. Rota, Reynolds operators. In *Proc. Sympos. Appl. Math., Vol. XVI,* Amer. Math. Soc., Providence, 1964, pp. 70–83.

[19] G.-C. Rota, Baxter algebras and combinatorial identities. I, II. *Bull. Amer. Math. Soc.* **75** (1969), 325–334.

Abel Ergodic Theory

Norton Starr

With his article "On the maximal ergodic theorem for Abel-limits", Gian-Carlo Rota opened up a new aspect of ergodic theory. Although there had been a prior history of Abel ergodic theory, it was largely concerned with limiting behavior in various topologies, as distinct from almost everywhere convergence (cf. D. A. Edwards [6]). In the present work, Rota expresses the view that Abel limits are the appropriate setting for ergodic theorems, an opinion only partially supported by subsequent developments in this area. However, his initiative did give rise to numerous and varied results in non-topological behavior for Abel limits. As a historical note, Gian-Carlo told this writer in July, 1962 that he had thought about the problem for some six months, then obtained his results in the last ten days (not an uncommon experience for mathematicians).

The first extension of Gian-Carlo's result was by Luis Báez-Duarte [1,2]. Making effective use of Rota's Abel approach, he obtained a simple proof of an a.e. limit ratio theorem which he said "justifies Rota's suggestion" ... "that Abel convergence is a more natural limiting process to be associated with the method of Chacon-Ornstein-Hopf". Subsequently, Chacon himself [4] showed that the Chacon-Ornstein [5] theorem can be extended to include, as a by-product, the Báez-Duarte theorem [1].

In the 1970's a variety of results inspired by Rota's approach appeared. At the first Midwestern Ergodic Theory Conference, M. M. Rao [9, p. 257] used Rota's Abel maximal theorem in deriving a.e. convergence for a generalized martingale. D. A. Edwards [6] made crucial use of Rota's result in proving a maximal ergodic theorem for Abel means of continuous-parameter operator semigroups, which he then used to obtain a.e. convergence when the semigroup is sub-Markovian. Edwards [7] also used Rota's result in developing an Abel sum version of Brunel's maximal ergodic theorem. From this last result Edwards derived the individual ergodic theorem of Báez-Duarte [2], as well as a general ergodic theorem for continuous-parameter semigroups, which, in turn, implied the corresponding result in his article [6]. Edwards' student Kopp [8] modified Rota's method to extend earlier Abel ergodic theorems of himself and of Edwards. In another direction, Sato [10] used Báez-Duarte's theorem in proving Abel individual ergodic theorems in the context of Brunel-Keane uniform sequences, acknowledging the influence of Rota's work with Abel limits. Indeed, Sato published a number of articles in Abel ergodic theory, giving maximal theory, ratio-limit theory, and vector valued results. See his *Studia* article [11] for a fairly extensive bibliography.

More recently, Rota's operator Abel-sum is reflected in the analysis of operator valued random processes by Blower [3].

References

[1] L. Báez-Duarte, An ergodic theorem of abelian type. *Notices Amer. Math. Soc.* **11**(1964), 467.

[2] L. Báez-Duarte, An ergodic theorem of abelian type. *J. Math. Mech.* **15**(1966), 599–607.

[3] G. Blower, Abel means of operator-valued processes. *Studia Math.* **115**(1995), 261–276.

[4] R. V. Chacon, Ordinary means imply recurrent means. *Bull. Amer. Math. Soc.* **70**(1964), 796–797.

[5] R. V. Chacon and D. S. Ornstein, A general ergodic theorem. *Illinois J. Math.* **4**(1960), 153–160.

[6] D. A. Edwards, On potentials and general ergodic theorems for resolvents. *Z. Wahrscheinlichkeitstheorie und Verw. Gebiete* **20**(1971), 1–8.

[7] D. A. Edwards, A maximal ergodic theorem for Abel means of continuous-parameter operator semigroups. *J. Functional Analysis* **7**(1971), 61–70.

[8] P. E. Kopp, Abelian ergodic theorems for vector-valued functions. *Glasgow Math. J.* **16**(1975), 57–60.

[9] M. M. Rao, Generalized Martingales. In *Contributions to Ergodic Theory and Probability*, Lecture Notes in Math. 160, Springer, Berlin and New York, 1970, pp. 241–261.

[10] Ryotaro Sato, Abel-ergodic theorems for subsequences. *Pacific J. Math.* **47**(1973), 233–242.

[11] Ryotaro Sato, Ratio limit theorems and applications to ergodic theory. *Studia Math.* **66**(1979/80), 237–245.

An "alternierende Verfahren" I

Radu Zaharopol

A very elegant, interesting, and difficult class of results in ergodic theory are the so-called almost everywhere (a.e.) convergence theorems; that is, theorems that offer sufficient conditions for the a.e. convergence of certain sequences of (equivalence classes of) measurable functions. In the paper "An "alternierende Verfahren" for general positive operators," Rota obtains a very general a.e. convergence theorem, and offers two applications of his result.

Since Rota's paper appeared in 1962 there has been a consistent interest in the ideas articulated in the work; this interest has been manifested by the multitude of papers that continue Rota's work. The unusual generality of the main result of Rota's paper has made this theorem into a tool that has been used many times, and in many scientific areas.

Let (S, Σ, μ) be a probability space. A positive contraction on $L^1(\mu)$ is called a *doubly stochastic operator* if $P1_S = 1_S$ and the adjoint $P^* : L^\infty(\mu) \to L^\infty(\mu)$ has the property $P^*1_S = 1_S$. (Note. If P is a doubly stochastic operator and if $1 \leq p \leq \infty$, then P can be restricted and P^* can be extended to $L^p(\mu)$. The restriction of P and the extension of P^* are positive contractions of $L^p(\mu)$ with respect to the L^p norm.)

The main result of the paper states that if $(P_n)_{n\in\mathbb{N}}$ is a sequence of doubly stochastic operators on $L^1(\mu)$, then the sequence

$$(P_1P_2\cdots P_nP_n^*P_{n-1}^*\cdots P_1^*f)_{n\in\mathbb{N}} \tag{$\dagger$}$$

converges μ-a.e. for every $f \in L^p(\mu), p > 1$. The proof is a tour de force in creativity: Rota constructs an AL-space, which, by a well-known representation theorem of Kakutani, can be thought of as an L^1 space of some measure space (S', Σ', μ'); next, a dilation of the sequence ($\dagger$) into a martingale is obtained; finally, by using the martingale convergence theorem the proof is completed. (As noted in Burkholder [15], the convergence also holds for $f \in L\log L$.)

The first application consists in proving that $(P^{2n}f)_{n\in\mathbb{N}}$ converges μ-a.e. for every $f \in \bigcup_{p>1} L^p(\mu)$ whenever P is doubly stochastic and selfadjoint on $L^2(\mu)$.

The second application was actually Rota's motivation for the paper: he proves that the sequence $((F_1F_2)^nf)_{n\in\mathbb{N}}$ converges μ-a.e. whenever $f \in \bigcup_{p>1} L^p(\mu)$ and F_1 and F_2 are conditional expectation operators. This second application of the main result is also the reason for the title of the paper: in the German literature a sequence of the form $((Q_1Q_2)^nf)_{n\in\mathbb{N}}$, where Q_1 and Q_2 are non-commuting projections, is known as an *alternierende Verfahren* or alternating process. The *alternierende Verfahren* application extends a result of Burkholder and Chow [16], and was obtained independently and by completely different methods by Stein [30]. Both Rota's and Stein's approaches have had a very significant influence on the areas of ergodic theory that are related to the *alternierende Verfahren* result. The line of research started by Burkholder and Chow is still being continued (for details, see Akcoglu and King [6]).

The main result of Rota's paper has been discussed in the monographs of Stein [31], Krengel [23], and Edgar and Sucheston [19]. However, because of the time when these works appeared and/or the specific goals of each volume, none of the above-mentioned books incorporated details about the lines of investigation generated by Rota's article and the numerous applications that the results of Rota's paper have had so far. Thus, it seems appropriate to point out here these lines of investigation and the applications.

The main result of Rota's paper has generated two lines of investigation: one of these consists (essentially, but not entirely) of extending the class of operators for which Rota's main result remains valid (we will call these extensions a.e. results); the second one consists of finding the proper setting and conditions for the norm convergence of sequences similar in spirit to the ones that appear in the main result of Rota's work (the results that belong to the second line of investigation will be called norm-convergence results).

The first of the a.e. results have been obtained by Doob [18], who reviewed Rota's paper in Mathematical Reviews, and by Starr [29], who extended both Rota's and Doob's results; these papers by Doob and Starr have created the framework on which many later extensions and applications of Rota's main result have been based. In a negative direction, Burkholder [15] and Ornstein [24] used conditional expectation operators to demonstrate that a.e. convergence may fail for functions in L^1. Later on, further a.e. results – a rather impressive body of work – were obtained by Akcoglu, Sucheston, and their collaborators. Thus, Frangos and Sucheston [22] obtained an extension of the main result of Rota's paper to Fava spaces (for details on Fava spaces see Fava [20] or Edgar and Sucheston [19]). Next, Akcoglu and Sucheston [8,9] obtained an extension to positive contractions on L^p-spaces, $1 < p < \infty$. These Akcoglu and Sucheston results were soon extended to non-positive operators by Akcoglu [2] (for $f \in L^p$, $1 < p < \infty$) and by Assani [11] (for $f \in L \log L$). These latter results were, in turn, followed by a sequence of three works by Akcoglu, Baxter and Lee [3], Akcoglu and Bradley [5], and Bradley [13] who used a famous dilation technique of Akcoglu [1] to further extend the main result of Rota's paper. More recently, a type of a.e.-convergence result which is known as a multiparameter version of the main result of Rota's paper was obtained by Sucheston [32] (see also [19]).

The first of the norm convergence results were obtained by Akcoglu and Sucheston [7] for contractions of L^p-spaces. Their pioneering work was followed by the paper of Assani [10], who extended their result to uniformly smooth Banach spaces. Assani's work has been continued by Xu and Roach [35], who obtained an estimate of the rate of convergence in Assani's result for certain duality maps.

We now turn our discussion to applications of the main result of Rota's paper. This result has been used in the study of convolution of measures by Akcoglu, Bellow, Jones, Losert, Reinhold-Larsson, and Wierdl [4], Fendler [21], and Pisier [25]; in the study of Littlewood-Paley functions by Shigekawa and Yoshida [28] (their result which uses Rota's main result was further used by Shigekawa in [27]); in the study of Gaussian distributions by Bryc [14]; in dynamical systems, in proving a.e.

and norm convergence by Tempel'man [33]; in deriving convergence for a nonlinear prediction problem by Denny [17]; in the study of algebraic numbers by Bourgain [12]; in the study of a.e. convergence of certain averages that involve Riemann sum operators by Ruch [26]; and in the study of the return to equilibrium in a stochastic lattice-gas model by Vanheuverzwijn [34].

We have tried here to outline the "success story" that Rota's paper has had so far, and which, we believe, will continue well into the future. There are two reasons for the interest awakened by Rota's work: first, the unusual generality of the main result (the operators P_n do not even have to be commuting); second, the beautiful image revealed by the proof of the main result. Although we have tried to gather as much information as possible, in all likelihood we have overlooked (hopefully not many) works for which Rota's paper was a source of inspiration; we sincerely regret any omission.

References

[1] M. A. Akcoglu, A pointwise ergodic theorem in L_p-spaces. *Canadian J. Math.* **27** (1975), 1075–1082.

[2] M. A. Akcoglu, Alternating sequences with nonpositive operators. *Proc. Amer. Math. Soc.* **104** (1988), 1124–1130.

[3] M. A. Akcoglu, J. R. Baxter, and W. M. F. Lee, Representation of positive operators and alternating sequences. *Adv. Math.* **87** (1991), 249–290.

[4] M. Akcoglu, A. Bellow, R. L. Jones, V. Losert, K. Reinhold-Larsson, and M. Wierdl, The strong sweeping out property for lacunary sequences, Riemann sums, convolution powers, and related matters. *Ergodic Theory Dynam. Systems* **16** (1996), 207–253.

[5] M. A. Akcoglu and R. E. Bradley, Alternating sequences and induced operators. *Trans. Amer. Math. Soc.* **325** (1991), 765–791.

[6] M. A. Akcoglu and J. L. King, An example of pointwise non-convergence of conditional expectation operators. *Israel J. Math.* **94** (1996), 179–188.

[7] M. A. Akcoglu and L. Sucheston, An alternating procedure for operators on L_p spaces. *Proc. Amer. Math. Soc.* **99** (1987), 555–558.

[8] M. A. Akcoglu and L. Sucheston, Convergence presque partout de procédés alternants. *C. R. Acad. Sci. Paris Sér. I Math.* **307** (1988), 271–272.

[9] M. A. Akcoglu and L. Sucheston, Pointwise convergence of alternating sequences. *Canadian J. Math.* **40** (1988), 610–632.

[10] I. Assani, Alternating procedures in uniformly smooth Banach spaces. *Proc. Amer. Math. Soc.* **104** (1988), 1131–1133.

[11] I. Assani, Rota's alternating procedure with non-positive operators. *Adv. Math.* **77** (1989), 183–188.

[12] J. Bourgain, Problems of almost everywhere convergence related to harmonic analysis and number theory. *Israel J. Math.* **71** (1990), 97–127.

[13] R. E. Bradley, On induced operators. *Canadian J. Math.* **43** (1991), 477–494.

[14] W. Bryc, Remarks on properties of probability distributions determined by conditional moments. *Probab. Theory Related Fields* **78** (1988), 51–62.

[15] D. L. Burkholder, Successive conditional expectations of an integrable function. *Ann. Math. Stat.* **33** (1962), 887–893.

[16] D. L. Burkholder and Y. S. Chow, Iterates of conditional expectation operators. *Proc. Amer. Math. Soc.* **12** (1961), 490–495.

[17] J. L. Denny, L_p functions. Their representation by sums of superpositions and their prediction using iterated conditional expectations. *Z. Wahrscheinlichkeitstheorie und Verw. Gebiete* **21** (1972), 225–232.

[18] J. L. Doob, A ratio operator limit theorem. *Z. Wahrscheinlichkeitstheorie und Verw. Gebiete* **1** (1962), 288–294.

[19] G. A. Edgar and L. Sucheston, *Stopping Times and Directed Processes.* Encyclopedia of Mathematics and its Applications, Vol. 47, Cambridge Univ. Press, Cambridge, 1992.

[20] N. A. Fava, Weak type inequalities for product operators. *Studia Math.* **42** (1972), 271–288.

[21] G. Fendler, An L^p version of a theorem of D. A. Raikov. *Ann. Inst. Fourier (Grenoble)* **35** (1985), 125–135.

[22] N. E. Frangos and L. Sucheston, On multiparameter ergodic and martingale theorems in infinite measure spaces. *Probab. Theory Related Fields* **71** (1986), 477–490.

[23] U. Krengel, *Ergodic Theorems.* Walter de Gruyter, Berlin, 1985.

[24] D. Ornstein, On the pointwise behavior of iterates of a self-adjoint operator. *J. Math. Mech.* **18** (1968), 473–477.

[25] G. Pisier, Holomorphic semigroups and the geometry of Banach spaces. *Ann. of Math. (2)* **115** (1982), 375–392.

[26] J.-J. Ruch, Convergence presque sûre de moyennes de sommes de Riemann. *Acta Arith.* **87** (1998), 1–12.

[27] I. Shigekawa, Sobolev spaces over the Wiener space based on an Ornstein-Uhlenbeck operator. *J. Math. Kyoto Univ.* **32** (1992), 731-748.

[28] I. Shigekawa and N. Yoshida, Littlewood-Paley-Stein inequality for a symmetric diffusion. *J. Math. Soc. Japan* **44** (1992), 251–280.

[29] N. Starr, Operator limit theorems. *Trans. Amer. Math. Soc.* **121** (1966), 90–115.

[30] E. M. Stein, On the maximal ergodic theorem. *Proc. Nat. Acad. Sci. U. S. A.* **47** (1961), 1894–1897.

[31] E. M. Stein, *Topics in Harmonic Analysis Related to the Littlewood-Paley Theory.* Princeton Univ. Press, Princeton, NJ, 1970.

[32] L. Sucheston, On one-parameter proofs of almost sure convergence of multiparameter processes. *Z. Wahrscheinlichkeitstheorie und Verw. Gebiete* **63** (1983), 43–49.

[33] A. A. Tempel'man, Ergodic theorems for general dynamical systems (Russian). *Dokl. Akad. Nauk SSSR* **176** (1967) 790–793. English translation: *Soviet Math. Dokl.* **8** (1967), 1213–1216.

[34] P. Vanheuverzwijn, A note on the stochastic lattice-gas model. *J. Phys. A* **14**(1981), 1149–1158.

[35] Z. B. Xu and G. F. Roach, An alternating procedure for operators on uniformly convex and uniformly smooth Banach spaces. *Proc. Amer. Math. Soc.* **111** (1991), 1067–1074.

An "alternierende Verfahren" II

Norton Starr

Gian-Carlo was justifiably proud of his paper "An "alternierende Verfahren" for general positive operators", even though it was not possible at the time to envision the ramifications discussed in Radu Zaharopol's essay in this chapter. Gian-Carlo's interest in the topic may well have been ignited by the Burkholder-Chow paper [2], which he reviewed for *Mathematical Reviews.* Whatever the source, he was inspired and his arguments in this Bulletin article evidence several leaps of the imagination leading to this seminal result. Shortly afterward his nascent involvement in combinatorial analysis led him to express the hope that extensions of his *alternierende Verfahren* could be developed by means of arguments using Galois connections. To date this approach has not borne fruit, and it was probably a disappointment that the initial strengthening of his result (Starr [5]) was based on the approach Doob [3] used in his own subsequent proof of the *alternierende Verfahren.*

As his interest and strength in combinatorial mathematics blossomed, some of Gian-Carlo's initiatives in operator theory were left incomplete. In Chapter 2 of this book are two preprints on the subject of dilations of positive operators, along with a Notices announcement in the same area. The dilation method was to have been the foundation for a sequel publication extending and generalizing the *alternierende Verfahren* article, but this did not come to pass. One proposed extension, a reverse index convergence result using

$$(P_n P_{n-1} \cdots P_1 P_1^* P_2^* \cdots P_n^* f)_{n \in \mathrm{N}},$$

turned out to be impossible (Starr [4]). However, the continuous parameter analog of the *alternierende Verfahren* was separately presented in Rota's "A Limit Theorem for the Time-Dependent Evolution Equation" (which appeared as notes of his summer institute lecture(s) at Varenna and is reprinted in this chapter) and in Starr [5]. The Varenna lecture notes also recapitulated the argument given in his Bulletin article, but with different details, assisting latter day readers. These notes are not well known and we hope their appearance here will be beneficial.

References

[1] D. L. Burkholder, Successive conditional expectations of an integrable function. *Ann. Math. Stat.* **33** (1962), 887–893.

[2] D. L. Burkholder and Y. S. Chow, Iterates of conditional expectation operators. *Proc. Amer. Math. Soc.* **12** (1961), 490–495.

[3] J. L. Doob, A ratio operator limit theorem. *Z. Wahrscheinlichkeitstheorie und Verw. Gebiete* **1** (1962), 288–294.

[4] N. Starr, On an operator limit theorem of Rota. *Ann. Math. Stat.* **36** (1965), 1864–1866.

[5] N. Starr, Operator limit theorems. *Trans. Amer. Math. Soc.* **121** (1966), 90–115.

CHAPTER 5

Inequalities

Reprinted from
INEQUALITIES, VOL. II

On the Inequality $\|f'\|^2 \leqslant 4\|f\| \cdot \|f''\|$

ROBERT R. KALLMAN* AND GIAN-CARLO ROTA†

Massachusetts Institute of Technology, Cambridge, Massachusetts

1. INTRODUCTION AND MAIN RESULT

If a function is small and its second derivative is small, then its first derivative must also be small. This general heuristic principle has been made precise in the literature in innumerable ways. Perhaps the best known statement of it is the inequality for functions on $[0, \infty)$,

$$\|f'\|_\infty^2 \leqslant 4\|f''\|_\infty \cdot \|f\|_\infty, \tag{1}$$

which was proved by Landau [*6*]. Hardy *et al.* give a discussion of this and a generalization to the L_2 case in their well-known book [*2*]; further discussion is given in the survey [*1*] by Beckenbach and Bellman. A particularly relevant paper is the one by Kolmogorov [*5*]. A great many generalizations of this inequality in various directions can be found in the literature.

Our present purpose is to give a generalization of this inequality which will keep the same bound, namely the number 4, and which will generalize, as much as possible, the Banach space norm which is used. Briefly, our result states that the inequality of the title extends to an arbitrary Banach space where it makes sense, and where the norm is independent of rearrangement of the function (see the precise definition of this class of Banach spaces below). Remarkably, the constant 4 remains unchanged in all these Banach spaces, although we do not establish that it is best (we conjecture, however, it is often best).

The class of Banach spaces we consider contains as special cases all the classical Banach spaces, including all the L_p-spaces, Orlicz spaces, Lorentz spaces, as well as the very general spaces recently introduced by Luxemburg and Zaanen [*7*]. The basic idea we use is extremely simple and depends only on Taylor's formula.

* This author's work was supported in part by NSF Grant GP-9141.
† This author's work was supported in part by the ONR.

2. The Inequality for Semi-Groups

For general information on semi-groups needed in the proof of the following theorem, consult Hille and Phillips [3].

Theorem 1. *Let $P(s)$ $(s > 0)$ be a C_0-contraction semi-group on the Banach space X. Let A be the infinitesimal generator of $P(s)$ and let x be an element of X which is in the domain of A^2. Then*

$$\| Ax \|^2 \leqslant 4 \| x \| \cdot \| A^2 x \|.$$

Proof. By [3, Formula (E_3), p. 354], Taylor's formula holds in the following form:

$$P(s)x = x + sAx + \int_0^s (s-t)\, P(t)\, A^2 x\, dt.$$

The integral in this formula is a Riemann integral. By the triangle inequality,

$$\begin{aligned} s \| Ax \| &\leqslant \| P(s)x \| + \| x \| + \int_0^s (s-t) \| P(t)\, A^2 x \|\, dt \\ &\leqslant 2 \| x \| + \tfrac{1}{2} s^2 \| A^2 x \|. \end{aligned}$$

Hence,

$$\| Ax \| \leqslant \frac{2}{s} \| x \| + \frac{s}{2} \| A^2 x \|,$$

for all $s > 0$. If $A^2 x \neq 0$, the theorem now follows in an elementary way, by minimizing the right side over all positive values of s. If $A^2 x = 0$, then letting $s \uparrow +\infty$ in the above inequality ,we obtain $Ax = 0$. Hence, if $A^2 x = 0$, the theorem holds trivially. q.e.d.

3. *LZ* Spaces

The preceding proof is not strong enough to give the inequality we want to prove in general enough Banach spaces. The reason for this failure lies in the difficulties in relating integration for Banach-space valued functions to ordinary Lebesgue integration for functions of two variables. This forces us to detour into a "weak" type integration technique suggested by (but not equivalent to) the Pettis integral [8].

We begin by defining the class of Banach spaces on which our inequality will hold. This class is suggested by the recent work of Lorentz, Luxemburg, and Zaanen (whence the name we give these spaces).

Recall that two real-valued Lebesgue measurable functions f and g on $[0, \infty)$ are said to be equidistributed if and only if $\mu(x \mid f(x) > t) = \mu(x \mid g(x) > t)$, where μ is Lebesgue measure. If f is such a function, let $(L(s) f)(x) = f(x + s)$.

If B is a Borel subset of the right half-axis, let χ_B be its characteristic function.

A normed linear space $X = (X, \| \cdots \|_X)$ consisting of equivalence classes (to within sets of measure zero) of complex-valued measurable functions defined on $[0, \infty)$ is said to be an *LZ-space* if it satisfies the following conditions:

(a) X does not contain only the zero element;

(b) if f is in X, then $|f|$ is in X and $\|f\|_X = \| |f| \|_X$;

(c) if f is in X, if g is complex-valued, measurable on $[0, +\infty)$, and $|g| \leqslant |f|$, then g is in X and $\|g\|_X \leqslant \|f\|_X$;

(d) if $f_n \geqslant 0$ $(n \geqslant 1)$, f_n is in $X (n \geqslant 1)$, $f_n \uparrow f$ (as $n \uparrow \infty$) almost everywhere, and $(\|f_n\|_X)_{n=1}^{\infty}$ is bounded, then f is in X and $\|f_n\|_X \uparrow \|f\|_X$ (the Fatou property);

(e) if f is in X, then f is integrable over every finite interval;

(f) if f is a real-valued element of X, and if f is equidistributed with g, then g is in X and $\|f\|_X = \|g\|_X$.

We now list some technical consequences of these axioms, to be used later. Most of them are closely related to the results of Luxemburg and Zaanen [7] on Banach function spaces.

Proposition 1. *If f is in X, then $L(s)f$ is in X and $\| L(s)f\|_X \leqslant \|f\|_X$, for all $s > 0$.*

Proof. If f is in X, then $|f|$ is in X. Also, $|f| = f_1 + f_2$, where $f_1 = \chi_{[0,s)} |f|$ and $f_2 = \chi_{[s,+\infty)} |f|$. Note that $L(s)|f| = L(s)f_2$. The function $L(s)f_2$ is equidistributed with f_2. Hence, $L(s)|f| = L(s)f_2$ is in X. Since $L(s)|f| = |L(s)f|$, $L(s)f$ is in X. Hence, $\| L(s)f\|_X = \| | L(s)f| \|_X = \| L(s)f_2\|_X = \|f_2\|_X \leqslant \| |f| \|_X = \|f\|_X$. q.e.d.

Proposition 2. *The normed linear space X is complete. Furthermore, if $\|f_n - f\|_X \to 0$ (as $n \uparrow +\infty$), then there exists a subsequence f_{n_k} such that $f_{n_k}(x) \to f(x)$ almost everywhere.*

In fact, Luxemburg and Zaanen show that assumption (d) implies this conclusion (cf. [*3*, Theorem 5.3, p. 149, Theorem 6.9, p. 244, and Theorem 4.8, p. 143]).

PROPOSITION 3. *Let K be a compact subset of the right half-axis. Then the mapping S: $X \to L^1(K)$, defined as $S(f) = f\chi_K$, is continuous.*

Proof. S has its range in $L^1(K)$, by assumption (e). X is complete, by Proposition 2. Hence, we may apply the Closed Graph Theorem. Let $f_n \to f$ and $Sf_n \to g$ in norm. We must show that $Sf = g$. By Proposition 2 and elementary measure theory, we can choose a subsequence f_{n_k} such that $f_{n_k}(x) \to g(x)$ a.e. and $f_{n_k}(x) \to f(x)$ a.e. Hence, $f(x) = g(x)$ a.e. on K, or $Sf = g$, so that S is continuous. q.e.d.

If φ is a bounded measurable function which vanishes outside of a compact subset of the right half axis, let $T(\varphi)$ be the linear functional, defined on X by $T(\varphi)(f) = \int_0^{+\infty} f(x)\, \varphi(x)\, dx$. Proposition 3 shows that $T(\varphi)$ is bounded. It is clear that all such $T(\varphi)$ form a linear set in X^*, the dual space of X, which separates the points of X.

The following proposition is close to a result of Luxemburg and Zaanen (cf. the proof of [3], Lemma 11.3, pp. 258–261). However, since it plays a key role in what follows, we shall sketch its proof.

PROPOSITION 4. $\|f\|_X = \sup_{\|T(\varphi)\| \leqslant 1} | T(\varphi)(f)|$, *where φ ranges over the set of bounded measurable functions on the right half-axis with compact support.*

Proof. If f is in X, let $\|f\|' = \sup_{\|T(\varphi)\| \leqslant 1} | T(\varphi)(f)|$. It is clear that $\|f\|' \leqslant \|f\|_X$. The proof will be complete if we show that $\|f\|' \geqslant \|f\|_X$.

First, assume that f has compact support, lying, say, in $[0, a]$. Without loss of generality we may suppose that $\|f\|_X = 1$. Let X_1 be the unit ball of X, and let S: $X \to L^1([0, a])$ be as in Proposition 3. $S(X_1)$ is a closed, balanced, convex subset of $L^1([0, a])$, by assumption (d). Let $f_0 = (1 + \epsilon)f$. Then $\|f_0\|_X = 1 + \epsilon$. Hence, $S(f_0)$ is not in $S(X_1)$, for if $f_n \in X_1$ and $S(f_n) \to S(f_0)$ (as $n \uparrow +\infty$), then $\|f_0\|_X \leqslant \underline{\lim}_{n\uparrow+\infty} \|f_n\|_X \leqslant 1$, by assumption (d). Therefore, by a standard separation theorem (cf. [4, Theorem 14.3, p. 118]), there exists a bounded measurable function φ, with support in $[0, a]$, such that $| \int_0^a g\varphi | \leqslant \| T(\varphi)\|$ for all g in X_1, and $\| T(\varphi)\| < | \int_0^a f_0\varphi | = (1 + \epsilon)| \int_0^a f\varphi|$. Hence, $\|f\|' \geqslant 1/(1 + \epsilon)$. Since $\epsilon > 0$ is arbitrary, $\|f\|' \geqslant 1 = \|f\|_X$.

Now let f be arbitrary in X. Note that $f_n = \chi_{[0,n]}f$ is in X and $\|f_n\|' \leqslant \|f\|'$. Hence,

$$
\begin{aligned}
\|f\|' \geqslant \|f_n\|' &= \|f_n\|_X \\
&= \| \, |f_n| \, \|_X \uparrow \| \, |f| \, \|_X \qquad (\text{as } n \uparrow +\infty) \\
&= \|f\|_X .
\end{aligned}
$$

Therefore, $\|f\|' \geqslant \|f\|_X$. q.e.d.

4. Main Inequality

Theorem 2. *Let f be in X. Suppose that f' and f'' exist as distributions and are elements of X. Then*

$$\|f'\|_X^2 \leqslant 4\|f\|_X \|f''\|_X .$$

Proof. Since f' and f'' are in X, they are integrable over finite intervals. A simple computation shows that Taylor's formula holds "weakly," that is, the identity

$$\int_0^{+\infty} L(s)f\varphi = \int_0^{+\infty} f\varphi + s\int_0^{+\infty} f'\varphi + \int_0^s (s-t)\left(\int_0^{+\infty} L(t)f''\varphi\right) dt$$

holds for all functions φ with compact support which are infinitely differentiable. But this means that

$$T(\varphi)(L(s)f) = T(\varphi)(f) + sT(\varphi)(f') + \int_0^s (s-t)\, T(\varphi)(L(t)f'')\, dt. \quad (*)$$

Fix $a \geqslant 0$. The set of all φ with support in $[0, a]$ such that (*) holds is closed under bounded pointwise convergence, by the Lebesgue dominated convergence theorem. Hence, (*) holds for every bounded measurable function φ with compact support.

Proceeding as in the proof of Theorem 1 and using Proposition 1, we obtain

$$s\,|\,T(\varphi)(f')| \leqslant 2\,\|\,T(\varphi)\|\,\|f\|_X + (s^2/2)\,\|\,T(\varphi)\|\,\|f''\|_X .$$

Hence, by Proposition 4, $\|f'\|_X \leqslant 2/s\,\|f\|_X + (s/2)\,\|f''\|_X$ for all $s > 0$. One may now finish the proof as in Theorem 1. q.e.d.

References

1. E. F. Beckenbach and R. Bellman, "Inequalities." Springer-Verlag, Berlin, 1961.
2. G. H. Hardy, J. E. Littlewood, and G. Pólya, "Inequalities." Cambridge, 1959.
3. E. Hille and R. S. Phillips, "Functional Analysis and Semi-groups." American Mathematical Society, Providence, Rhode Island, 1957.
4. J. L. Kelley and I. Namioka, "Linear Topological Spaces." Van Nostrand, Princeton, New Jersey, 1963.
5. A. Kolmogorov, On inequalities between Upper Bounds of Consecutive Derivatives of an Arbitrary Function Defined on an Infinite Interval. *Učhen. Zap. Moskov. Gos. Univ. Mat.* **30** (1939), 3–16.
6. E. Landau, Einige ungleichungen fur zweimal differentiierbare Functionen. *Proc. London Math. Soc. Ser. 2* **13** (1913), 43–49.

7. W. A. J. Luxemburg and A. C. Zaanen, Notes on Banach Function spaces I-IV. *Indag. Math.* *66* (1963), 135–153, 239–263, 496–504.
8. B. J. Pettis, On integration in vector spaces . *Trans. Am. Math. Soc.* **44** (1938), 277–304.
9. L. Schwartz, Theorie des distributions I, II. Paris, 1950.

The Kallman-Rota inequality; a survey

W. N. Everitt and Anton Zettl

1. Introduction

This survey paper is concerned with the inequality now called the Kallman-Rota inequality. This remarkable and significant result was first discovered and proved by Robert Kallman and Gian-Carlo Rota in 1967, and published in 1970; for details see [42]. The title of the Kallman-Rota paper [42],

$$\text{(1.1)} \qquad \text{'On the inequality } \|f'\|^2 \leq 4\,\|f\|\,\|f''\| \text{'},$$

whilst accurate in one sense, is otherwise misleading. The inequality is concerned in general with the infinitesimal generator A of a one parameter contraction semi-group $\{T(s) : s \in [0,\infty)\}$ of bounded operators acting in a Banach space X; in this setting the inequality takes the global form

$$\text{(1.2)} \qquad \|Af\|^2 \leq 4\,\|f\|\,\|A^2 f\| \text{ for all } f \in D(A^2).$$

Not only is the general setting for this inequality significant, but it is quite remarkable that the number 4 in (1.2) is a global bound for all Banach spaces.

However, the important special case (1.1) has proved to be of significance when the Banach space X is a function space defined on some unbounded interval I of the real line $\mathbb{R}$, and $\{T(s) : s \in [0,\infty)\}$ is the contraction semi-group of translation operators with $(T(s)f)(x) = f(x+s)$ for $x \in I$ and $s \in [0,\infty)$; in these circumstances the infinitesimal generator is the differentiation operator and the general inequality (1.2) reduces to the form (1.1).

In Section 2, we outline our approach to the decision making process of preparing the extensive list of references for this survey paper. In Section 3, we state and prove the original inequality of the Kallman-Rota theorem as given in [42], and in Section 4, its Hilbert space offspring, the Kato theorem [43]. In Section 5, we state and prove the Landau and Hadamard inequalities [57] and [32], and in Section 6, the Hardy and Littlewood inequality [33], in order to provide examples that show the remarkable best possible global properties of the Kallman-Rota and Kato result; it is clear that these earlier results played a significant role in the ideas that led to the formulation of the Kallman-Rota inequality, and to the Kato inequality. In Section 7, we consider various extensions of, and then in Section 8, special cases of these extensions to, the Kallman-Rota inequality. Finally, in Section 9, we consider a number of inequalities indirectly related to the original Kallman-Rota inequality.

2. THE REFERENCES

We have found it no easy task to decide on the inclusion of references for this survey paper; the early results in 1913 and 1914 of Landau [57] and Hadamard [32] led to large number of results concerning the connection between the norm of a function, defined on an interval I of the real line, and the norms of its derivatives. Here the norms are taken in a variety of function spaces, the original results involving the space $L^\infty(I)$ of bounded functions on I. Consideration of such inequalities was intensified with the Hardy-Littlewood result [33] of 1932, involving the space $L^2(I)$, and subsequently considered in greater detail in the now classic text [34] of 1934, by Hardy, Littlewood and Pólya.

Many of these results were then brought together and unified in the remarkable Kallman-Rota theorem [42] of 1967, together with the significant observation on this result by Kato [43].

We have, then, made an assessment of the inequality results following these early theorems; the extraordinary variety and number of these later results has made it essential to exercise a choice in compiling the list of references for this paper.

The major references consulted for this survey are:

(*i*) *Inequalities* by Hardy, Littlewood and Pólya; see [34].

(*ii*) *Inequalities involving functions and their integrals and derivatives* by Mitrinović, Pečarić and Fink; see [59].

(*iii*) *Norm inequalities for derivatives and differences* by Kwong and Zettl; see [56].

(*iv*) *A dictionary of inequalities* by Bullen; see [13].

Both the books [59] and [56] contain extensive collections of references and we have drawn on these sources in compiling the list of references for this paper.

Other references are given a mention, which in many cases is all too brief, in the text that follows; however, some of the cited references are not mentioned in the text but are quoted in view of their direct or indirect connection with the Kallman-Rota inequality.

3. THE KALLMAN-ROTA INEQUALITY

The Kallman-Rota inequality is based on the properties of contraction semi-groups of operators defined within a Banach space, say X. For reference to the following definitions and properties see the texts of Yosida [72, Chapter IX, Sections 2 and 3] and Hille [38, Chapter 5, Section 5.3].

Let X be a Banach space over the complex scalar field $\mathbb{C}$, with norm $\|\cdot\|$, and let $\{T(s) : s \in [0,\infty)\}$ be a family of bounded linear operators $T(s) : X \to X$ for all $s \in [0,\infty)$. The family $\{T(s) : s \in [0,\infty)\}$ is said to satisfy the semi-group property if

$$T(s)T(t) = T(s+t) \text{ for all } s,t \in [0,\infty). \tag{3.1}$$

A semi-group $\{T(s) : s \in [0,\infty)\}$ on X is said to be a strongly continuous semi-group of class C_0 if, where I is the identity operator on X,

$$T_0 = I \text{ and } \lim_{t \to s+} \|T(t)f - T(s)f\| = 0 \text{ for all } s,t \in [0,\infty) \text{ and } f \in X. \tag{3.2}$$

If a semi-group of class C_0 satisfies the condition, where $M > 0$,

$$\|T(s)\| \le M \text{ for all } s \in [0,\infty) \tag{3.3}$$

then the semi-group is said to be bounded. If $M \le 1$ then $\{T(s) : s \in [0,\infty)\}$ is a contraction semi-group of class C_0.

Let $\{T(s) : s \in [0,\infty)\}$ be a strongly continuous semi-group of class C_0; then there exists an infinitesimal generator A of the semi-group, where $A : D(A) \subseteq X \to X$ is in general an unbounded operator, defined by, where the strong limit in X is implied,

$$D(A) := \{f \in X : \lim_{h\to 0+} h^{-1}(T(h) - I)\, f \text{ exists in } X\} \tag{3.4}$$

and

$$Af := \lim_{h\to 0+} h^{-1}(T(h) - I)\, f \quad \text{for all } f \in D(A). \tag{3.5}$$

It is known that the domain $D(A)$ is dense in X; also that all the iterated domains $D(A^n)$, for $n \ge 2$, are dense in X.

As an example of these definitions and properties let X be one of the Banach Lebesgue integration spaces $L^p(0,\infty)$, with $p \in [1,\infty]$ and norm $\|\cdot\|_p$ defined by

$$\|f\|_p := \left\{\int_0^\infty |f(x)|^p \, dx\right\}^{1/p} \quad \text{for all } p \in [1,\infty)$$

$$\|f\|_\infty := \text{ess sup}\{|f(x)| : x \in [0,\infty)\}. \tag{3.6}$$

Let the family $\{T(s) : s \in [0,\infty)\}$ of operators be defined by the translation operator

$$(T(s)f)(x) := f(x+s) \text{ for all } s, x \in [0,\infty). \tag{3.7}$$

Then it may be shown that this definition gives, for all $p \in [1,\infty]$, a strongly continuous semi-group of class C_0; the infinitesimal generator for this example is given by

$$D(A) := \{f \in L^p[0,\infty) : f \in AC_{\text{loc}}[0,\infty) \text{ and } f' \in L^p[0,\infty); \tag{3.8}$$

$$(Af)(x) := f'(x) \text{ for almost all } x \in [0,\infty). \tag{3.9}$$

For some details of these results see [72, Chapter 9, Section 5].

We can now state the Kallman-Rota theorem leading to the Kallman-Rota inequality.

Theorem 3.1. *Let X be a Banach space over the complex scalar field $\mathbb{C}$, with norm $\|\cdot\|$; let $\{T(s) : s \in [0,\infty)\}$ be a strongly continuous, contraction semi-group of class C_0, with infinitesimal generator A. Then*

$$\|Af\|^2 \le 4\,\|f\|\,\|A^2 f\| \quad \textit{for all } f \in D(A^2). \tag{3.10}$$

There are particular cases of the inequality (3.10) *when the number* 4 *is best possible; there are particular cases of the inequality when the result is valid but with a number strictly less than* 4.

Proof. The proof, which here follows the proof in the original Kallman-Rota paper [42], requires the use of abstract Riemann-Stieltjes integrals in Banach spaces; for the general theory see the text of Hille [38, Chapter 7, Section 7.4].

Let $f \in D(A^2)$; then we have the integral identity in X

$$T(s)f = f + sAf + \int_0^s (s-u)T(u)A^2 f\, du \text{ for all } s \in [0,\infty). \tag{3.11}$$

The existence and proof of this identity requires abstract integration by parts, see [38, Chapter 7, Section 4, Theorem 7.4.1]. Thus we have

$$\begin{aligned} s\,\|Af\| &\le \|f\| + \|T(s)f\| + \int_0^s (s-u)\,\|T(u)A^2 f\|\, du \\ &\le 2\,\|f\| + \tfrac{1}{2}s^2\,\|A^2 f\| \quad \text{for all } s \in (0,\infty) \text{ and all } f \in D(A^2) \end{aligned} \tag{3.12}$$

where we have used integration results from [38, Chapter 7, Section 4, Theorem 7.4.1] and the condition $\|T(s)\| \le 1$ for all $s \in [0,\infty)$. If $A^2 f \ne 0$ then, on dividing (3.12) by $s > 0$, the right-hand side is minimised by choosing $s = 2\left\{\|f\| / \|A^2 f\|\right\}^{1/2}$ and substitution yields

$$\|Af\| \le 2\left\{\|f\|\,\|A^2 f\|\right\}^{1/2}.$$

If $A^2 f = 0$ then on letting $s \to +\infty$ in (3.12) it follows that $Af = 0$ and the inequality (3.10) follows.

Thus for all $f \in D(A^2)$ we have, as required for the theorem,

$$\|Af\|^2 \le 4\,\|f\|\,\|A^2 f\|.$$

The statement concerning particular cases is seen to hold in the light of the Landau-Hadamard inequality (5.7), when 4 is the best possible number, and the Hardy-Littlewood inequality (6.3) when 2 is the best possible number. See also Remarks 3.1, 5.1 and 6.1 below. □

The classical examples of the Kallman-Rota inequality are given in

Corollary 3.1. *Let the Banach space $L^p[0,\infty)$, the family of translation operators $\{T(s) : s \in [0,\infty)\}$ and the infinitesimal generator A be given by* (3.6) *to* (3.9) *above; thus $Af = f'$. Then for $p \in [1,\infty]$*

$$\|f'\|_p \le 4\,\|f\|_p\,\|f''\|_p \tag{3.13}$$

for all $f \in L^p[0,\infty)$ with $f, f' \in AC_{loc}[0,\infty)$ and $f'' \in L^p[0,\infty)$.

Remark 3.1. (1) Although the proof of Theorem 3.1 given above appears to be elementary it has to be recognised that it requires the initial study of contraction semi-groups, infinitesimal generators and abstract integration in Banach spaces.

(2) The proof follows the main idea of the Landau paper [57], but with the use of the classical Taylor theorem replaced by its analogue from the theory of semi-groups of operators.

(3) The global coverage of the Kallman-Rota inequality leads to the consequence that the proof gives no information concerning the best possible number $K \leq 4$ in the inequality

$$\|Af\|^2 \leq K \|f\| \, \|A^2 f\| \text{ for all } f \in D(A^2), \tag{3.14}$$

nor the existence of cases of equality when K is known or characterised.

(4) There exist cases of the general inequality (3.14) when the best possible number K is strictly less than 4; see Theorem 4.1 below for a class of cases when the best possible number is 2.

(5) The title of the original Kallman-Rota paper [42], see (1.1), is clearly connected with the inequalities (3.13) as given by the classical Banach spaces $L^p[0, \infty)$, with $p \in [1, \infty]$.

4. The Kato inequality

We have the following theorem due to Kato in 1971, see [43]; for the theory of semi-groups of operators and their associated infinitesimal generators see [45, Chapter IX, Sections 1.1 to 1.4]; for the theory of semi-bounded and accretive operators in Hilbert space see [45, Chapter V, Section 3.10].

Theorem 4.1. *Let A be the infinitesimal generator of a contraction semi-group of class C_0 on a Hilbert space H. Then*

$$\|Af\|^2 \leq 2 \|f\| \, \|A^2 f\| \text{ for all } f \in D(A^2). \tag{4.1}$$

The inequality (4.1) *is globally best possible; there is equality if and only if for some $\gamma \in [0, \infty)$ and for some $f \in D(A^2)$*

$$A^2 f + \gamma A f + \gamma^2 f = 0 \text{ and } \operatorname{Re}(f, A^2 f) = 0, \tag{4.2}$$

in which case, necessarily, $\|A^2 f\| = \gamma^2 \|f\|$.

Proof. We follow the proof given in [43].

The assumed properties of the operator A are equivalent to assuming that the operator $-A$ is m-accretive, see [45, Chapter V, Section 3.10, Page 279]. In other words, $(\gamma - A)^{-1}$ is defined everywhere on the space H if $\gamma > 0$ and $\operatorname{Re}(Ag, g) \leq 0$ for all $g \in D(A)$. It follows that if we define the operator $B_\gamma := (\gamma + A)(\gamma - A)^{-1}$, the Cayley transform of A, then $\|B_\gamma\| \leq 1$.

A computation now shows that, with $\|f\| > 0$ and $\gamma \in (0, \infty)$,

$$-2\gamma A f = A^2 f + \gamma^2 f + B_\gamma (A^2 f - \gamma^2 f) \text{ for all } f \in D(A^2). \tag{4.3}$$

Hence

$$\begin{aligned} \|Af\|^2 &\leq \left(\|A^2 f + \gamma^2 f\| + \|B_\gamma (A^2 f - \gamma^2 f)\|\right)^2 / 4\gamma^2 \\ &\leq \left(\|A^2 f + \gamma^2 f\| + \|A^2 f - \gamma^2 f\|\right)^2 / 4\gamma^2 \\ &\leq \left(\|A^2 f + \gamma^2 f\|^2 + \|A^2 f - \gamma^2 f\|^2\right) / 2\gamma^2 \\ &= \gamma^{-2} \|A^2 f\|^2 + \gamma^2 \|f\|^2 \text{ for all } \gamma \in (0, \infty). \end{aligned} \tag{4.4}$$

If $A^2 f = 0$ then on letting $\gamma \to 0^+$ in (4.4) it follows that $Af = 0$ and (4.1) follows. If $A^2 f \neq 0$ then the inequality (4.1) now follows by minimising the right-hand side of (4.4) with respect to $\gamma \in (0, \infty)$; this process determines $\gamma^2 = \|A^2 f\| / \|f\|$.

Now suppose that $\|f\| > 0$ and equality holds in (4.1). If $A^2 f = 0$, then $Af = 0$ and (4.3) is satisfied with $\gamma = 0$. Otherwise set $\gamma = (\|A^2 f\| / \|f\|)^{1/2} > 0$; then

$$\|Af\|^2 = 2\|f\| \|A^2 f\|;$$

thus equality holds everywhere in the display (4.4); for this to hold it is necessary that

$$\|A^2 f + \gamma^2 f\| = \|A^2 f - \gamma^2 f\| = \|B_\gamma(A^2 f - \gamma^2 f)\| \tag{4.5}$$

and

$$B_\gamma(A^2 f - \gamma^2 f) = A^2 f + \gamma^2 f. \tag{4.6}$$

The equality (4.5) implies that $\mathrm{Re}(f, A^2 f) = 0$, whilst the equality (4.6), together with (4.3), implies that $-2\gamma A f = 2(A^2 f + \gamma^2 f)$.

Taken all together these results imply that (4.2) holds.

Conversely, we now show that (4.2) implies that equality holds in (4.1). To see this result, first suppose that (4.2) holds with $\gamma > 0$; then

$$\|Af\|^2 = \gamma^{-2} \|A^2 f\|^2 + \gamma^2 \|f\|^2;$$

from (4.1) it follows that

$$(\gamma^{-1}\|A^2 f\| - \gamma\|f\|)^2 \leq 0.$$

Hence, $\|A^2 f\| = \gamma^2 \|f\|$ and $\|Af\|^2 = 2\|f\| \|A^2 f\|$ as required. Finally if (4.2) is true with $\gamma = 0$, then $A^2 f = 0$ and the equality (4.1) is satisfied; in this case it also follows that $\|A^2 f\| = 0 = \gamma^2 \|f\|$.

This completes the proof of Theorem 4.1. □

We now apply the results of this theorem to establish the Hardy-Littlewood integral inequality [33], as discussed in classical form below, in Section 6.

Let the Hilbert space H be the Lebesgue space $L^2(0, \infty)$, with the usual norm $\|\cdot\|_2$; in this space the family of translation operators $\{T(s) : s = [0, \infty)\}$ defined by

$$(T(s)f)(x) := f(x + s) \text{ for all } x, s \in [0, \infty)\}, \tag{4.7}$$

is a C_0 contraction semi-group; the infinitesimal generator A of this semi-group is defined by

$$D(A) := \{f : [0, \infty) \to \mathbb{C} : f \in AC_{\mathrm{loc}}[0, \infty) \text{ and } f, f' \in L^2[0, \infty)\} \tag{4.8}$$

$$Af := f' \text{ for all } f \in D(A). \tag{4.9}$$

Conversely, the operator A, given by (4.8) and (4.9), generates the translation semi-group determined by (4.7).

Corollary 4.1. *Let A in $H = L^2[0, \infty)$ be determined by* (4.8) *and* (4.9); *then A is the infinitesimal generator of the translation C_0 semi-group determined by* (4.7). *The operator A^2 is determined by*

$$D(A^2) := \{f : [0, \infty) \to \mathbb{C} : f, f' \in AC_{\mathrm{loc}}[0, \infty) \text{ and } f, f'' \in L^2[0, \infty)\} \tag{4.10}$$

$$A^2 f := f'' \textit{ for all } f \in D(A^2).$$

The application of Theorem 4.1 *is justified and results in the Hardy-Littlewood inequality*

$$\|f'\|_2^2 \le 2\,\|f\|_2\,\|f''\|_2 \textit{ for all } f \in D_2([0,\infty)) \equiv D(A^2). \tag{4.11}$$

The general condition (4.2) *reduces to the requirement that f satisfies the linear differential equation, for some $\gamma > 0$,*

$$f''(x) + \gamma f'(x) + \gamma^2 f(x) = 0 \textit{ for all } x \in [0,\infty)$$

and $\mathrm{Re}(f, f'')_2 = 0$. *A straightforward calculation then shows that, for some number $K \in \mathbb{C}$,*

$$f(x) = K \exp(-x/(2\gamma)) \sin\left(x\sqrt{3}/(2\gamma) - \pi/3\right) \textit{ for all } x \in [0,\infty). \tag{4.12}$$

Proof. This is an immediate application of Theorem 4.1 and the indicated calculation. □

Remark 4.1. (1) It is clear from Remark 2 in [43] that Kato was inspired to conjecture and then prove Theorem 4.1, as a result of observing the Kallman-Rota inequality of Theorem 3.1.

(2) We note that this Kato theorem has a restricted application but more definite consequences, in comparison with the Kallman-Rota theorem; the reason here is, of course, the reduction of the Banach space condition to the Hilbert space setting.

(3) The number 2 in the Kato inequality is globally best possible, as is the number 4 in the Kallman-Rota inequality.

(4) The wide generality of the Kallman-Rota inequality prevents the determination of any explicit form of cases of equality, in any particular example when the number 4 is best possible; the Kato inequality does give explicit information for the cases of equality when the number 2 is best possible.

(5) There are cases when the conditions for the Kato inequality are satisfied and yet the best possible number is strictly less than 2. If A is a self-adjoint operator in a Hilbert space H such that the spectrum $\sigma(A)$ of A satisfies $\sigma(A) \subseteq [0,\infty)$, then $-A$ generates a C_0-contraction semi-group in H; for this result see [72, The Corollary, Chapter IX, Section 8]. From the self-adjoint properties of A we then have

$$\|Af\|_H^2 = (Af, Af)_H = (f, A^2 f)_H \le \|f\|_H \left\|A^2 f\right\|_H \text{ for all } f \in D(A^2).$$

5. The Landau-Hadamard inequality

We state and prove one form of the Landau-Hadamard inequality, see [57] and [32]. However to make a good comparison with the Hardy-Littlewood inequality, considered in Section 6 below, we extend the domain of the original Landau-Hadamard inequality to include functions with locally integrable second derivatives, rather that. the restriction to continuous second derivatives. This extension allows of the definition of interesting non-null extremals (cases of equality) which show that the numbers in the Landau-Hadamard inequalities are best possible.

For any interval $I \subseteq \mathbb{R}$ define

$$(5.1) \qquad C(I) := \{f : I \to \mathbb{C} : f \text{ is continuous at all points of } I\}.$$

Let $f \in L^\infty(I)$ if $f : I \to \mathbb{C}$, and if f satisfies the two properties:

$$(5.2) \qquad (i)\ f \text{ is Lebesgue measurable on } I$$

$$(5.3) \qquad (ii)\ \operatorname{ess\,sup}\{|f(x)| : \text{ for all } x \in I\} < \infty;$$

then $L^\infty(I)$ is a Banach space over the complex field $\mathbb{C}$ with the norm defined by

$$(5.4) \qquad \|f\|_\infty := \operatorname{ess\,sup}\{|f(x)| : \text{ for all } x \in I\}.$$

Theorem 5.1. *Define the domains*

$$(5.5) \qquad D_\infty([0,\infty)) := \{f : [0,\infty) \to \mathbb{C} : \begin{array}{ll} (i) & f \in C([0,\infty)) \\ (ii) & f' \in AC_{loc}([0,\infty)) \\ (iii) & f, f'' \in L^\infty([0,\infty))\} \end{array}$$

$$(5.6) \qquad D_\infty((-\infty,\infty)) := \{f : (-\infty,\infty) \to \mathbb{C} : \begin{array}{ll} (i) & f \in C((-\infty,\infty)) \\ (ii) & f' \in AC_{loc}((-\infty,\infty)) \\ (iii) & f, f'' \in L^\infty((-\infty,\infty))\}. \end{array}$$

The following results hold:

(*a*) *Let* $f \in D_\infty([0,\infty))$; *then* $f' \in L^\infty([0,\infty)))$ *and*

$$(5.7) \qquad \|f'\|_\infty^2 \le 4\,\|f\|_\infty\,\|f''\|_\infty \text{ for all } f \in D_\infty([0,\infty)).$$

(*b*) *Let* $f \in D_\infty((-\infty,\infty))$; *then* $f' \in L^\infty((-\infty,\infty)))$ *and*

$$(5.8) \qquad \|f'\|_\infty^2 \le 2\,\|f\|_\infty\,\|f''\|_\infty \text{ for all } f \in D_\infty((-\infty,\infty)).$$

The inequalities (5.7) *and* (5.8) *are best possible, and there are non-null cases of equality.*

Remark 5.1. We note that the Landau-Hadamard inequality (5.7) is a special case of the Kallman-Rota inequality, when the Banach space involved is $L^\infty([0,\infty))$ and the semi-group is translation on the interval $[0,\infty)$, with the number 4 as best possible.

Proof. We give here the extension of the original proofs of the Landau and Hadamard inequalities; for an extensive and thorough development of these inequalities see the text by Kwong and Zettl [56, Chapters 1 and 2].

Proof of (*a*).

Recall that if $f' \in AC_{loc}([0,\infty))$ then f'' exists almost everywhere (Lebesgue) and $f'' \in L_{loc}([0,\infty))$. Let $f \in D_\infty([0,\infty))$; then for all $x \in [0,\infty)$ and all $s \in [0,\infty)$ it may be verified, on using the properties of the Lebesgue integral, that the following identity holds

$$(5.9) \quad f(x+s) = f(x) + sf'(x) + \int_0^s (s-u)f''(x+u)\,du \text{ for all } f \in D_\infty([0,\infty)).$$

With the norm definition given in (5.4) and $I = [0, \infty)$ we obtain, for $s \in (0, \infty)$,

$$|sf'(x)| \le 2\,\|f\|_\infty + \int_0^s (s-u)\,|f''(x+u)|\,du$$

$$\le 2\,\|f\|_\infty + \tfrac{1}{2}s^2\,\|f''\|_\infty \text{ for all } x \in [0,\infty). \tag{5.10}$$

Thus

$$\|f'\|_\infty \le \frac{2}{s}\,\|f\|_\infty + \frac{s}{2}\,\|f''\|_\infty \text{ for all } s \in (0,\infty). \tag{5.11}$$

If $\|f''\|_\infty = 0$ then on letting $s \to \infty$ it follows that $\|f'\|_\infty = 0$ and the inequality (5.7) holds.

If $\|f''\|_\infty > 0$ then the right-hand side of (5.11) is minimised by choosing

$$s = \{\|f\|_\infty / \|f''\|_\infty\}^{1/2}$$

and substitution yields

$$\|f'\|_\infty \le 2\,\{\|f\|_\infty\,\|f''\|_\infty\}^{1/2}$$

and then (5.7) follows.

To show that the inequality (5.7) is best possible we follow [56, Chapter 2, Theorem 2.1] and define $f \in D_\infty([0,\infty))$ by

$$f(x) := \begin{cases} -\frac{1}{4} + x - \frac{1}{2}x^2 & \text{for all } x \in [0,1] \\ +\frac{1}{4} & \text{for all } x \in [1,\infty) \end{cases};$$

substitution in (5.7) gives equality. It is remarked in [56, Chapter 2, Theorem 2.1] that such cases of equality are not unique and that there is no explicit formula to describe all such cases.

Proof of (b).

Let $f \in D_\infty((-\infty,+\infty))$; then for all $x \in (-\infty,+\infty)$ and all $s \in [0,\infty)$ it may be verified that the following identity holds

$$f(x+s)-f(x-s) = 2sf'(x)+\int_0^s (s-u)f''(x+u)\,du-\int_{-s}^0 (s+u)f''(x+u)\,du. \tag{5.12}$$

With the norm definition given in (5.4) and $I = (-\infty,+\infty)$ we obtain, for $s \in (0,\infty)$,

$$|2sf'(x)| \le 2\,\|f\|_\infty + \int_0^s (s-u)\,|f''(x+u)|\,du + \int_{-s}^0 (s+u)\,|f''(x+u)|\,du$$

$$\le 2\,\|f\|_\infty + s^2\,\|f''\|_\infty \text{ for all } x \in (-\infty,\infty). \tag{5.13}$$

Thus

$$\|f'\|_\infty \le \frac{1}{s}\,\|f\|_\infty + \frac{s}{2}\,\|f''\|_\infty \text{ for all } s \in (0,\infty). \tag{5.14}$$

If $\|f''\|_\infty = 0$ then on letting $s \to \infty$ it follows that $\|f'\|_\infty = 0$ and the inequality (5.8) holds good.

If $\|f''\|_\infty > 0$ then the right-hand side of (5.14) is minimised by choosing

$$s = \sqrt{2}\,\{\|f\|_\infty / \|f''\|_\infty\}^{1/2}$$

and substitution yields

$$\|f'\|_\infty \le \sqrt{2}\{\|f\|_\infty \|f''\|_\infty\}^{1/2};$$

the inequality (5.8) now follows.

To show that the inequality (5.8) is best possible we follow [56, Chapter 2, Theorem 2.1] and define $f \in D_\infty((-\infty,+\infty))$ by

$$f(x) := \begin{cases} x(2-x) & \text{for all } x \in [0,2] \\ (x-2)(x-4) & \text{for all } x \in [2,4] \end{cases},$$

and then extend the definition of f to the whole of $\mathbb{R}$ by periodicity of period 4. It is remarked in [56, Chapter 2, Theorem 2.1] that such cases of equality are not unique and that there is no explicit formula to describe all such cases. □

Remark 5.2. We have given the proofs of both the Kallman-Rota and the classical Landau-Hadamard inequalities.

The proof of the Landau-Hadamard inequality is not "strictly elementary" in the sense given by Hardy, Littlewood and Pólya [34, Chapter 1, Section 1.7], but can be counted as "elementary" in that only the basic properties of mathematical analysis are required.

Although the original proof of the Kallman-Rota inequality follows the same lines as the Landau-Hadamard inequality the Kallman-Rota proof is not elementary in the above sense, but requires a knowledge of the methods of functional analysis and operator theory.

In the Landau-Hadamard inequalities (5.7) and (5.8) there are known non-null cases of equality, as given above; however the generality of the Kallman-Rota inequality seems to imply that it is difficult to provide information about the existence and form of cases of equality.

6. The Hardy-Littlewood inequality

We state and prove, see [33] and [34, Chapter VII, Theorem 259],

Theorem 6.1. *For any interval I of the real line $\mathbb{R}$ let $L^2(I)$ denote the Hilbert space of Lebesgue measurable, complex-valued functions defined on I, with the finite norm*

$$\|f\|_2^2 := \int_I |f(x)|^2\,dx < +\infty. \tag{6.1}$$

Define the domain

$$D_2(I) := \{f : I \to \mathbb{R} : (i)\ f, f' \in AC_{loc}(I),\ (ii)\ f, f'' \in L^2(I)\}. \tag{6.2}$$

Then:

(*a*) *If $I = [0,\infty)$ and $f \in D_2([0,\infty))$ then $f' \in L^2([0,\infty)))$ and*

$$\|f'\|_2^2 \le 2\,\|f\|_2\,\|f''\|_2 \text{ for all } f \in D_2([0,\infty)). \tag{6.3}$$

(*b*) *If $I = (-\infty,\infty)$ and $f \in D_2((-\infty,\infty))$ then $f' \in L^2((-\infty,\infty)))$ and*

$$\|f'\|_2^2 \le \|f\|_2\,\|f''\|_2 \text{ for all } f \in D_2((-\infty,\infty)). \tag{6.4}$$

The inequalities (6.3) *and* (6.4) *are best possible; the cases of equality are determined as follows:*

(c) *There is equality in* (6.3) *if and only* $f(x) = \alpha F(\beta x)$ *for all* $x \in [0, \infty)$ *where* $\alpha \in \mathbb{C}$, $\beta \in (0, \infty)$ *and the extremal function* F *is defined by*

$$F(x) := \exp(-x/2)\sin(x\sqrt{3}/2 - \pi/3) \text{ for all } x \in [0, \infty). \tag{6.5}$$

(d) *There is equality in* (6.4) *if and only if* f *is the null function on* $(-\infty, \infty)$.

Remark 6.1. We note that the Hardy-Littlewood inequality (6.3) is a special case of the Kallman-Rota inequality, when the Banach (Hilbert) space involved is $L^2([0, \infty))$ and the semi-group is translation on the interval $[0, \infty)$, but now with the number 2 as best possible; see also Remark 3.1 (3).

To prove (*a*) and (*c*) of this theorem we first state and prove the following lemma:

Lemma 6.1. *Under the statement and conditions of Theorem* 6.1*:*

(*i*) $f' \in L^2([0, \infty))$ *for all* $f \in D_2([0, \infty))$.

(*ii*) *Define the functional* $J : D_2([0, \infty)) \to \mathbb{R}$ *by*

$$J(f) := \int_0^\infty \{f(x)^2 - f'(x)^2 + f''(x)^2\}\, dx \text{ for all } f \in D_2([0, \infty)) \tag{6.6}$$

then

$$J(f) > 0 \text{ for all } f \in D_2([0, \infty)) \tag{6.7}$$

unless $f(x) = \alpha F(\beta x)$ *for all* $x \in [0, \infty)$, *as in Theorem* 6.1.

Proof. If $f \in D_2([0, \infty))$ then, on integration by parts and following [34, Chapter VII, Section 7.8],

$$\int_0^X f'(x)^2 dx = [f(x)f'(x)]_0^X - \int_0^X f(x)f''(x)\, dx \text{ for all } X \in (0, X).$$

As $X \to \infty$ the integral on the right-hand side is convergent, since both f and $f'' \in L^2([0, \infty))$. If the integral on the left-hand side is divergent, as $x \to \infty$, then

$$\lim_{x \to \infty} f(x)f'(x) = +\infty \tag{6.8}$$

and this implies that $\lim_{X \to \infty} |f(x)| = \infty$, a contradiction on $f \in L^2([0, \infty))$. Hence $\lim_{x \to \infty} f(x)f'(x)$ exists and is finite; this limit value has to be zero, otherwise again there is a contradiction on $f \in L^2([0, \infty))$. Thus $f' \in L^2([0, \infty))$, (*i*) above holds and

$$\int_0^\infty f'(x)^2 dx = -f(0)f'(0) - \int_0^\infty f(x)f''(x)\, dx \text{ for all } f \in D_2([0, \infty)). \tag{6.9}$$

We now note that, with F defined as in (6.5), a calculation shows that

$$F(x) + F'(x) + F''(x) = 0 \text{ for all } x \in [0, \infty), \tag{6.10}$$

and

$$F(0) + F'(0) = 0 \text{ and } F''(0) = 0. \tag{6.11}$$

An additional, and more lengthy, calculation shows that

$$(6.12) \qquad J(F) = \int_0^\infty \{F(x)^2 - F'(x)^2 + F''(x)^2\}\, dx = 0.$$

Now write, for any $f \in D_2([0,\infty))$ and for $\gamma \in \mathbb{R}$,

$$(6.13) \qquad f(x) = g(x) + \gamma F(x) \text{ for all } x \in [0,\infty)$$

thus defining $g \in D_2([0,\infty))$; choose $\gamma \in \mathbb{R}$ so that $g(0) = 0$. Note that, see [34, Chapter VI, Theorem 223],

$$(6.14) \qquad \lim_{x\to\infty} x^{-1/2} g(x) = \lim_{x\to\infty} x^{-1/2} g'(x) = \lim_{x\to\infty} g(x)g'(x) = 0.$$

We have now, from (6.13),

$$J(f) = J(g) + 2\gamma K(F,g) + \gamma^2 J(F)$$

where, using integration by parts,

$$\begin{aligned} K(F,g) &= \int_0^\infty (F(x)g(x) - F'(x)g'(x) + F''(x)g''(x))\ dx \\ &= -\int_0^\infty (F'(x) + F'''(x))\, g(x)\ dx - \int_0^\infty (F(x) + F'(x))\, g''(x)\ dx - \int_0^\infty F'(x)g'(x)\ d \\ &= \int_0^\infty \{F(x) + F'(x) + F'''(x)\}\, g'(x)\ dx = 0. \end{aligned}$$

Thus, from (6.10) and (6.12),

$$(6.15) \qquad J(f) = J(g),$$

and it is enough to prove that $J(g) > 0$ unless g is the null function on $[0,\infty)$. However, since $g(0) = 0$ and $\lim_{x\to\infty} g(x)g'(x) = 0$,

$$\int_0^\infty g'(x)^2 dx = -\int_0^\infty g(x)g''(x) dx$$

and so

$$\begin{aligned} J(g) &= \int_0^\infty \{g(x)^2 + g(x)g''(x) + g''(x)^2\}\, dx \\ (6.16) \qquad &= \int_0^\infty (g(x) + g''(x))^2\, dx + \int_0^\infty g'(x)^2\ dx > 0 \end{aligned}$$

unless g is the null function on $[0,\infty)$.

From the representation (6.13), the equality (6.15) and the inequality (6.16) a straightforward calculation shows that the inequality (6.7) is satisfied and that there is equality if and only if (6.5) holds good.

This completes the proof of Lemma 6.1. □

We now consider the proof of parts (a) and (c) of Theorem 6.1.

Proof. The proof that $f' \in L^2(0,\infty)$ is given in the proof of the lemma. The proof of (6.3) follows from the inequality (6.7) using a scaling argument; details are given in [34, Chapter VII, End of Section 7.8].

Taken together these results complete the proof of parts (a) and (c) of the theorem. □

Remark 6.2. In [34, Chapter VII] three proofs of Lemma 6.1 are to be found; the proof given above is the second of these proofs. The third proof is based on the calculus of variations and is given only in part; the difficulties here are due to the singular nature of this problem in the calculus of variations. For a complete and illuminating proof of Lemma 6.1 using the calculus of variations see the paper of Ahlbrandt [1].

We now consider the proof of parts (*b*) and (*d*) of Theorem 6.1.

Proof. To prove parts (*b*) and (*d*) of the theorem we use the result in [34, Chapter VII, Section 7.9] and note that, on integration by parts and use of (6.8) at both $\pm\infty$, we obtain

$$\int_{-\infty}^{+\infty} f'(x)^2\, dx = -\int_{-\infty}^{+\infty} f(x)f''(x)\, dx \text{ for all } f \in D_2((-\infty,+\infty)).$$

To this result apply the Cauchy-Schwarz integral inequality to yield

$$\int_{-\infty}^{+\infty} f'(x)^2\, dx \le \left\{\int_{-\infty}^{+\infty} f(x)^2\, dx \int_{-\infty}^{+\infty} f''(x)^2\, dx\right\}^{1/2} \quad \text{for all } f \in D_2((-\infty,+\infty))$$

which gives (6.4) and so proves part (*b*).

The proof of part (*d*) is omitted here but is given in [34, Chapter VII, Section 7.9]. □

Remark 6.3. We have given one proof of the Hardy-Littlewood inequality (for other proofs see [34, Chapter VII, Section 7.8]) in Theorem 6.1, in order to illustrate two points:

(*a*) Firstly, the proof is "elementary", in the sense of Remark 5.2, in that the proof uses only methods of classical analysis.

(*b*) Secondly, the shorter proof of the Hardy-Littlewood inequality, including the cases of equality, given in Corollary 4.1 follows from the Kato inequality; however the proof of Theorem 4.1 is clearly not "elementary" but again depends upon a knowledge of the methods of functional analysis and operator theory.

7. Extensions of the Kallman-Rota inequality

The concepts of the infinitesimal generator A of semi-groups and groups of operators, and the properties of dissipative and m-dissipative operators in Banach and Hilbert space have led to a large number of generalisations and extensions of the original Kallman-Rota inequality. To present some of these results we have called on the references given in the list at the end of this paper: references of particular importance are Certain and Kurtz [14], Chernoff [15], Ditzian [17] and [16], Gindler and Goldstein [28] and [29], Goldstein [30], Goldstein and Räbinger [31], Hille [37], Kwong and Zettl [55] and [56], Mitrinović, Pečarić and Fink [59].

7.1. **Semi-groups.** (1) Suppose that $\{T(s) : s \in [0,\infty)\}$ is a semi-group of operators satisfying the conditions (3.1), (3.2) and (3.3) within a Banach space X; let the number M be defined by

$$M := \sup\{\|T(s)\| : s \in [0,\infty)\}$$

and suppose that $M \in (0,\infty)$. Then with A as the infinitesimal generator of $\{T(s)\}$ we obtain the inequality

$$\|Af\|^2 \le 2M(M+1)\,\|f\|\,\|A^2f\| \text{ for all } f \in D(A^2). \tag{7.1}$$

This inequality, which reduces to the Kallman-Rota inequality (3.10) when $M = 1$, is due to Kurepa [49], and Kraljević and Kurepa [47].

(2) Suppose now that, in addition, $\{T(s) : s \in [0,\infty)\}$ is a strongly continuous semi-group of operators satisfying the conditions (3.1), (3.2) and (3.3) within a Banach space X; let A be the infinitesimal generator of $\{T(s)\}$. Then for any $m \in \mathbb{N}$ with $m \ge 2$, and any $k \in \mathbb{N}$ with $1 \le k < m$ there exist positive numbers $C_{m,k}$ such that

$$\|A^kf\|^m \le C_{m,k}^m\,\|f\|^{m-k}\,\|A^mf\|^k \text{ for all } f \in D(A^m). \tag{7.2}$$

This result is due to Hille [36]; it reduces to the Kallman-Rota inequality (3.10) when $m = 2$ and $k = 1$.

In the special case when $n \in \mathbb{N}$, $m = 2n$ and $k = n$ we obtain

$$\|A^nf\|^2 \le C_{2n,n}^2\,\|f\|\,\|A^{2n}f\| \text{ for all } f \in D(A^{2n}), \tag{7.3}$$

a result also due to Hille [36]; this result can be applied to the cases when $X = C[0,\infty]$ or $L^p[0,\infty)$ for $p \in [1,\infty]$ and A is classical differentiation on $[0,\infty)$ to give

$$\left\|f^{(n)}\right\|^2 \le C_{2n,n}^2\,\|f\|\,\left\|f^{(2n)}\right\| \tag{7.4}$$

on the domain where both f and $f^{(2n)}$ are in the space concerned.

(3) Following the Landau-Kallman-Rota path Hille [35] showed, but using different methods, that (7.4) can also be considered in the cases when $X = C[-\infty,\infty]$ or $L^p(-\infty,\infty)$ with $p \in [1,\infty]$; the result is the inequality

$$\left\|f^{(n)}\right\|^2 \le 4\,\|f\|\,\left\|f^{(2n)}\right\| \tag{7.5}$$

holding on the domain where both f and $f^{(2n)}$ are in the space concerned. For a precise result of the inequality (7.5) when $p = 2$ and $n = 2$ see the result in [8].

(4) Let the weight function $w : [0,\infty) \to [0,\infty)$ satisfy

$$\begin{cases} (i) & w(x) \ge 0 \text{ for all } x \in [0,\infty) \\ (ii) & w \text{ is Lebesgue measurable on } [0,\infty) \\ (iii) & w \text{ is monotone non-decreasing on } [0,\infty). \end{cases} \tag{7.6}$$

Define the spaces $L^p((0,\infty);w)$ for all $p \in [1,\infty)$ as the set of all $f : [0,\infty) \to \mathbb{C}$ such that

$$\|f\|_{p,w} := \left\{ \int_0^\infty w(x)\,|f(x)|^p\,dx \right\}^{1/p} < \infty. \tag{7.7}$$

Then if w satisfies conditions (i) and (ii) of (7.6), with elements as equivalence classes, $L^p((0,\infty);w)$ is a Banach space over the scalar field $\mathbb{C}$.

If now w further satisfies condition (iii) of (7.6), and this condition is essential, then it may be shown that the translation family $\{T(s) : s \in [0,\infty)\}$ of Section 3 is a contraction semi-group of class C_0 for the Banach space $L^p((0,\infty);w)$; here again, as in Section 3, the action of the infinitesimal generator A is classical differentiation.

These results lead to an application of the Kallman-Rota inequality to give, with f and $A^2 f = f'' \in L^p((0,\infty);w)$, the inequality

$$\|f'\|_{p,w}^2 \le 4\,\|f\|_{p,w}\,\|f''\|_{p,w}\,. \tag{7.8}$$

There are extensions of the inequality (7.8) to inequalities of the form (7.2).

The inequality (7.8) is due to Kwong and Zettl; see [52] and [56, Chapter 3, Section 3.5, Theorem 3.17].

Remark 7.1. (1) There is a review of the results in this Section 7.1 in the reference book by Mitrinović, Pečarić and Fink [59, Chapter I, Section 34].

(2) There is an extensive coverage of inequalities of the form (7.2) in the book of Kwong and Zettl [56, Chapter 3].

(3) The inequality (7.2) is also considered in Section 8 below.

(4) An alternative analysis of the inequality (7.8), under conditions (7.6), is given in the paper [25]; this paper uses only classical analytic methods and follows the techniques of Hardy-Littlewood in [33]. In this paper the example $L^2((0,\infty);x)$ with the translation family for $\{T(s) : s \in [0,\infty)\}$, is considered in detail; the best possible number K in the corresponding inequality

$$\int_0^\infty x\,|f'(x)|^2\,dx \le K \left\{ \int_0^\infty x\,|f(x)|^2\,dx \int_0^\infty x\,|f''(x)|^2\,dx \right\}^{1/2}$$

satisfies $0 < 2 < K < 4$ and there is a continuum of non-null cases of equality.

(5) Examples show that if the condition (iii) of (7.6) is not satisfied by the weight w then the corresponding inequality to (7.8) may not hold; moreover that the conditions for an application of the Kallman-Rota inequality may not be satisfied; the space $L^p((0,\infty);\exp(-x))$ is such an example with the translation family $\{T(s) : s \in [0,\infty)\}$ of Section 3.

7.2. Cosine functions.

Let X be a Banach space over the complex scalar field $\mathbb{C}$, with norm $\|\cdot\|$. A one-parameter family $\{C(s) : s \in [0,\infty)\}$ of bounded linear operators in X is said to be a strongly continuous cosine function if the following conditions are satisfied

(i) $C(t+s) + C(t-s) = 2C(t)C(s)$ for all $t, s \in \mathbb{R}$

(ii) $C(0) = I$ the identity operator in X

(*iii*) $\lim_{t\to s}\|C(t)f - C(s)f\| = 0$ for all $t, s \in \mathbb{R}$ and all $f \in X$.

Every cosine function in X has an infinitesimal generator A defined by

(*a*)

$$D(A) := \left\{ f \in X : \lim_{h\to 0} \frac{C(t+h)f - C(t-h)f}{2h} \text{ exists in } X \right\} \tag{7.9}$$

(*b*)

$$Af := \lim_{h\to 0} \frac{C(t+h)f - C(t-h)f}{2h} \text{ for all } f \in D(A). \tag{7.10}$$

If in addition to the conditions (*i*), (*ii*) and (*iii*) above

$$\sup\{\|C(s)\| : s \in [0,\infty)\} = M < \infty \tag{7.11}$$

then the cosine function $\{C(s)\}$ is said to bounded; if $M \le 1$ then $\{C(s)\}$ is a contraction cosine function.

If the cosine function $\{C(s)\}$ is bounded then the domain $D(A)$ is dense in the Banach space X; indeed the domain $D(A^n)$ of the power A^n is dense in X for all $n \in \mathbf{N}$.

For these results concerning cosine functions see Goldstein [30, Page 14, Section 2 and Page 118, Section 8] and Hille [38, Chapter 5, Pages 167 to 169].

The general Kallman-Rota inequality (7.1) inspired an extension to bounded cosine functions:

Theorem 7.1. *Let the cosine function* $\{C(s) : s \in [0,\infty)\}$ *satisfy the conditions* (*i*), (*ii*) *and* (*iii*) *above, and be bounded by the positive number* M *as in* (7.11); *then*

$$\|Af\|^2 \le \frac{2}{3}M(M+1)\,\|f\|\,\left\|A^2 f\right\| \text{ for all } f \in D(A^2). \tag{7.12}$$

The proof of this cosine inequality is given in the papers [47] and [49].

In the cases when the Banach space X is taken to be $C[-\infty,\infty]$, or $L^p(-\infty,\infty)$ with $p \in [1,\infty]$ we can define a cosine function by

$$(C(s)f)(t) := \frac{f(t+s) + f(t-s)}{2} \text{ for all } t \in (-\infty,\infty) \text{ and } s \in [0,\infty). \tag{7.13}$$

In the spaces concerned it may be shown that this definition yields a bounded cosine function with $M = 1$. It follows that the infinitesimal generator A is then given by $Af = f''$, for all $f \in D(A)$, with classical differentiation on $(-\infty,\infty)$. The application of the inequality (7.12) to the cosine function (7.13) for these function Banach spaces, taking the appropriate domains as in Theorem 5.1, yields the inequalities

$$\|f''\|^2 \le \frac{4}{3}\,\|f\|\,\left\|f^{(4)}\right\|.$$

We note that this last result is a considerable improvement on the general result (7.5) in the case when $n = 2$.

7.3. Additional extensions. For additional information on general and function inequalities inspired by the original Kallman-Rota inequality, reference should be made to the extensive collection of inequalities given in the book [59]; see in particular [59, Chapter I, Sections 1 to 34, and Section 94].

8. Special cases of the Kallman-Rota inequality

Many of the extensions of the Kallman-Rota inequality involve results from the theory of m-dissipative operators. According to a well-known characterisation a linear operator A (bounded or unbounded) acting on a Banach space X, is the infinitesimal generator of a strongly continuous semi-group C_0 if and only if A is m-dissipative. For a discussion of these general results see the texts [39], [45] and the more recent work of [4].

It is straightforward to show that in the classical Banach function spaces, such as considered in this paper, the norm of a function f and and the norm of the first derivative f' are not, in general, related. In particular for the Banach spaces $L^p(I)$, for some interval I of the real line, it may be shown that for given positive numbers $\mu_0 > 0, \mu_1 > 0$ there is a real-valued function $f \in L^p(I)$ satisfying $\|f\|_p = \mu_0$ and $\|f'\|_p = \mu_1$. However, this freedom for these first two norms does not extend to the three norms $\|f\|_p, \|f'\|_p$ and $\|f''\|_p$ as has been seen in earlier sections of this paper; indeed the whole family of Kallman-Rota type inequalities show that the values of these three norms cannot be chosen arbitrarily. In particular, this lack of numerical freedom for the three norms certainly holds for the spaces $L^p(0,\infty)$ and $L^p(-\infty,\infty)$ as is seen in Theorems 5.1 and 6.1; see also [56, Chapter 1].

Such inequalities have higher order extensions as is seen in Section 7.1. In a widely quoted paper [46] Kolmogorov showed that if k, n are positive integers with $1 \le k < n$ and $n \ge 2$, then there is a positive number $K(\mathbb{R}, \infty, k, n)$, depending upon the variables shown, such that for all real-valued functions $f \in L^\infty(\mathbb{R})$ with the classical derivative $f^{(n)} \in L^\infty(\mathbb{R})$, it follows that $f^{(k)} \in L^\infty(\mathbb{R})$ and

$$\left\|f^{(k)}\right\|_\infty^n \le K(\mathbb{R}, \infty, k, n) \, \|f\|_\infty^{n-k} \left\|f^{(n)}\right\|_\infty^k . \tag{8.1}$$

Kolmogorov gives a remarkable explicit formula for the best possible number $K(\mathbb{R}, \infty, k, n)$.

The positive half-line version of this result is given by Schoenberg and Cavaretta [69]:

$$\left\|f^{(k)}\right\|_\infty^n \le K(\mathbb{R}^+, \infty, k, n) \, \|f\|_\infty^{n-k} \left\|f^{(n)}\right\|_\infty^k . \tag{8.2}$$

These authors characterised the best possible numbers $K(\mathbb{R}^+, \infty, k, n)$ in terms of Euler splines thus making it possible to compute approximations to these numbers; however, there is still no closed form formula known to give explicit values for these best possible numbers. The following given extension of the original Kallman-Rota inequality was discovered independently by Ditzian [17] and Certain and Kurtz [14], see also the paper by Chernoff [15]:

Theorem 8.1. *Suppose that A is an m-dissipative operator, bounded or unbounded, on a Banach space X, with norm $\|\cdot\|$. Let k,n be positive integers satisfying $1 \le k < n$ with $n \ge 2$. Then*

$$\|A^k f\|^n \le K(\mathbb{R}^+,\infty,k,n)\,\|f\|^{n-k}\,\|A^n f\|^k \quad \text{for all } f \in D(A^n) \tag{8.3}$$

where the numbers $K(\mathbb{R}^+,\infty,k,n)$ are the best possible numbers the inequality in (8.2).

Remark 8.1. The inequality (8.3) can be established for some positive number K directly from the Kallman-Rota inequality, as given in Theorem 3.1, on using arguments based on mathematical induction. This induction argument yields a number K which is larger than the optimal result, even when the optimal number for $n = 2$ is known. Theorem 8.1 gives the remarkable result that the numbers $K(\mathbb{R}^+,\infty,k,n)$ obtained from the inequality (8.2) are universal for all m-dissipative operators in any Banach space, just as in the second order case first established by Kallman and Rota. However, there is no direct application of the original Kallman-Rota method to yield the inequality (8.3); Certain and Kurtz [14] used a different method of analysis.

Certain and Kurtz [14], and later Chernoff [15] also proved a remarkable result to match the inequality (8.3) but now utilising the original Kolmogorov inequalities (8.1):

Theorem 8.2. *Suppose that each of A and $-A$ is a m-dissipative operator, bounded or unbounded, on a Banach space X. (Thus A generates a group of isometries.) Let k,n be positive integers satisfying $1 \le k < n$ and $n \ge 2$. Then*

$$\|A^k f\|^n \le K(\mathbb{R},\infty,k,n)\,\|f\|^{n-k}\,\|A^n f\|^k \quad \text{for all } f \in D(A^n). \tag{8.4}$$

Remark 8.2. This theorem shows that the Kolmogorov numbers $K(\mathbb{R},\infty,k,n)$ are universal for the inequality (8.4) when both A and $-A$ are m-dissipative; since the Kolmogorov numbers $K(\mathbb{R},\infty,k,n)$ are smaller than the Schoenberg-Cavaretta numbers $K(\mathbb{R}^+,\infty,k,n)$, Theorem 8.2 is a significant improvement over Theorem 8.1, provided that the conditions of Theorem 8.2 are satisfied.

Since the differentiation operator $D \equiv d/dt$ on the real line $\mathbb{R}$ is known to be m-dissipative on each of the classical Banach spaces

$$X_p := L^p(0,\infty) \text{ for all } p \in [1,\infty]$$

we have as a special case of Theorem 8.1 the inequalities, valid for all $p \in [1,\infty]$, and for all $f \in L^p(0.\infty)$ with $f^{(n)} \equiv D^n f \in L^p(0,\infty)$,

$$\left\|f^{(k)}\right\|_p^n \le K(\mathbb{R}^+,\infty,k,n)\,\|f\|_p^{n-k}\left\|f^{(n)}\right\|_p^k. \tag{8.5}$$

Let $K(D,\mathbb{R}^+,p,k,n)$ denote the best possible numbers in the inequalities (8.5), then the Kallman-Rota-Certain-Kurtz-Ditzian-Chernoff results show that

$$K(D,\mathbb{R}^+,p,k,n) \le K(\mathbb{R}^+,\infty,k,n) \text{ for all } p \in [1,\infty]. \tag{8.6}$$

Similarly, since both D and $-D$ are m-dissipative on the Banach spaces X_p, it follows from Theorem 8.2 that

$$K(D,\mathbb{R},p,k,n) \leq K(\mathbb{R},\infty,k,n) \text{ for all } p \in [1,\infty]. \tag{8.7}$$

The Kallman-Rota inequality has inspired the search for results analogous to the inequalities given above but now replacing the differentiation operator by the difference operator. For an extensive discussion of the properties of difference operators and Kallman-Rota type inequalities see the results in Kwong and Zettl [56, Chapter 4].

Let Δ denote the classical difference operator; for any sequence of real numbers $\mathbf{x} = \{x_n : n \in \mathbb{Z}^+ \text{ or } \mathbb{Z}\}$, where $\mathbb{Z}^+ = \{0,1,2,\ldots\}$ and $\mathbb{Z} = \{\ldots,-2,-1,0,1,2,\ldots\}$, the sequence $\Delta\mathbf{x}$ is defined by

$$\Delta\mathbf{x} := \{x_n - x_{n-1} : n \in \mathbb{Z}^+ \text{ or } \mathbb{Z}\}. \tag{8.8}$$

The product difference operator Δ^r is defined inductively by $\Delta^r := \Delta\left(\Delta^{r-1}\right)$ for $r \geq 2$.

In contrast with the differentiation operator $D \equiv d/dt$, considered above, Δ is a bounded operator on the sequence spaces ℓ^p, considered on $\mathbb{Z}^+$ or $\mathbb{Z}$, for all $p \in [1,\infty]$.

There are results analogous to Theorems 8.1 and 8.2, respectively, for the difference operator Δ; let k, n be positive integers satisfying $1 \leq k < n$ with $n \geq 2$: then for all $p \in [1,\infty]$,

$$\left\|\Delta^k\mathbf{x}\right\|_p^n \leq C(\Delta,\mathbb{Z}^+,p,k,n) \|\mathbf{x}\|_p^{n-k} \|\Delta^n\mathbf{x}\|_p^k \text{ for all } \mathbf{x} \in \ell^p(\mathbb{Z}^+) \tag{8.9}$$

and

$$\left\|\Delta^k\mathbf{x}\right\|_p^n \leq C(\Delta,\mathbb{Z},p,k,n) \|\mathbf{x}\|_p^{n-k} \|\Delta^n\mathbf{x}\|_p^k \text{ for all } \mathbf{x} \in \ell^p(\mathbb{Z}). \tag{8.10}$$

Here the symbols $C(\Delta,\mathbb{Z}^+,p,k,n)$ and $C(\Delta,\mathbb{Z},p,k,n)$ denote the optimal numbers for the given inequalities.

For all $p \in [1,\infty]$ it is shown in [56, Chapter 4] that the numbers involved in the derivative and difference inequalities, given above, satisfy

$$C(\Delta,\mathbb{Z}^+,p,k,n) \leq K(D,\mathbb{R}^+,\infty,k,n) \text{ for all } p \in [1,\infty]$$

and

$$C(\Delta,\mathbb{Z},p,k,n) \leq K(D,\mathbb{R},\infty,k,n) \text{ for all } p \in [1,\infty].$$

Further results for both the differentiation and difference inequalities are given in the text [56].

9. Inequalities associated with the Kallman-Rota inequality

In addition to the impressive array of inequalities that are directly related to the original Kallman-Rota inequality there is a number of inequalities indirectly related to Kallman-Rota.

9.1. HELP inequalities. Probably the most extensive collection of such inequalities is to be found under the heading of the so-called HELP inequalities; see the catalogue [3].

The original HELP inequality was inspired by the Hardy-Littlewood integral inequality of 1932, see Theorem 6.1, and the 1967 Kallman-Rota inequality, see Theorem 3.1, when considered in the environment of Hilbert spaces. The early HELP references are the 1972 paper of Everitt [20] and the extension in the 1982 paper of Evans and Everitt [18].

These early papers are concerned with the generation of HELP integral inequalities associated with the Sturm-Liouville differential equation

$$-(py')' + qy = \lambda wy \text{ on } [a,b) \subset \mathbb{R} \tag{9.1}$$

where the coefficients p, q, w satisfy the minimal conditions

$$\begin{cases} (i) & p, q, w : [a,b) \to \mathbb{R} \\ (ii) & p^{-1}, q, w \in L^1_{\text{loc}}[a,b) \\ (iii) & w > 0 \text{ almost everywhere on } [a,b). \end{cases} \tag{9.2}$$

The HELP inequality takes the form

$$\left(\int_a^b \left\{p\,|f'|^2 + q\,|f|^2\right\}\right)^2 \leq K \int_a^b w\,|f|^2 \int_a^b w\,\left|w^{-1}\left((pf')' - qf\right)\right|^2 \tag{9.3}$$

where the function f is restricted so that both terms on the right-hand side of (9.3) exist and are finite. The problem then is to decide if there is a positive number K, which depends only on the coefficients p, q, w, such that the inequality (9.3) holds for all such functions f.

The connection with the Hardy-Littlewood and Kallman-Rota inequalities is seen in the example

$$\begin{array}{ll} (i) & a = 0 \text{ and } b = \infty \\ (ii) & p = w = 1 \text{ and } q = 0 \text{ on } [0, \infty), \end{array}$$

when (9.3) reduces to

$$\left(\int_0^\infty |f'|^2\right)^2 \leq K \int_0^\infty |f|^2 \int_0^\infty |f''|^2. \tag{9.4}$$

With these restrictions it is seen, from the Hardy-Littlewood inequality, that the inequality (9.4) is valid with the best possible number 4 for K, together with the cases of equality given in Theorem 6.1.

In this example the HELP inequality is a special case of the Kallman-Rota inequality, and is also the driving example behind the Kato inequality. However, in general, the HELP inequality (9.3) cannot be seen as lying within the scope of the Kallman-Rota and Kato inequalities, even in the special circumstance when $q = 0$ and $w = 1$ on $[a, b)$; in this respect see the inequality (9.8) below.

9.2. Separation inequalities. For a recent survey of separation properties of ordinary differential expressions defined on the real line $\mathbb{R}$ see the paper by Brown and

Hinton [12]; earlier work in this area is due to Everitt, Giertz and Weidmann, see the references in [12].

For some inequalities associated with separation results see the paper [23]; in this paper the results arise from consideration of the Sturm-Liouville differential expression M defined by

$$M[f](x) := -((p(x)f'(x))' + q(x)f(x) \text{ for all } x \in [a,\infty), \tag{9.5}$$

where the coefficients p and q satisfy the conditions (9.2), together with the conditions $p(x)$ and $q(x) \geq 0$, and $w(x) = 1$ for all $x \in [a,\infty)$. Let the domain $D \subset L^2[a,\infty)$ be defined by

$$D := \{f : [a,\infty) \to \mathbb{C} : f \text{ and } M[f] \in L^2[a,\infty)\}. \tag{9.6}$$

One of the inequalities considered in [23, Section 8, Theorem 3] is of the form, for all $f \in D$,

$$A\int_a^\infty |(pf')'|^2 + B\int_a^\infty pq|f'|^2 + C\int_a^\infty q^2|f|^2 \leq \int_a^\infty |M[f]|^2 + D\int_a^\infty |f|^2 \tag{9.7}$$

where A, B, C, D are positive constants.

The interest for this inequality, in respect of the Kallman-Rota inequality, is that an additional condition on M is required for (9.7) to hold, *i.e.* there exists a positive number K such that

$$\left(\int_a^\infty p|f'|^2\right)^2 \leq K\int_a^\infty |f|^2 \int_a^\infty |(pf')'|^2 \tag{9.8}$$

for f and $(pf')' \in L^2[a,\infty)$. The connection between this required inequality (9.8) and the Kallman-Rota inequality proved to be illusive.

Acknowledgements

Both authors thank Professor J. A. Goldstein and Professor Dr. Michael Plum for advice and help at a number of stages in the preparation of this survey paper.

Norrie Everitt gives his special thanks to Professor Norton Starr, an editor for this volume, for his patience and understanding over a period of some months during the writing of this contribution to the volume, and for his careful reading of a number of drafts of the manuscript.

References

[1] C. D. Ahlbrandt, Variational inequalities. In [21], pp. 1–19.

[2] P. R. Beesack, A simpler proof of two inequalities of Brodlie and Everitt. *Proc. Roy. Soc. Edinburgh Sect. A* **84** (1979), 259–261.

[3] M. Benammar, C. Bennewitz, M. J. Beynon, B. M. Brown, N. G. J. Dias, W. D. Evans, W. N. Everitt, V. G. Kirby, and L. L. Littlejohn, A catalogue of HELP and HELP-type integral and series inequalities. In G. V. Milovanović, ed., *Recent Progress in Inequalities*, Kluwer, Dordrecht, 1998, pp. 127–160.

[4] P. Böhm, *Die Landau-Ungleichung für Generatoren stark stetiger Halbgruppen.* Diplomarbeit der Facultät für Mathematik an der Eberhard-Karls-Universität Tübingen, März, 1999.

[5] B. Bollobás, The spatial numerical range and powers of an operator. *J. London Math. Soc.* (2) **7** (1974), 435–440.

[6] B. Bollobás and J. R. Partington, Inequalities for quadratic polynomials in Hermitian and dissipative operators. *Adv. in Math.* **51** (1984), 271–280.
[7] J. S. Bradley and W. N. Everitt, Inequalities associated with regular and singular problems in the calculus of variations. *Trans. Amer. Math. Soc.* **182** (1973), 303–321.
[8] J. S. Bradley and W. N. Everitt, On the inequality $\|f''\|^2 \leq K\|f\|\,\|f^{(4)}\|$. *Quart. J. Math. Oxford Ser.* (2) **25** (1974), 241–252.
[9] J. S. Bradley and W. N. Everitt, A singular integral inequality on a bounded interval. *Proc. Amer. Math. Soc.* **61** (1976), 29–35.
[10] K. W. Brodlie and W. N. Everitt, On an inequality of Hardy and Littlewood. *Proc. Roy. Soc. Edinburgh Sect. A* **72** (1975), 179–186.
[11] R. C. Brown and D. B. Hinton, Interpolation inequalities with power weights for functions of one variable. *J. Math. Anal. Appl.* **172** (1993), 233–242.
[12] R. C. Brown and D. B. Hinton, Two separation criteria for second order ordinary or partial differential operators. *Mathematica Bohemica.* **124** (1999), 273–292.
[13] P. S. Bullen, *A Dictionary of Inequalities*. Longman, Harlow, 1998.
[14] M. W. Certain and T. G. Kurtz, Landau-Kolmogorov inequalities for semi-groups and groups. *Proc. Amer. Math. Soc.* **63** (1977), 226–230.
[15] P. R. Chernoff, Optimal Landau-Kolmogorov inequalities for dissipative operators in Hilbert and Banach spaces. *Adv. in Math.* **34** (1979), 137–144.
[16] Z. Ditzian, Some remarks on inequalities of Landau and Kolmogorov. *Aequationes Math.* **12** (1975), 145–151.
[17] Z. Ditzian, A new fractional form of the Landau-Kolmogorov inequality. *J. Math. Anal. Appl.* **148** (1990), 528–537.
[18] W. D. Evans and W. N. Everitt, A return to the Hardy-Littlewood integral inequality. *Proc. Roy. Soc. London Ser. A* **380** (1982), 447–486.
[19] W. D. Evans, W. N. Everitt, W. K. Hayman, and D. S. Jones, Five integral inequalities; an inheritance from Hardy and Littlewood. *J. Inequal. Appl.* **2** (1998), 1–36.
[20] W. N. Everitt, On an extension to an integro-differential inequality of Hardy, Littlewood and Pólya. *Proc. Roy. Soc. Edinburgh Sect. A* **69** (1971/1972), 295–333.
[21] W. N. Everitt, ed., *Inequalities: Fifty Years on from Hardy, Littlewood and Pólya*. Marcel Dekker, New York, 1991.
[22] W. N. Everitt, Analytic and numerical aspects of the HELP integral inequality. In B. D. Sleeman and R. J. Jarvis, eds., *Ordinary and Partial Differential Equations IV; Proceedings of the Twelfth Dundee Conference, 1992, in honour of Professor D. S. Jones*, Longman, Harlow, 1993, pp. 19–48.
[23] W. N. Everitt and M. Giertz, Some inequalities associated with certain ordinary differential operators. *Math. Zeit.* **126** (1972), 308–326.
[24] W. N. Everitt and M. Giertz, On the integro-differential inequality $\|f'\|_2^2 \leq K\|f\|_p\,\|f''\|_q$. *J. Math. Anal. Appl.* **45** (1974), 639–653.
[25] W. N. Everitt and A. P. Guinand, On a Hardy-Littlewood type integral inequality with a monotonic weight function. In W. Walter, ed., *General Inequalities 5 (Oberwolfach, 1986)*, Internat. Schriftenreihe Numer. Math., 80, Birkhäuser, Basel, 1987, pp. 29–63.
[26] W. N. Everitt and D. S. Jones, On an integral inequality. *Proc. Roy. Soc. London Ser. A* **357** (1977), 271–288.
[27] Z. M. Franco, H. G. Kaper, M. K. Kwong, and A. Zettl, Best constants in norm inequalities for derivatives on a half-line. *Proc. Roy. Soc. Edinburgh Sect. A* **100** (1985), 67–84.
[28] H. A. Gindler and J. A. Goldstein, Dissipative operator versions of some classical inequalities. *J. Analyse Math.* **28** (1975), 213–238.
[29] H. A. Gindler and J. A. Goldstein, Dissipative operators and series inequalities. *Bull. Austral. Math. Soc.* **23** (1981), 429–442.
[30] J. A. Goldstein, *Semigroups of Linear Operators and Applications*. Oxford University Press, New York, 1985.
[31] J. A. Goldstein and F. Räbinger, On Hardy-Landau-Littlewood inequalities. In *Semesterbericht Funktionanalysis: Operator Semi-groups and Evolution Equations*, Mathematisches Institut, Eberhard-Karls-Universität, Tübingen, 1989, 61–75.

[32] J. S. Hadamard, Sur le module maximum d'une fonction et de ses dérivées. *Comptes Rendus des Séances de la Société Mathematique de France*, (1914), 68–72.

[33] G. H. Hardy and J. E. Littlewood, Some integral inequalities connected with the calculus of variations. *Quart. J. Math. Oxford* **3** (1932), 241–252.

[34] G. H. Hardy, J. E. Littlewood, and G. Pólya, *Inequalities*. Cambridge University Press, Cambridge, 1934.

[35] E. Hille, Remarks on the Landau-Kallman-Rota inequality. *Aequationes Math.* **4** (1970), 239–240.

[36] E. Hille, Generalizations of Landau's inequality to linear operators. In *Linear Operators and Approximation, Proc. Conf., Oberwolfach, 1971*, Internat. Ser. Numer. Math., 20, *Birkhäuser, Basel*, 1972, 20–32.

[37] E. Hille, On the Landau-Kallman-Rota inequality. *J. Approx. Theory* **6** (1972), 117–122.

[38] E. Hille, *Methods in Classical and Functional Analysis*, Addison-Wesley, London, 1972.

[39] E. Hille and R. S. Phillips, *Functional Analysis and Semi-Groups*. Revised edition, Amer. Math. Soc., Providence, RI, 1957.

[40] J. A. R. Holbrook, A Kallman-Rota inequality for nearly Euclidean spaces. *Adv. in Math.* **14** (1974), 335–345.

[41] R. J. Hughes, Hardy-Landau-Littlewood inequalities for fractional derivatives in weighted L^p spaces. *J. London Math. Soc.* (2) **35** (1987), 489–498.

[42] R. R. Kallman, and G.-C. Rota, On the inequality $\|f'\|^2 \leq 4\|f\|\,\|f''\|$. In O. Shisha, ed., *Inequalities, II*, Academic Press, New York, 1970, pp. 187–192. Reprinted in this chapter.

[43] T. Kato, On an inequality of Hardy, Littlewood and Pólya. *Adv. in Math.* **7** (1971), 217–218.

[44] T. Kato, The Cauchy problem for quasi-linear symmetric hyperbolic systems. *Arch. Rational Mech. Anal.* **58** (1975), 181–205.

[45] T. Kato, *Perturbation Theory for Linear Operators*. Corrected printing of the second edition, Springer-Verlag, Heidelberg, 1980.

[46] A. N. Kolmogorov, On inequalities between the upper bounds of the successive derivatives of an arbitrary function on an infinite interval. *Amer. Math. Soc. Translations* (1) **2** (1962), 233–243.

[47] H. Kraljević and S. Kurepa, Semigroups on Banach spaces. *Glasnik Mat. Ser.* III **5(25)** (1970), 109–117.

[48] N. P. Kupcov, Kolmogorov estimates for derivatives in $L_2(0,\infty)$. *Proc. Steklov Inst. Math.* **138** (1975), 94–117. *Amer. Math. Soc. Translations* (1977), 101–125.

[49] S. Kurepa, Remarks on the Landau inequality. *Aequationes Math.* **4** (1970), 240–241.

[50] M. K. Kwong and A. Zettl, Remarks on best constants for norm inequalities among powers of an operator. *J. Approx. Theory* **26** (1979), 249–258.

[51] M. K. Kwong and A. Zettl, Norm inequalities for dissipative operators on inner product spaces. *Houston J. Math.* **5** (1979), 543–557.

[52] M. K. Kwong and A. Zettl, An extension of the Hardy-Littlewood inequality. *Proc. Amer. Math. Soc.* **77** (1979), 117–118.

[53] M. K. Kwong and A. Zettl, Ramifications of Landau's inequality. *Proc. Roy. Soc. Edinburgh Sect. A* **86** (1980), 175–212.

[54] M. K. Kwong and A. Zettl, Norm inequalities of product form in weighted L^p spaces. *Proc. Roy. Soc. Edinburgh Sect. A* **89** (1981), 293–307.

[55] M. K. Kwong and A. Zettl, Norm inequalities for derivatives and differences. In [21], pp. 91–121.

[56] M. K. Kwong and A. Zettl, *Norm Inequalities for Derivatives and Differences*. Springer, Berlin, 1992.

[57] E. Landau, Einige Ungleichungen für zweimal differenzierbare Funktionen. *Proc. London Math. Soc.* **13** (1913), 43–49.

[58] Ju. I. Ljubič, On inequalities between powers of a linear operator. *Izv. Akad. Nauk SSSR Ser. Mat.* **24** (1960), 825–864. *Amer. Math. Soc. Translations* **40** 2 (1964), 39–84.

[59] D. S. Mitrinović, J. E. Pečarić, and A. M. Fink, *Inequalities Involving Functions and their Integrals and Derivatives*. Kluwer, Dordrecht, 1991.

[60] L. Nirenberg, Remarks on strongly elliptic partial differential equations. *Comm. Pure Appl. Math.* **8** (1955), 649–675.

[61] S. V. Parter and M. Steuerwalt, Block iterative methods for elliptic finite element equations. *SIAM J. Numer. Anal.* **22** (1985), 146–179.

[62] J. R. Partington, Hadamard-Landau inequalities in uniformly convex spaces. *Math. Proc. Cambridge Philos. Soc.* **90** (1981), 259–264.

[63] V. Q. Phóng, On inequalities for powers of linear operators and for quadratic forms. *Proc. Roy. Soc. Edinburgh Sect. A* **89** (1981), 25–50.

[64] M. H. Protter, Inequalities for powers of unbounded operators. *Manuscripta Math.* **28** (1979), 71–79.

[65] J. M. Rassias, Landau's type inequalities. *J. Math. Anal. Appl.* **202** (1996), 280–301.

[66] R. M. Redheffer, Über eine beste Ungleichung zwischen den Normen von f, f', f''. *Math. Zeit.* **80** (1963), 390–397.

[67] R. M. Redheffer and W. Walter, Inequalities involving derivatives. *Pacific J. Math.* **85** (1979), 165–178.

[68] I. J. Schoenberg, The elementary cases of Landau's problem of inequalities between derivatives. *Amer. Math. Monthly* **80** (1973), 125–158.

[69] I. J. Schoenberg and A. Cavaretta, Solution of Landau's problem concerning higher derivatives on the halfline. *Constructive Theory of Functions (Proc. Internat. Conf., Varna, 1970)*, Izdat. Bolgar. Akad. Nauk, Sofia, 1972, pp. 297–308.

[70] J. A. Siddiqi and A. El Koutri, Norm inequalities for generators of analytic semigroups and cosine operator functions. *Canad. Math. Bull.* **32** (1989), 47–53.

[71] W. Trebels and U. Westphal, A note on the Landau-Kallman-Rota-Hille inequality. In *Linear Operators and Approximation (Proc. Conf., Oberwohlfach, 1971)*, Internat. Ser. Numer. Math., 20, Birkhäuser, Basel, 1972, pp. 115-119.

[72] K. Yoshida. *Functional Analysis*. Sixth edition, Springer, Berlin and New York, 1980.

CHAPTER 6

Geometry Probability and Profinite Combinatorics

Totally Invariant Set Functions of Polynomial Type

BEIFANG CHEN

AND

GIAN-CARLO ROTA

Massachusetts Institute of Technology

Dedicated to Henry McKean on his sixtieth birthday.

1. Introduction

It has long been known (though not well known) that volume is not the only measure in ordinary n-dimensional Euclidean space that is invariant under rigid motions. A celebrated result of Hadwiger states that the space of measures invariant under rigid motions, defined on the family of all polyconvex sets (= finite unions of compact convex sets) and satisfying a mild continuity condition (v. below) is of dimension $n + 1$.

Of all bases of the vector space of invariant measures, the best known, and probably the most natural, is provided by the elementary mixed volumes (quermassintegrale). Following the exposition of the theory given by one of us (see [5]), these invariant measures can be defined as follows. For any integer k between 0 and n, define the measure μ_k of a rectangular parallelotope P with sides $x_1, x_2, \ldots, x_n$ to be the k-th elementary symmetric function of the numbers $x_1, x_2, \ldots, x_n$, that is,

$$\mu_k(P) = e_k(x_1, x_2, \ldots, x_n) = \sum_{i_1 < i_2 \cdots < i_k} x_{i_1} x_{i_2} \cdots x_{i_k} .$$

It is then shown, by a very indirect argument, that this set function defined on all rectangular parallelotopes has a unique extension to the family of all polyconvex sets in $\mathbb{R}^n$, and one thereby obtains the elementary mixed volumes. For k = 0, one obtains the Euler characteristic.

These measures have been defined in several other ways in the literature, but the present definition shows the strict link between the elementary mixed volumes and the elementary symmetric functions.

From this point of view, it is natural to ask whether symmetric functions other than the elementary symmetric functions can be associated with set functions defined on the family of polyconvex sets. Such set functions need not be additive; that is, they need not be measures.

The literature on non-additive set functions is particularly wanting; not only are the basic definitions missing, but no clearly focused set of problems has been put forward, to the best of our knowledge. The purpose of this note is to single out a class of non-additive real-valued set functions defined on polyconvex sets which is the natural generalization of the elementary mixed volumes. Our objec-

Communications on Pure and Applied Mathematics, Vol. XLVII, 187–197 (1994)

tive is twofold: First, we wish to define a natural concept of a "polynomial" set function, one which, by suitable polarization, can be made to correspond to a sum of multilinear set functions. Second, we define a concept of invariance for such non-additive set functions (which we call total invariance) which turns out to be the correct analog of invariance under rigid motions. Having thus defined the two basic concepts, we proceed to prove that the class of polynomial totally invariant set functions is naturally projected to the vector space of symmetric polynomials in n variables. More precisely, given a symmetric function $f(x_1, x_2, \ldots, x_n)$ of the real variables $x_1, x_2, \ldots, x_n$, we show that the set function taking the value $f(x_1, x_2, \ldots, x_n)$ on a rectangular parallelotope of sides $x_1, x_2, \ldots, x_n$ has extensions to the family of all polyconvex sets which is a totally invariant set function, and, conversely, given a totally invariant continuous set function of polynomial type, there is a unique symmetric function which is obtained by evaluating it on rectangular parallelotopes.

In closing, we show by an example that the obvious notion of set function which is merely invariant under rigid motions (rather than totally invariant in our sense) is insufficient for present purposes.

We hope the present note will stimulate interest in the study of non-additive set functions and their applications in integral geometry and geometric combinatorics.

2. Linear Forms

We recall some well-known facts about multilinear functionals and linear forms, for which we have found no good reference in the literature.

Definition 2.1. Let V be a vector space over the field $\mathbb{R}$ of real numbers. A continuous nonzero functional f on V is called a linear n-form ($n \geqq 1$) if

(i) $f(\lambda v) = \lambda^n f(v)$,

(ii) $f(v_1 + \cdots + v_{n+1}) = \sum_{k=1}^{n}(-1)^{n+k} \sum_{1 \leqq i_1 < \cdots < i_k \leqq n+1} f(v_{i_1} + \cdots + v_{i_k})$,

where $v, v_1, \ldots, v_n, v_{n+1} \in V$, and $\lambda \in \mathbb{R}$.

For each functional f on a vector space V, we associate a symmetric functional φ_f defined on $\underbrace{V \times \cdots \times V}_{m}$ by

$$\varphi_f(v_1, \ldots, v_m) = \frac{(-1)^m}{m!} \sum_{k=1}^{m} (-1)^k \sum_{i_1 < \cdots < i_k} f(v_{i_1} + \cdots + v_{i_k}) . \tag{2.1}$$

With this association, linear forms are closely related to symmetric multilinear functionals. The following propositions reveal such relationships.

Proposition 2.2. *A continuous functional f on a vector space V satisfies condition* (ii) *in Definition* 2.1 *if and only if its associated symmetric functional φ_f,* *which is defined by* (2.1), *is multilinear when* $m = n$ *and* φ_f *is identically zero when* $m > n$.

Proof: It can be routinely checked by using the definition of φ_f that

$$\begin{aligned}(m+1)\varphi_f(v_0, v_1, \ldots, v_m) &= \varphi_f(v_0 + v_1, v_2, \ldots, v_m) \\ &\quad -\varphi_f(v_0, v_2, \ldots, v_m) \\ &\quad -\varphi_f(v_1, v_2, \ldots, v_m)\,.\end{aligned}$$

Note that condition (ii) in Definition 2.1 is equivalent to

$$\varphi_f(v_1, \ldots, v_{n+1}) = 0\,.$$

Now it is easily followed from the above relations that (ii) is equivalent to the additivity of $\varphi_f(v_1, \ldots, v_n)$ in each of its variables and $\varphi_f(v_1, \ldots, v_m) = 0$ for $m > n$. Note that φ_f is also continuous. Hence the multilinearity of $\varphi_f(v_1, \ldots, v_n)$ is equivalent to its additivity in each of its variables.

PROPOSITION 2.3. *For each nonzero symmetric multilinear functional* φ *in* n *variables on a vector space* V, *define the functional* f_φ *on* V *by*

$$f_\varphi(v) = \varphi(\underbrace{v, \ldots, v}_{n})\,. \tag{2.2}$$

Then f_φ *is a linear* n*-form, and* φ *is the same as* φ_{f_φ} *which is defined by* (2.1).

Proof: It is obvious that $f_\varphi(\lambda v) = \lambda^n f_\varphi(v)$ for $v \in V$ and $\lambda \in \mathbb{R}$. Let $v_1, \ldots, v_r \in V$, and let $v = v_1 + \cdots + v_r$. Then

$$\begin{aligned}f_\varphi(v) &= \sum_{i_1, \ldots, i_n = 1}^{r} \varphi(v_{i_1}, \ldots, v_{i_n}) \\ &= \sum_{|(k)|=n} \frac{n!}{(k)!}\varphi(v_1, k_1; \cdots; v_r, k_r)\,,\end{aligned} \tag{2.3}$$

where $(k) = (k_1, \ldots, k_r)$, $k_i \geqq 0$, $|(k)| = k_1 + \cdots + k_r$, $(k)! = k_1! \cdots k_r!$. Set $u = v_1 + \cdots + v_n$, we have

$$\begin{aligned}n!\varphi(v_1, \ldots, v_n) &= f_\varphi(u) - \sum_{\substack{|(k)|=n \\ (k)\neq(1,\ldots,1)}} \frac{n!}{(k)!}\varphi(v_1, k_1; \cdots; v_n, k_n) \\ &= f_\varphi(u) - \sum_i \sum_{\substack{|(k)|=n \\ k_i=0}} \frac{n!}{(k)!}\varphi(v_1, k_1; \cdots; v_n, k_n) \\ &\quad + \sum_{i<j} \sum_{\substack{|(k)|=n \\ k_i, k_j=0}} \frac{n!}{(k)!}\varphi(v_1, k_1; \cdots; v_n, k_n)\end{aligned}$$

$$-\sum_{i<j<k}\sum_{\substack{|(k)|=n\\ k_i,k_j,k_k=0}}\frac{n!}{(k)!}\varphi(v_1,k_1;\cdots;v_n,k_n)+\cdots$$

Therefore

$$\varphi(v_1,\ldots,v_n)=\frac{1}{n!}f_\varphi(v_1+\cdots+v_n)$$
$$+\frac{1}{n!}\sum_{k=1}^{n-1}(-1)^k\sum_{i_1<\cdots<i_k}f_\varphi(v_1+\cdots+\widehat{v_{i_1}}+\cdots+\widehat{v_{i_k}}+\cdots+v_n)\,.$$

This identity and (2.1) show that $\varphi=\varphi_{f_\varphi}$ when $m=n$ and $\varphi_{f_\varphi}=0$ when $m>n$. So f_φ is a linear n-form.

PROPOSITION 2.4. *For each nonzero linear n-form f on a vector space V, its associated symmetric functional φ_f, which is defined by* (2.1), *is multilinear when $m=n$, and f is the same as f_{φ_f}. Moreover, for all $v\in V$,*

$$\varphi_f(\underbrace{v,\ldots,v}_{n})=f(v)\,.$$

Proof: It follows from Proposition 2.2 that φ_f is a symmetric multilinear functional. Let $v_1=\cdots=v_n=v$. We have by (i) and (2.1)

$$\varphi_f(\underbrace{v,\ldots,v}_{n})=\frac{(-1)^n}{n!}\sum_{k=1}^{n}(-1)^k\binom{n}{k}k^nf(v)\,.$$

It suffices to show that

$$\frac{(-1)^n}{n!}\sum_{k=1}^{n}(-1)^k\binom{n}{k}k^n=1\,. \tag{2.4}$$

Note that the symmetric multilinear functional φ_f defines another linear n-form f_{φ_f} by Proposition 2.3. Thus (2.1) is also satisfied by φ_f and f_{φ_f}. Similarly, for all $v\in V$,

$$\varphi_f(\underbrace{v,\ldots,v}_{n})=\frac{(-1)^n}{n!}\sum_{k=1}^{n}(-1)^k\binom{n}{k}k^nf_{\varphi_f}(v)\,,$$

$$f_{\varphi_f}(v)=\varphi_f(v,\ldots,v)\,.$$

Identity (2.4) follows immediately. Obviously, $f=f_{\varphi_f}$.

Identities (2.1) and (2.2) are basic relations between linear n-forms and symmetric multilinear functionals. A clearer relation between them can be demonstrated by Möbius function of Boolean algebras. Let X be a multiset on V. Suppose X

is labeled, then X is a finite set. Consider the incidence algebra of the power set $\mathcal{P}(X)$ of X, ordered by inclusion. For each $S \subset X$, define

$$f(S) = f\left(\sum_{v \in S} v\right).$$

If f is a linear n-form, then by Propositions 2.3 and 2.4, $f(S)$ can be written as

$$f(S) = \sum_{T \subset S} \sum_{\substack{T \leq M \\ |M|=n}} \binom{n}{M} \varphi(M), \tag{2.5}$$

where the first sum runs over all subsets T of S, the second sum means that M runs over all multisets on T such that each element of T appears in M and $|M| = n$, i.e., M is a positive integer-valued function on T such that $\sum_{v \in T} M(v) = n$, $\binom{n}{M}$ denotes the multinomial coefficient $\binom{n}{M(v_1),\ldots,M(v_t)}$ whenever $T = \{v_1, \ldots, v_t\}$, and $\varphi(M) = \varphi(v_1, M(v_1); \cdots; v_t, M(v_t))$. We now define for each $S \subset X$

$$g(S) = \sum_{\substack{S \leq M \\ |M|=n}} \binom{n}{M} \varphi(M). \tag{2.6}$$

By the Möbius function $\mu(T, S) = (-1)^{|S-T|}$ of $\mathcal{P}(X)$, we have

$$g(S) = \sum_{T \subset S} (-1)^{|S-T|} f(T). \tag{2.7}$$

The combination of identities (2.6) and (2.7) gives the following proposition.

PROPOSITION 2.5. *Let f be an n-form on a vector space V and φ_f its associated symmetric multilinear function. Then for any $v_1, \ldots, v_s \in V$,*

$$\begin{aligned} &\sum_{k=1}^{s} (-1)^{s-k} \sum_{i_1 < \cdots < i_k} f(v_{i_1} + \cdots + v_{i_k}) \\ &\qquad = \sum_{\substack{l_1 + \cdots + l_s = n \\ l_1, \ldots, l_s \geq 1}} \frac{n!}{l_1! \cdots l_s!} \varphi_f(v_1, l_1; \cdots; v_s, l_s). \end{aligned} \tag{2.8}$$

Let $v_i = x_i v$ in (2.8), $1 \leqq i \leqq s$, we have

COROLLARY 2.6.

$$\sum_{k=1}^{s} (-1)^{s-k} \sum_{i_1 < \cdots < i_k} (x_{i_1} + \cdots + x_{i_k})^n = \sum_{\substack{l_1 + \cdots + l_s = n \\ l_1, \ldots, l_s \geq 1}} \frac{n!}{l_1! \cdots l_s!} x_1^{l_1} \cdots x_s^{l_s}.$$

3. Totally Invariant Set Functions

Let $\mathscr{K}^n$ be the set of all compact convex sets of $\mathbb{R}^n$. Let $L(\mathscr{K}^n)$ be the lattice generated by $\mathscr{K}^n$ by taking finite unions. The elements of $L(\mathscr{K}^n)$ are said to be *polyconvex sets*. By a *set function* on $\mathbb{R}^n$ we mean a real-valued function defined on $L(\mathscr{K}^n)$. A set function ν is said to be *totally invariant* if $\nu(\emptyset) = 0$ and

$$\nu(K_1 \cup K_2 \cup g(K_1 \cap K_2) \cup g_3K_3) = \nu(g_1K_1 \cup g_2K_2 \cup g_3K_3)$$

for all $K_1, K_2, K_3 \in L(\mathscr{K}^n)$, where g, g_1, g_2, g_3 are rigid motions of $\mathbb{R}^n$ such that $K_1 \cup K_2, g(K_1 \cap K_2), g_1K_1, g_2K_2, g_3K_3$ are pairwise disjoint. The idea of total invariance is that the value $\nu(K_1 \cup K_2 \cup g(K_1 \cap K_2) \cup g_3K_3)$ shall be independent of the positions where the polyconvex sets K_1, K_2, and K_3 are placed in $\mathbb{R}^n$, provided that the sets K_1, K_2, and K_3 are pairwise disjoint, or their overlaped part is placed disjointly.

An *invariant measure* μ on $\mathbb{R}^n$ is a set function such that

$$\mu(gK_1) = \mu(K_1), \qquad \mu(\emptyset) = 0,$$
$$\mu(K_1 \cup K_2) = \mu(K_1) + \mu(K_2) - \mu(K_1 \cap K_2)$$

for $K_1, K_2 \in L(\mathscr{K}^n)$, and for any rigid motion g. Invariant measures are examples of totally invariant set functions. An invariant *r-multimeasure* ν on $\mathbb{R}^n$ is a real-valued function on $\underbrace{L(\mathscr{K}^n) \times \cdots \times L(\mathscr{K}^n)}_{r}$ such that for any fixed polyconvex sets $K_1, \ldots, K_r$, the set function $\nu(K_1, \ldots, K_{i-1}, \cdot, K_{i+1}, \ldots, K_r)$ is an invariant measure on $L(\mathscr{K}^n)$.

It is well known that there is a metric d on $\mathscr{K}^n$, called the *Hausdorff metric*, defined by

$$d(K, K') = \max\left\{\sup_{x\in K} d(x, K'), \sup_{x'\in K'} d(x', K)\right\},$$

where $d(x, K) = \inf_{y\in K} d(x, y)$. A set function ν on $L(\mathscr{K}^n)$ is said to be *continuous* if $\nu(K_m)$ converges to $\nu(K)$ whenever the sequence $\{K_m\}$ of compact convex sets converges to the compact convex set K under the Hausdorff metric.

We shall consider totally invariant real-valued continuous set functions of *polynomial type* defined as the following.

Definition 3.1. A totally invariant, continuous set function ν on $\mathbb{R}^n$ is said to be homogeneous of degree r ($r \geqq 1$) if for each positive integer k,

$$\nu(g_1K \cup \cdots \cup g_kK) = k^r \nu(K) \tag{3.1}$$

for any $K \in \mathscr{K}^n$ and for any choice of rigid motions $g_1, \ldots, g_k$ of $\mathbb{R}^n$ such that the sets $g_1K, \ldots, g_kK$ are pairwise disjoint. Moreover,

$$\nu(K_1 \cup \cdots \cup K_{r+1}) = \sum_{k=1}^{r}(-1)^{r+k} \sum_{1\leqq i_1<\cdots<i_k\leqq r+1} \nu(K_{i_1} \cup \cdots \cup K_{i_k}) \tag{3.2}$$

if $K_1, \dots, K_{r+1}$ are pairwise disjoint compact convex sets.

A continuous invariant set function ν is said to be of polynomial type if it is a linear combination of homogeneous set functions.

By a *parallelotope* P of sides $x_1, \dots, x_n$ we mean the image of the set

$$\{(a_1, a_2, \dots, a_n) : 0 \leqq a_i \leqq x_i, i = 1, 2, \dots, n\}$$

under a rigid motion of $\mathbb{R}^n$. For any parallelotope P of sides $x_1, x_2, \dots, x_n$, set

$$\mu_k(P) = e_k(x_1, x_2, \dots, x_n)\,,$$

where e_k $(k = 0, 1, \dots, n)$ is the k-th elementary symmetric polynomial and $e_0 \equiv 1$. It can be shown (e.g., Rota, [5]) that μ_k can be uniquely extended to a continuous invariant measure on $\mathscr{K}^n$. A constructive definition of μ_k up to a constant for bounded polyhedra by polyhedral curvatures can be found in [1]. It is interesting that the 0-dimensional measure μ_0 is just the Euler characteristic which is a topological invariant. Hadwiger characterized continuous invariant measures on $\mathscr{K}^n$ by showing that every such a measure is a linear combination of μ_k $(k = 0, 1, \dots, n)$.

We come now to our main result.

THEOREM 3.2. *For every totally invariant continuous set function of polynomial type ν on $L(\mathscr{K}^n)$, there exists a unique symmetric polynomial f of n variables such that for any parallelotope P of sides $x_1, x_2, \dots, x_n$,*

$$\nu(P) = f(x_1, x_2, \dots, x_n)\,. \tag{3.3}$$

Conversely, for every symmetric polynomial f of n variables, there exist totally invariant continuous set functions ν of polynomial type such that (3.3) *is satisfied.*

Proof: We first prove that our condition is sufficient. We may assume that ν is of degree r. Let φ_ν be the symmetric set function defined on

$$\underbrace{L(\mathscr{K}^n) \times \cdots \times L(\mathscr{K}^n)}_{r}$$

as

$$\varphi_\nu(K_1, \dots, K_r) = \frac{(-1)^r}{r!} \sum_{k=1}^{r} (-1)^k \sum_{i_1 < \cdots < i_k} \nu(g_{i_1} K_{i_1} \cup \cdots \cup g_{i_k} K_{i_k})\,, \tag{3.4}$$

where $g_1, \dots, g_r$ are rigid motions of $\mathbb{R}^n$ such that the sets $g_1 K_1, \dots, g_r K_r$ are pairwise disjoint. If K_0 and K_1 are disjoint, then it can be shown that

$$\begin{aligned}(s+1)\varphi_\nu(K_0, K_1, K_2, \dots, K_s) &= \varphi_\nu(K_0 \cup K_1, K_2, \dots, K_s) \\ &\quad - \varphi_\nu(K_0, K_2, \dots, K_s) \\ &\quad - \varphi_\nu(K_1, K_2, \dots, K_s)\end{aligned}$$

for $s \geqq r$. Thus it follows from (3.2) that

$$\varphi_\nu(K_0 \cup K_1, K_2, \ldots, K_r) = \varphi_\nu(K_0, K_2, \ldots, K_r) + \varphi_\nu(K_1, K_2, \ldots, K_r) .$$

Let g be the rigid motion of $\mathbb{R}^n$ such that $g(K_0 \cap K_1)$ is disjoint from $K_0 \cup K_1$. Therefore the total invariance condition of φ_ν implies that

$$\varphi_\nu(K_0 \cup K_1 \cup g(K_0 \cap K_1), K_2, \ldots, K_r) = \varphi_\nu(g_0 K_0 \cup g_1 K_1, K_2, \ldots, K_r) ,$$

where g_0 and g_1 are rigid motions such that $g_0 K_0$ and $g_1 K_1$ are disjoint. Hence

$$\begin{aligned}&\varphi_\nu(K_0 \cup K_1, K_2, \ldots, K_r)\\ &\quad = \varphi_\nu(K_0, K_2, \ldots, K_r) + \varphi_\nu(K_0, K_2, \ldots, K_r) - \varphi_\nu(K_0 \cap K_1, K_2, \ldots, K_r) .\end{aligned}$$

This shows that the associated set function φ_ν is a symmetric r-multimeasure. It is obvious that φ_ν is invariant and continuous in each of its variables. By Hadwiger's theorem characterizing invariant continuous measures on $\mathscr{K}^n$ (see [2] and [5]), there exist unique real constants $a_{i_1 \cdots i_r}$ such that for $K_1, \ldots, K_r \in \mathscr{K}^n$,

$$\varphi_\nu(K_1, \ldots, K_r) = \sum_{i_1, \ldots, i_r = 0}^{n} a_{i_1 \cdots i_r} \mu_{i_1}(K_1) \cdots \mu_{i_r}(K_r) .$$

Let f be the symmetric polynomial of $x = (x_1, \ldots, x_n)$ defined as

$$f(x) = \sum_{i_1, \ldots, i_r = 0}^{n} a_{i_1 \cdots i_r} e_{i_1}(x) \cdots e_{i_r}(x) .$$

It follows from (3.4) that

$$\varphi_\nu(P, \ldots, P) = \frac{(-1)^n}{n!} \sum_{k=1}^{n} \binom{n}{k} k^n \nu(P) = \nu(P) .$$

Thus

$$\begin{aligned}\nu(P) &= \sum_{i_1, \ldots, i_r = 0}^{n} a_{i_1 \cdots i_r} \mu_{i_1}(P) \cdots \mu_{i_r}(P)\\ &= \sum_{i_1, \ldots, i_r = 0}^{n} a_{i_1 \cdots i_r} e_{i_1}(x) \cdots e_{i_r}(x) .\end{aligned}$$

The uniqueness of f is obvious.

Conversely, given any symmetric polynomial $f(x)$ of n variables, it can be written as a polynomial $p(e_1, \ldots, e_n)$ of the elementary symmetric polynomials e_k ($k = 0, 1, \ldots, n$). Let $p(e_1, \ldots, e_n)$ be written as a linear combination $\sum_r a_r p_r$ of homogeneous polynomials p_r of degree r, where $p_0 = e_0$. For each p_r ($r > 0$), define a set function ν_r by

$$\nu_r(K) = p_r(\mu_1(K), \ldots, \mu_n(K))$$

for $K \in L(\mathcal{K}^n)$. Let p_r be written as the following form

$$p_r = \sum_{j_1,\ldots,j_r=0}^{n} b_{j_1\cdots j_r} e_{j_1} \cdots e_{j_r} \,,$$

where the coefficients $b_{j_1\cdots j_r}$ are chosen to be invariant under permutations of $(1, \ldots, r)$. It is obvious that the set function

$$\varphi_r(K_1, \ldots, K_r) = \sum_{j_1,\ldots,j_r=0}^{n} b_{j_1\cdots j_r} \mu_{j_1}(K_1) \cdots \mu_{j_r}(K_r)$$

is an invariant, continuous symmetric r-multimeasure on $L(\mathcal{K}^n)$. Let $g_1, \cdots, g_k$ be rigid motions such that the sets $g_1K, \ldots, g_kK$ are pairwise disjoint. Then

$$\nu_r(g_1K \cup \cdots \cup g_kK) = k^r \nu_r(K) \,.$$

Note that φ_r can be extended to a multilinear functional on the vector space generated by indicator functions of polyconvex sets, so ν_r is a symmetric r-form by Proposition 2.3. Hence

$$\varphi_r(1_{K_1}, \ldots, 1_{K_r}) = \frac{(-1)^r}{r!} \sum_{k=1}^{r} (-1)^k \sum_{i_1<\cdots<i_k} \nu_r(1_{g_{i_1}K_{i_1}} + \cdots + 1_{g_{i_k}K_{i_k}}) \,.$$

When φ is written as a multimeasure, we have

$$\varphi_r(K_1, \ldots, K_r) = \frac{(-1)^r}{r!} \sum_{k=1}^{r} (-1)^k \sum_{i_1<\cdots<i_k} \nu_r(g_{i_1}K_{i_1} \cup \cdots \cup g_{i_k}K_{i_k}) \,.$$

This shows that ν_r is an invariant continuous homogeneous set function of degree r. For polyconvex sets K_1, K_2, K_3, and rigid motions g, g_1, g_2, g_3 such that g_1K_1, g_2K_2, g_3K_3 are pairwise disjoint and $g(K_1 \cap K_2)$ is disjoint from $K_1 \cup K_2$ and g_3K_3, we have

$$\begin{aligned}
&\nu_r(K_1 \cup K_2 \cup g(K_1 \cap K_2) \cup g_3K_3) \\
&\quad = p_r(\mu_1(K_1 \cup K_2 \cup g(K_1 \cap K_2) \cup g_3K_3), \\
&\qquad \ldots, \mu_n(K_1 \cup K_2 \cup g(K_1 \cap K_2) \cup g_3K_3)) \\
&\quad = p_r(\mu_1(g_1K_1 \cup g_2K_2 \cup g_3K_3), \ldots, \mu_n(g_1K_1 \cup g_2K_2 \cup g_3K_3)) \\
&\quad = \nu_r(g_1K_1 \cup g_2K_2 \cup g_3K_3) \,.
\end{aligned}$$

So ν_r is totally invariant. Therefore $\nu = \sum_r a_r \nu_r$ is a totally invariant continuous set function of polynomial type such that $\nu(P) = f(x_1, x_2, \ldots, x_n)$.

It is clear that p can also be written as

$$a_0 p_0 e_0^m + \sum_r a_r \sum_{j_1,\dots,j_r=1}^{n} b_{j_1\cdots j_r} e_{j_1} \cdots e_{j_r} e_0^{m(b)} ,$$

where m and $m(b) = m(b_{j_1\cdots j_r})$ are non-negative integers. Thus the totally invariant, continuous set functions of polynomial type with the form

$$a_0 \mu_0^{m+1} + \sum_r a_r \sum_{j_1,\dots,j_r=1}^{n} b_{j_1\cdots j_r} \mu_{j_1} \cdots \mu_{j_r} \mu_0^{m(b)}$$

are all such set functions which satisfy (3.3).

Theorem 3.2 gives a surjective map from totally invariant continuous set functions of polynomial type on $\mathbb{R}^n$ onto symmetric polynomials of n-variables.

It should be noticed that invariant set functions need not be totally invariant in our sense. A nontrivial example is the set function μ associated with Minkowski's mixed volume $V(K_1,\dots,K_n)$ (see [4]), i.e.,

$$\mu(K) = V(\underbrace{K,\dots,K}_{n})$$

for $K \in L(\mathcal{K}^n)$. Note that $V(K_1,\dots,K_n)$ is a multimeasure on the lattice $L(\mathcal{K}^n)$. If K is a compact convex set, $\mu(K)$ is just the volume vol(K) of K. In general $\mu(K)$ is not the volume of K if K is not convex. It is easy to see that μ is an invariant, continuous, homogeneous set function of degree n on $L(\mathcal{K}^n)$, but it is not totally invariant when $n > 1$ because $V(K_1,\dots,K_n)$ is not a rigid motion invariant multimeasure on $L(\mathcal{K}^n)$; it is only a translation invariant multimeasure.

One can consider totally invariant continuous set functions of polynomial type on $\mathbb{R}^\infty$ which is the direct sum of countably many copies of $\mathbb{R}$. Here we assume the canonical embedding $\mathbb{R}^0 \subset \mathbb{R}^1 \subset \cdots \subset \mathbb{R}^n \subset \cdots \subset \mathbb{R}^\infty$. In this case, $L(\mathcal{K}^n)$ is replaced by the lattice $L(\mathcal{K})$ which is generated by members of $\mathcal{K} = \bigcup_{n=0}^{\infty} \mathcal{K}^n$ by taking finite unions. A measure μ on $L(\mathcal{K})$ is said to be of *polynomial growth* if for any $a > 1$,

$$\lim_{n\to\infty} \frac{\mu([0,a]^n)}{\mathrm{vol}([0,a]^n)} = 0 . \tag{3.5}$$

When considering $L(\mathcal{K})$, if the condition (3.5) is added to the totally invariant set functions in Definition 3.1, we will have a similar result of Theorem 3.2 in which symmetric polynomials are replaced by symmetric functions of infinitely many variables $x_1, x_2, \dots$ in the sense of Macdonald; see [3]. More precisely, we have

PROPOSITION 3.3. *For every totally invariant continuous set function of polynomial type ν on $L(\mathcal{K})$, there exists a unique symmetric function f of countably many variables such that for any parallelotope P of sides $x_1, x_2, \dots, x_n$,*

$$\nu(P) = f(x_1, x_2, \dots, x_n, 0, 0, \dots) . \tag{3.6}$$

Conversely, for every symmetric function f of countably many variables, there exist totally invariant continuous set functions ν of polynomial type such that (3.6) *is satisfied.*

Remark. It is well known that the ring of symmetric functions is a free λ-ring with one generator e_1. It induces a λ-ring structure on the ring of totally invariant set functions of polynomial type. In fact, the λ-operations are generated by μ_1 as follows:

$$\lambda^n(\mu_1) = \mu_n\,, \qquad n = 0, 1, 2, \dots\,,$$

$$\lambda^k(\mu\nu) = P_k(\lambda^1(\mu), \dots, \lambda^k(\mu), \lambda^1(\nu), \dots, \lambda^k(\nu))\,,$$

$$\lambda^k(\lambda^j(\mu)) = P_{k,j}(\lambda^1(\mu), \dots, \lambda^{kj}(\mu))\,,$$

where μ and ν are any totally invariant continuous set functions of polynomial type, P_k and $P_{k,j}$ are universal polynomials defined by

$$\sum_{k=0}^{\infty} P_k(e_1, \dots, e_k, e'_1, \dots, e'_k)t^k = \prod_{i,j=1}^{m,n} (1 + x_i y_j t)\,,$$

$$\sum_{k=0}^{\infty} P_{k,j}(e_1, \dots, e_{kj})t^k = \prod_{i_1 < \cdots < i_j} (1 + x_{i_1} \dots x_{i_j} t)\,,$$

and e_i is the i-th elementary symmetric polynomial of variables $x_1, x_2, \dots, x_m$, and e'_j the j-th elementary symmetric polynomial of $y_1, y_2, \dots, y_n$. P_k is uniquely determined whenever $m \geqq k, n \geqq k$, and so is $P_{k,j}$ if $m \geqq kj$.

Bibliography

[1] Chen, B., *The Gram-Sommerville and Gauss-Bonnet theorems and combinatorial geometric measures for noncompact polyhedra,* Adv. Math. 91, 1992, pp. 269–291.

[2] Hadwiger, H., *Vorlesungen über Inhalt, Oberfläche und Isoperimetrie,* Springer-Verlag, Berlin, 1957.

[3] Macdonald, I. G., *Symmetric Functions and Hall Polynomials,* Clarendon Press, Oxford, 1979.

[4] Minkowski, H., *Theorie der konvexen Körper, insbesondere Begründung ihres Oberflächenbegriffs,* Ges. Abh., Leipzig-Berlin 2, 1911, pp. 131–229.

[5] Rota, G.-C., *Introduction to Geometric Probability,* Lezioni Lincee held at the Scuola Normale Superiore, Pisa, December 2–22, 1986.

Received October 1992.

A Continuous Analogue of Sperner's Theorem

DANIEL A. KLAIN
Georgia Institute of Technology

AND

GIAN-CARLO ROTA
Massachusetts Institute of Technology

Abstract

One of the best-known results of extremal combinatorics is Sperner's theorem, which asserts that the maximum size of an antichain of subsets of an n-element set equals the binomial coefficient $\binom{n}{\langle n/2 \rangle}$, that is, the maximum of the binomial coefficients. In the last twenty years, Sperner's theorem has been generalized to wide classes of partially ordered sets.

It is the purpose of the present paper to propose yet another generalization that strikes in a different direction. We consider the lattice $\mathrm{Mod}(n)$ of linear subspaces (through the origin) of the vector space $\mathbb{R}^n$. Because this lattice is infinite, the usual methods of extremal set theory do not apply to it. It turns out, however, that the set of elements of rank k of the lattice $\mathrm{Mod}(n)$, that is, the set of all subspaces of dimension k of $\mathbb{R}^n$, or Grassmannian, possesses an invariant measure that is unique up to a multiplicative constant. Can this multiplicative constant be chosen in such a way that an analogue of Sperner's theorem holds for $\mathrm{Mod}(n)$, with measures on Grassmannians replacing binomial coefficients? We show that there is a way of choosing such constants for each level of the lattice $\mathrm{Mod}(n)$ that is natural and unique in the sense defined below and for which an analogue of Sperner's theorem can be proven.

The methods of the present note indicate that other results of extremal set theory may be generalized to the lattice $\mathrm{Mod}(n)$ by similar reasoning. ©1997 John Wiley & Sons, Inc.

1 The Lattice of Subspaces

Let $\mathrm{Mod}(n)$ denote the set of all linear subspaces of $\mathbb{R}^n$, that is, the set of all linear varieties passing through the origin (having fixed an origin once and for all). The set $\mathrm{Mod}(n)$ is a partially ordered set under the relation of inclusion of linear subspaces. Moreover, it is a lattice, where the join $x \vee y$ and the meet $x \wedge y$ of two elements $x, y \in \mathrm{Mod}(n)$ are defined, respectively, as the linear subspaces spanned by x and y and as the intersection of x and y. The lattice $\mathrm{Mod}(n)$ may be viewed as a continuous analogue of the lattice $P(S)$ of subsets of a set S with n elements. Note, however, that this analogy is only a partial one, since the distributive law governing unions and intersections of subsets of S does not hold in the lattice $\mathrm{Mod}(n)$. Nonetheless, this analogy shall carry us as far as we need to go.

We shall apply to the lattice $\mathrm{Mod}(n)$ the notions of *chain*, *flag*, and *antichain*, which are defined for any partially ordered set. An element x of

Communications on Pure and Applied Mathematics, Vol. L, 0205–0223 (1997)

$\mathrm{Mod}(n)$ has rank k (that is, $r(x) = k$) whenever x is a linear subspace of dimension k. The subspace $\{0\}$ is the minimal element of the lattice $\mathrm{Mod}(n)$.

Much as the group of permutations of the set S acts naturally on $P(S)$, the *orthogonal group* $O(n)$ acts naturally on the lattice $\mathrm{Mod}(n)$.

Recall that set of all elements of $\mathrm{Mod}(n)$ of dimension (rank) k, denoted $G(n,k)$, is called the *Grassmannian*. There exists an invariant (Haar) measure acting on $G(n,k)$ that is unique up to a common factor (see [15]). In the appropriate normalization, the total measure of $G(n,k)$ should be in some sense an analogue of the binomial coefficient.

Let τ_n denote the invariant measure on $G(n,1)$, that is, on the set of all straight lines through the origin. Denote

$$[n] = \tau_n(G(n,1)) . \tag{1.1}$$

The value of $[n]$ depends on the normalization for the measure τ_n, to be determined later.

Let $\mathrm{Flag}(n)$ be the set of flags in $\mathrm{Mod}(n)$. For $x \in \mathrm{Mod}(n)$, denote by $\mathrm{Flag}(x)$ the set of all flags that contain x, that is, the set of all sequences $(x_0, x_1, \ldots, x_n)$ of $x_i \in \mathrm{Mod}(n)$ where $\dim(x_i) = i$ such that $x_0 \leq x_1 \leq \cdots \leq x_n$ and such that one of the x_i equals x. Note that $x_0 = \{0\}$ and $x_n = \mathbb{R}^n$.

For fixed x_k, the set of all sequences $(x_k, x_{k+1}, \ldots, x_n)$ with $x_i \in \mathrm{Mod}(n)$ such that $\dim(x_i) = i$ and $x_i \leq x_{i+1}$ is isomorphic to $\mathrm{Flag}(n-k)$. Similarly, the set of sequences $(x_0, x_1, \ldots, x_k)$ is isomorphic to $\mathrm{Flag}(k)$.

Denote by ϕ_n the invariant measure on the set $\mathrm{Flag}(n)$, which is computed as follows: The measure τ_n on $G(n,1)$ induces a measure $\widetilde{\tau}_n$ on the set $G(n, n-1)$ via orthogonal duality. If $f(x_0, x_1, \ldots, x_n)$ is a simple function on the set $\mathrm{Flag}(n)$, then define

$$\int f \, d\phi_n = \iint f(x_0, x_1, \ldots x_n) \, d\phi_{n-1}(x_0, \ldots, x_{n-1}) \, d\widetilde{\tau}_n(x_n) , \tag{1.2}$$

so that the measure ϕ_n is inductively defined. It is clearly invariant.

A more explicit form for ϕ_n can be obtained by the following argument: If $(x_0, x_1, \ldots, x_n)$ is a flag in $\mathrm{Mod}(n)$, let

$$y_1 = x_1 , \quad y_2 = x_1^{\perp} \cap x_2 , \quad y_3 = x_2^{\perp} \cap x_3, \quad \ldots, \quad y_n = x_{n-1}^{\perp} \cap x_n .$$

The sequence $(y_1, y_2, \ldots, y_n)$ is a sequence of orthogonal straight lines, a frame. Conversely, given a frame $(y_1, y_2, \ldots, y_n)$, we obtain a flag by setting

$$x_0 = \{0\} , \quad x_1 = y_1 , \quad x_2 = y_1 \vee y_2 , \quad x_3 = y_1 \vee y_2 \vee y_3 , \quad \ldots .$$

This gives a one-to-one correspondence between flags and frames. If

$$f(x_0, x_1, \ldots, x_n)$$

is a real-valued (measurable) function on flags, let $\bar{f}(y_1, y_2, \ldots, y_n)$ be the corresponding function on frames. Then

$$\int f\, d\phi_n = \int\int \cdots \int \bar{f}(y_1, y_2, \ldots, y_n)\, d\tau_1(y_n)\, d\tau_2(y_{n-1}) \cdots d\tau_n(y_1).$$

This result can be read in the simpler language of combinatorics. Having chosen the line $x_1 = y_1$, which can be done in τ_n ways, the subspace x_2 is determined by the choice of a straight line y_2 in the space orthogonal to x_1 (through the origin), which can be done in τ_{n-1} ways, and so on.

The measure of Flag(n) turns out to be

$$\phi_n(\text{Flag}(n)) = [n][n-1] \cdots [2][1],$$

which is also written $[n]!$, where $[1] = 1$.

We now define an invariant measure on $G(n,k)$. For $A \subseteq G(n,k)$, let Flag(A) be the set of all flags $(x_0, x_1, \ldots, x_n)$ such that $x_k \in A$. Set

$$\nu_k^n(A) = \frac{1}{[k]![n-k]!}\phi_n(\text{Flag}(A)). \tag{1.3}$$

The measure ν_k^n is evidently invariant under rotation, and we have

$$\nu_k^n(G(n,k)) = \frac{[n]!}{[k]!\,[n-k]!} = \begin{bmatrix} n \\ k \end{bmatrix}. \tag{1.4}$$

These values, called *flag coefficients*, are continuous analogues of the binomial coefficients of the discrete lattice of subsets of a finite set.

Define a measure ν_n on Mod(n) by taking the direct sum of the measures ν_k^n. That is, for any measurable subset $A \subseteq \text{Mod}(n)$, define

$$\nu_n(A) = \sum_{k=0}^{n} \nu_k^n(A \cap G(n,k)).$$

Henceforth, we make the assumption that subsets of Mod(n) under discussion are measurable with respect to ν_n.

The measure ν_n satisfies the following analogue of the classical Lubell-Yamamoto-Meshalkin (LYM) inequality:

THEOREM 1.1 (The Continuous LYM Inequality) *Let* $A \subseteq \text{Mod}(n)$ *be an antichain. For* $0 \leq k \leq n$ *let*

$$A_k = A \cap G(n,k)\,,$$

so that

$$A = \bigcup_k A_k$$

is a disjoint union. Then

$$\sum_k \frac{\nu_k^n(A_k)}{\begin{bmatrix} n \\ k \end{bmatrix}} \leq 1\,. \tag{1.5}$$

PROOF: For each $0 \leq k \leq n$, the measure of flags meeting A_k is given by

$$\phi_n(\text{Flag}(A_k)) = \nu_k^n(A_k)\,[k]!\,[n-k]!$$

by the definition (1.3) of ν_k^n. Since every flag in Flag(n) meets A in at most one point, we have

$$\sum_k \nu_k^n(A_k)\,[k]!\,[n-k]! = \sum_k \phi_n(\text{Flag}(A_k)) = \phi_n(\text{Flag}(A)) \leq [n]!$$

It follows that

$$\sum_k \frac{\nu_k^n(A_k)}{\begin{bmatrix} n \\ k \end{bmatrix}} \leq 1\,.$$

∎

The preceding results hold independently of the normalization chosen for the measure τ_n, that is, the value of $[n]$. For example, in [4], Fisk developed a similar construction, in which the total measure of $G(n,1)$ is taken to be $n\omega_n/2$, the measure suggested by the usual two-to-one quotient map from the unit sphere in $\mathbb{R}^n$. Unfortunately, this normalization does not permit an extension of the lattice analogy, nor does it agree with a related and fundamental normalization for the rigid motion invariant measures on $\mathbb{R}^n$ and its subspaces. To see this and to determine the appropriate choice of normalization for τ_n, we turn briefly to the theory of convex bodies and intrinsic volumes.

2 The Intrinsic Volumes

Denote by $\mathcal{K}^n$ the collection of all compact convex subsets of $\mathbb{R}^n$, that is, n-dimensional Euclidean space. The elements of $\mathcal{K}^n$ are also known as *convex bodies*.

A function $\mu : \mathcal{K}^n \longrightarrow \mathbb{R}$ is called a *valuation* on $\mathcal{K}^n$ if $\mu(\varnothing) = 0$, where $\varnothing$ is the empty set and

$$\mu(K \cup L) = \mu(K) + \mu(L) - \mu(K \cap L)$$

for all $K, L \in \mathcal{K}^n$ such that $K \cup L \in \mathcal{K}^n$ as well.

A valuation μ on $\mathcal{K}^n$ is said to be *continuous* if

$$\lim_{i \to \infty} \mu(K_i) = \mu(K)$$

for any convergent sequence $K_i \longrightarrow K$ in the Hausdorff topology on $\mathcal{K}^n$ (see [17, p. 47]).

Let E_n denote the group of Euclidean (or rigid) motions, that is, the group generated by all translations and rotations in $\mathbb{R}^n$. A valuation μ on $\mathcal{K}^n$ is *invariant under rigid motions*, or simply *invariant*, if $\mu(\phi K) = \mu(K)$ for all $\phi \in E_n$.

A well-known example of a continuous rigid motion invariant valuation on $\mathcal{K}^n$ is the *volume*, denoted V_n. Another example is *surface area*, denoted S. Notice that if P is a rectilinear box with edge lengths $a_1, \ldots, a_n$, then

$$V_n(P) = a_1 a_2 \cdots a_n\,,$$

and

$$S(P) = 2 \sum_{i=1}^{n} a_1 \cdots \hat{a}_i \cdots a_n\,,$$

where $\hat{a}_i$ denotes the deletion of a_i from the indicated product. Setting $V_{n-1} = \frac{1}{2}S$, it follows that V_n and V_{n-1} evaluate on a rectilinear box P as the nth and $(n-1)^{\text{th}}$ elementary symmetric polynomials of the edge lengths $a_1, \ldots, a_n$ of P. In fact, it can easily be seen that each of the elementary symmetric polynomials $e_1, \ldots, e_n$ in n variables defines a continuous valuation on the collection of *rectilinear boxes* (with respect to a given ordered orthogonal frame) by evaluating on the edge lengths of such boxes. Moreover, these valuations are evidently invariant under permutations of the edges (or basis elements of the given frame). Therefore, define

$$V_i(P) = e_i(a_1, \ldots, a_n)$$

for all rectilinear boxes P and all $1 \leq i \leq n$. Define $V_0(P)$ to equal 1 if P is nonempty and 0 if P is empty. It turns out that, as is true for the volume V_n and for V_{n-1}, each of the valuations V_i extends to a continuous invariant valuation on all of $\mathcal{K}^n$. In particular, V_0 extends to the Euler characteristic. These (extended) valuations are known as McMullen's *intrinsic volumes* [2, 12, 13][17, p. 210] and have numerous expressions in terms of integral formulae, some of which are described in the following sections.

The intrinsic volume V_i is equal to the i-dimensional volume when restricted to any i-dimensional subspace of $\mathbb{R}^n$. In fact, the intrinsic volumes are normalizations of special cases of *Minkowski mixed volumes*, also known as the *quermassintegrals*. Consequently, the intrinsic volumes satisfy Steiner's formula [17, p. 210] for the volume of Minkowski sums.

Let B_n denote the unit ball in $\mathbb{R}^n$. It follows easily from Steiner's formula that

$$V_i(B_n) = \binom{n}{i} \frac{\omega_n}{\omega_{n-i}} \tag{2.1}$$

where ω_k denotes the k-dimensional volume of the unit ball in $\mathbb{R}^k$.

Hadwiger's characterization theorem states that the intrinsic volumes span the vector space of *all* continuous valuations on $\mathcal{K}^n$ that are invariant under rigid motions (see [2, 5, 8, 13][17, p. 210]).

A valuation μ on $\mathcal{K}^n$ is said to be *homogeneous of degree i* if

$$\mu(\alpha K) = \alpha^i \mu(K)$$

for all $\alpha \geq 0$ and all $K \in \mathcal{K}^n$. It is well-known that, for $0 \leq i \leq n$, the valuation V_i is homogeneous of degree i. Hadwiger's characterization theorem implies that, up to a constant factor, V_i is the *unique* continuous invariant valuation that is invariant under rigid motions and is homogeneous of degree i. In particular, the intrinsic volume V_i is the only continuous and invariant valuation extending i-dimensional volume on $\mathbb{R}^i$ to convex bodies in $\mathbb{R}^n$ for $n > i$.

The intrinsic volumes are related to the invariant measures ν_k^n on the Grassmannians by the *mean projection formulae* [16, p. 221][17, p. 295]). If $K \in \mathcal{K}^n$ and if ξ is a k-dimensional subspace of $\mathbb{R}^n$, denote by $K \mid \xi$ the orthogonal projection of K onto ξ. The mean projection formula (for dimension $k \in \{0, 1, \ldots, n\}$) states that

$$V_k(K) = c_k \int_{G(n,k)} V_k(K \mid \xi)\, d\nu_k^n(\xi)\,, \tag{2.2}$$

for all $K \in \mathcal{K}^n$, where the value of the constant c_k depends on the normalization chosen for the measure ν_k^n. The mean projection formula (2.2) forms a crucial link between measures on the lattice $\mathrm{Mod}(n)$ and the intrinsic volumes on compact convex sets.

For the special case of $k = n - 1$, the formula (2.2) can be expressed in the following form, known as Cauchy's formula:

$$S(K) = \frac{1}{\omega_{n-1}} \int_{\mathbb{S}^{n-1}} V_{n-1}(K \mid u^{\perp})\, du.$$

Note that $V_{n-1}(K \mid u^{\perp})$ is just the $(n-1)$-dimensional volume of the projection of K onto the subspace $u^{\perp}$. Since $S(K) = 2V_{n-1}(K)$, we have

$$V_{n-1}(K) = \frac{1}{2\omega_{n-1}} \int_{\mathbb{S}^{n-1}} V_{n-1}(K \mid u^{\perp})\, du. \tag{2.3}$$

Alternatively, this result can be expressed as an integral over the projective space $G(n,1)$ (i.e., the set of all lines ℓ through the origin in $\mathbb{R}^n$) rather than an integral over the sphere. Integrating with respect to the Haar *probability* measure on $G(n,1)$, we have

$$V_{n-1}(K) = \alpha \int_{G(n,1)} V_{n-1}(K \mid \ell^{\perp})\, d\ell$$

where α is a constant independent of K. To compute α, set $K = B_n$ to obtain

$$\frac{n\omega_n}{2} = V_{n-1}(B_n) = \alpha \int_{G(n,1)} V_{n-1}(B_n \mid \ell^{\perp})\, d\ell = \alpha\omega_{n-1}$$

so that $\alpha = \frac{n\omega_n}{2\omega_{n-1}}$. We come now to the normalization of the Haar measure on $G(n,1)$, which is the central point of the present work. Denote by τ_n the Haar measure on $G(n,1)$ normalized so that

$$[n] = \tau_n(G(n,1)) = \frac{n\omega_n}{2\omega_{n-1}}, \tag{2.4}$$

where ω_n denotes the volume of the unit ball B_n in $\mathbb{R}^n$:

$$\omega_n = \frac{\pi^{n/2}}{\Gamma((n/2)+1)}. \tag{2.5}$$

The choice of normalization just described is crucial. For example, (2.3) now becomes

$$V_{n-1}(K) = \int_{G(n,1)} V_{n-1}(K \mid \ell^{\perp})\, d\tau_n. \tag{2.6}$$

This is the first of several confirmations that our choice of normalization is the "right" one.

Having chosen a normalization for the measure τ_n on $G(n,1)$, the normalizations for the measures ϕ_n and ν_k^n on $\mathrm{Flag}(n)$ and $G(n,k)$ are then precisely determined by (1.2) and (1.3). Specifically, we have

$$[n]! = \frac{n!\,\omega_n\omega_{n-1}\cdots\omega_1}{2^n\omega_{n-1}\omega_{n-2}\cdots\omega_0} = \frac{n!\,\omega_n}{2^n}. \tag{2.7}$$

The flag coefficients now assume a particularly pleasing form. From (1.4) and (2.7) we obtain

$$\begin{bmatrix} n \\ k \end{bmatrix} = \frac{n!}{k!(n-k)!}\frac{\omega_n}{\omega_k\omega_{n-k}} = \binom{n}{k}\frac{\omega_n}{\omega_k\omega_{n-k}}. \tag{2.8}$$

Once again our computations of the flag coefficients differ from the calculations of Fisk [4] due to our special choice of normalization (2.4) for the measure τ_n on $G(n,1)$.

With our choice of normalization (2.4), we have set the factor c_{n-1} from (2.2) equal to 1. This single assignment then determines all other normalizations (2.8) for the measures ν_k^n on $G(n,k)$, where $k \in \{1,\ldots,n\}$.

Moreover, the normalizations (2.8) are compatible not only with the intrinsic volume V_{n-1} (see formula (2.6)) but with *all* of the intrinsic volumes. To see this, we compute the rest of the constants c_k in (2.2) by evaluating at $K = B_n$:

$$\begin{aligned} V_k(B_n) = c_k\int_{G(n,k)} V_k(B_n \mid \xi)\,d\nu_k^n(\xi) &= c_k V_k(B_k)\int_{G(n,k)} d\nu_k^n(\xi) \\ &= c_k\omega_k\begin{bmatrix} n \\ k \end{bmatrix}. \end{aligned}$$

It now follows from (2.1) and (2.8) that

$$c_k = \frac{V_k(B_n)}{\omega_k}\begin{bmatrix} n \\ k \end{bmatrix}^{-1} = \binom{n}{k}\frac{\omega_n}{\omega_{n-k}\omega_k}\begin{bmatrix} n \\ k \end{bmatrix}^{-1} = 1,$$

so that in fact, the formula (2.2) becomes

$$V_k(K) = \int_{G(n,k)} V_k(K \mid \xi)\,d\nu_k^n(\xi), \tag{2.9}$$

for all $k \in \{1,\ldots,n\}$.

Recall that V_k is the *unique* continuous invariant valuation defined simultaneously on $\mathcal{K}^n$, for all $n > 0$, so that V_k restricts to the k-dimensional volume on all k-dimensional subspaces of $\mathbb{R}^n$. Recall also that the intrinsic volumes V_k are normalized independently of n; that is, $V_k(L)$ of an l-dimensional convex body L is the same regardless of the dimension $n \geq l$ of the ambient space $\mathbb{R}^n$. The absence of an additional normalizing factor in any of the mean projection formulae (2.9) suggests that our choice of normalization for the measures ν_k^n is indeed the correct one.

When expressed in the language of this section, Kubota's generalization of the mean projection formula [9][17, p. 295] takes the following combinatorially suggestive form:

$$\int_{G(n,l)} V_k(K \mid \xi)\, d\nu_l^n(\xi) = \begin{bmatrix} n-k \\ l-k \end{bmatrix} V_k(K)$$

for $0 \leq k \leq l \leq n$ and all $K \in \mathcal{K}^n$.

The flag coefficients also appear in the principal kinematic formula for convex bodies (see [9], also [16, p. 262] and [17, p. 253]) when expressed in terms of the intrinsic volumes. Let E_n denote the group of Euclidean motions in $\mathbb{R}^n$, the semidirect product of the orthogonal group $O(n)$ (equipped with the Haar probability measure) with the group of translations. The principal kinematic formula then takes the form

$$\int_{E_n} V_0(K \cap gL)\, dg = \sum_{i=0}^{n} \begin{bmatrix} n \\ i \end{bmatrix}^{-1} V_i(K) V_{n-i}(L)$$

for all $K, L \in \mathcal{K}^n$. Here V_0 denotes zeroth intrinsic volume, usually called the Euler characteristic, which takes the value 0 on the null set and the value 1 on all nonempty, compact, convex sets.

More generally, we have

$$\int_{E_n} V_k(A \cap gK) dg = \sum_{i=0}^{n-k} \begin{bmatrix} i+k \\ k \end{bmatrix} \begin{bmatrix} n \\ i \end{bmatrix}^{-1} V_{k+i}(A) V_{n-i}(K)$$

for $0 \leq k \leq n$ and all $K, L \in \mathcal{K}^n$. Similarly clean formulations also exist for Crofton's formulae (see [9], also [16] and [17, p. 235]).

3 A Sperner Theorem for Subspaces

From now on, all flag coefficients will be normalized as specified in the previous section. Thanks to our choice of normalization, the flag coefficients satisfy a number of properties analogous to those of the binomial coefficients. For example,

$$\begin{bmatrix} n \\ k \end{bmatrix} = \begin{bmatrix} n \\ n-k \end{bmatrix}.$$

In analogy to Pascal's triangle relations, we have the following:

PROPOSITION 3.1 *For* $1 \leq k \leq n-1$,

$$\frac{\omega_{k-1}\omega_{n-k}}{\omega_{n-1}} \begin{bmatrix} n-1 \\ k-1 \end{bmatrix} + \frac{\omega_k \omega_{n-k-1}}{\omega_{n-1}} \begin{bmatrix} n-1 \\ k \end{bmatrix} = \frac{\omega_k \omega_{n-k}}{\omega_n} \begin{bmatrix} n \\ k \end{bmatrix}.$$

PROOF: By direct computation, we have

$$\frac{\omega_{k-1}\omega_{n-k}}{\omega_{n-1}} \begin{bmatrix} n-1 \\ k-1 \end{bmatrix} + \frac{\omega_k \omega_{n-k-1}}{\omega_{n-1}} \begin{bmatrix} n-1 \\ k \end{bmatrix}$$
$$= \binom{n-1}{k-1} + \binom{n-1}{k} = \binom{n}{k} = \frac{\omega_k \omega_{n-k}}{\omega_n} \begin{bmatrix} n \\ k \end{bmatrix}.$$

∎

In constructing the flag coefficients, the real numbers $[n]$ play a role analogous to that of the positive integers n in the discrete case of the classical binomial coefficients.

The following proposition demonstrates once again the expedience of our choice of normalization for the measure τ_n.

PROPOSITION 3.2 *The map* $n \longmapsto [n]$ *is an increasing function.*

PROOF: From (2.5) and (1.1) we have

$$[n] = \frac{n\omega_n}{2\omega_{n-1}} = \frac{n\sqrt{\pi}}{2} \frac{\Gamma\left(\frac{n+1}{2}\right)}{\Gamma\left(\frac{n+2}{2}\right)} = \frac{n\sqrt{\pi}}{2} \frac{\Gamma\left(\frac{n+1}{2}\right)}{\frac{n}{2}\Gamma\left(\frac{n}{2}\right)} = \sqrt{\pi}\frac{\Gamma\left(\frac{n}{2}+\frac{1}{2}\right)}{\Gamma\left(\frac{n}{2}\right)}.$$

Define a function f on the positive real numbers by

$$f(t) = \sqrt{\pi}\frac{\Gamma\left(t+\frac{1}{2}\right)}{\Gamma(t)}.$$

Since $f(\frac{n}{2}) = [n]$, it is sufficient to show that f is an increasing function of t. Recall that [1, p. 15]

$$\Gamma(t) = \lim_{k\to\infty} \frac{k^t k!}{t(t+1)\cdots(t+k)}.$$

This implies that

$$
\begin{aligned}
f(t) &= \sqrt{\pi}\left(\lim_{k\to\infty}\frac{k^{t+\frac{1}{2}}k!}{(t+\frac{1}{2})(t+\frac{1}{2}+1)\cdots(t+\frac{1}{2}+k)}\right) \\
&\quad\times\left(\lim_{k\to\infty}\frac{k^{t}k!}{t(t+1)\cdots(t+k)}\right)^{-1} \qquad (3.1)\\
&= \sqrt{\pi}\lim_{k\to\infty}\frac{t(t+1)\cdots(t+k)\sqrt{k}}{(t+\frac{1}{2})(t+\frac{1}{2}+1)\cdots(t+\frac{1}{2}+k)}
\end{aligned}
$$

Since the function

$$\frac{t}{t+\frac{1}{2}}$$

is increasing with respect to $t > 0$, so is the product on the right-hand side of (3.1). It follows that f is increasing for $t > 0$, and we conclude that $[n]$ is an increasing function of the positive integers. ∎

The generalized factorial $[n]!$ also satisfies the following property:

PROPOSITION 3.3 *For* $0 \leq k \leq l \leq \frac{n}{2}$,

$$[k]!\,[n-k]! \geq [l]!\,[n-l]!$$

PROOF: To begin, note that

$$0 \leq k \leq l \leq \frac{n}{2} \leq n-l \leq n-k \leq n\,.$$

It follows from Proposition 3.2 that

$$[n-k]\cdots[n-l+1] \geq [l]\cdots[k+1]\,.$$

(Note that there are $l-k$ factors on each side of this identity.) Multiplying on both sides, we obtain

$$[n-k]!\,[k]! \geq [n-l]!\,[l]!$$

∎

The flag coefficients in turn satisfy the following property, in evident analogy to the classical binomial coefficients:

PROPOSITION 3.4 *For* $0 \leq k \leq n$,

$$\begin{bmatrix} n \\ k \end{bmatrix} \leq \begin{bmatrix} n \\ \langle \frac{n}{2} \rangle \end{bmatrix}.$$

Here $\langle \frac{n}{2} \rangle$ *denotes the largest integer less than or equal to* $\frac{n}{2}$.

PROOF: Since

$$\begin{bmatrix} n \\ k \end{bmatrix} = \frac{[n]!}{[k]!\,[n-k]!} = \begin{bmatrix} n \\ n-k \end{bmatrix},$$

it is sufficient to consider the case where $k \leq \langle \frac{n}{2} \rangle$. The result then follows immediately from Proposition 3.3 after setting $l = \langle \frac{n}{2} \rangle$. ∎

Our choice of normalization for the measures ν_k^n now enables us to prove a continuous analogue of Sperner's theorem on lattice antichains [6, 11, 18][10, p. 542].

THEOREM 3.5 (Continuous Sperner Theorem) *Assume that* A *is an antichain in* $\mathrm{Mod}(n)$. *Then*

$$\nu_n(A) \leq \begin{bmatrix} n \\ \langle \frac{n}{2} \rangle \end{bmatrix}.$$

PROOF: For $0 \leq k \leq n$, let $A_k = A \cap G(n,k)$. Combining (1.5) with Proposition 3.4, we obtain

$$\sum_k \frac{\nu_k^n(A_k)}{\begin{bmatrix} n \\ \langle \frac{n}{2} \rangle \end{bmatrix}} \leq \sum_k \frac{\nu_k^n(A_k)}{\begin{bmatrix} n \\ k \end{bmatrix}} \leq 1$$

so that

$$\nu_n(A) = \sum_k \nu_k^n(A_k) \leq \begin{bmatrix} n \\ \langle \frac{n}{2} \rangle \end{bmatrix}.$$

∎

A subset $F \subseteq \mathrm{Mod}(n)$ is called an *r-family* if chains in F contain no more than r elements. For example, an antichain is a 1-family. Given an r-family F in $\mathrm{Mod}(n)$, let $F_k = F \cap G(n,k)$. Since every flag in $\mathrm{Mod}(n)$ meets F in at most r elements, we have

$$\sum_{k=0}^{n} \phi_n(\mathrm{Flag}(F_k)) \leq [n]! \cdot r.$$

From (1.3) we then obtain the following generalization of (1.5):

$$\sum_k \frac{\nu_k^n(F_k)}{\begin{bmatrix} n \\ k \end{bmatrix}} \leq r. \tag{3.2}$$

This inequality leads in turn to a continuous analogue of Sperner's theorem for r-families. (For the classical result, see [3, 6] and [10, p. 543].)

THEOREM 3.6 *Let F be an r-family in* $\mathrm{Mod}(n)$. *Then*

$$\nu_n(F) \leq \begin{bmatrix} n \\ \langle \frac{n+1}{2} \rangle \end{bmatrix} + \begin{bmatrix} n \\ \langle \frac{n+2}{2} \rangle \end{bmatrix} + \cdots + \begin{bmatrix} n \\ \langle \frac{n+r}{2} \rangle \end{bmatrix}.$$

Once again $\langle x \rangle$ denotes the largest integer less than or equal to a real number x. In order to prove Theorem 3.6 we make use of the following lemma from [6] (see also [9]):

LEMMA 3.7 *Suppose that* $c_0 \geq c_1 \geq \cdots \geq c_n > 0$. *If* $c_i \geq x_i \geq 0$ *for* $0 \leq i \leq n$, *and if*

$$x_0 + x_1 + \cdots + x_n \geq c_0 + c_1 + \cdots + c_{r-1},$$

then

$$\sum_{k=0}^{n} \frac{x_k}{c_k} \geq r.$$

If $c_0 > \cdots > c_n > 0$, *then equality holds if and only if* $x_i = c_i$ *for* $0 \leq i \leq r-1$ *and* $x_i = 0$ *for* $r \leq i \leq n$.

PROOF OF THEOREM 3.6 Relabel the flag coefficients $c_0, c_1, \ldots, c_n$ in *descending* order, so that $c_0 \geq c_1 \geq \cdots \geq c_n$; then relabel the numerators $\nu_k^n(F_k)$ in (3.2) by $x_0, x_1, \ldots, x_n$, so that each x_k is the numerator of that term of (3.2) having c_k as denominator. The inequality (3.2) now becomes

$$\sum_{k=0}^{n} \frac{x_k}{c_k} \leq r.$$

It then follows from Lemma 3.7 that

$$x_0 + x_1 + \cdots + x_n \leq c_0 + c_1 + \cdots + c_{r-1}.$$

In other words,

$$\nu_n(F) = \sum_{k=0}^{n} \nu_k^n(F_k) \leq \begin{bmatrix} n \\ \langle \frac{n+1}{2} \rangle \end{bmatrix} + \begin{bmatrix} n \\ \langle \frac{n+2}{2} \rangle \end{bmatrix} + \cdots + \begin{bmatrix} n \\ \langle \frac{n+r}{2} \rangle \end{bmatrix}.$$

∎

4 A Multinomial Generalization

In [14] (see also [7]) Meshalkin generalizes the notion of antichain in the finite lattice $P(S)$, defining an *s-system* σ to be a collection of ordered partitions (called *decompositions*) of a finite set S into r blocks, such that the set σ_j, whose elements are the jth blocks of the ordered partitions in σ, is an antichain for each $1 \leq j \leq r$. Meshalkin has shown that the maximum size of an s-system σ of r-decompositions (for fixed r) is given by the maximal multinomial coefficient in r parameters, namely,

$$\binom{n}{\underbrace{\langle n/r\rangle, \ldots, \langle n/r\rangle}_{r-b}, \underbrace{\langle n/r\rangle + 1, \ldots, \langle n/r\rangle + 1}_{b}}$$

where $n \cong r \mod b$ and $0 \leq b \leq r-1$. We now define analogous notions for the lattice $\mathrm{Mod}(n)$.

A map $\delta : \{1, \ldots, r\} \longrightarrow \mathrm{Mod}(n)$ is called an *r-decomposition of* $\mathbb{R}^n$ if

1. $\delta(i) \perp \delta(j)$ for $i \neq j$, and
2. $\delta(1) \oplus \cdots \oplus \delta(r) = \mathbb{R}^n$.

Denote by $\mathrm{Dec}(n, r)$ the set of all r-decompositions of $\mathbb{R}^n$. Note that for each $\delta \in \mathrm{Dec}(n, r)$

$$\dim \delta(1) + \cdots + \dim \delta(r) = n\,.$$

Given positive integers $a_1, a_2, \ldots, a_r$ such that $a_1 + \cdots + a_r = n$ we denote by $\mathrm{Mult}(n; a_1, \ldots, a_r)$ the set of all r-decompositions δ such that $\dim \delta(i) = a_i$ for $i = 1, \ldots, r$. In other words, $\mathrm{Mult}(n; a_1, \ldots, a_r)$ is the set of all (ordered) decompositions of $\mathbb{R}^n$ into direct sums of subspaces having dimensions $a_1, \ldots, a_r$. Evidently the set $\mathrm{Dec}(n, r)$ can be expressed as the finite disjoint union

$$\mathrm{Dec}(n, r) = \biguplus_{a_1 + \cdots + a_r = n} \mathrm{Mult}(n; a_1, \ldots, a_r)\,.$$

An s-system of order r is a subset $\sigma \subseteq \mathrm{Dec}(n, r)$ such that the set

$$\{\delta(i) \mid \delta \in \sigma\} \tag{4.1}$$

is an antichain in $\mathrm{Mod}(n)$ for each $1 \leq i \leq r$.

An obvious example of an s-system of order r is $\mathrm{Mult}(n; a_1, \ldots, a_r)$ for some admissible selection of $a_1, \ldots, a_r$. If $\delta, \zeta \in \mathrm{Mult}(n; a_1, \ldots, a_r)$, then

$\delta(i)$ and $\zeta(i)$ both have dimension a_i, so that either $\delta(i) = \zeta(i)$ or the two subspaces are incomparable in the subset partial ordering on $\mathrm{Mod}(n)$. This holds for $i = 1, \ldots, r$, and so the antichain condition on (4.1) is satisfied.

Other examples with which we have already worked are the s-systems of order 2. Let A be an antichain in $\mathrm{Mod}(n)$. For each $V \in A$, express $\mathbb{R}^n$ as the direct sum $V \oplus V^\perp$, so that the pair $(V, V^\perp)$ is a 2-decomposition in $\mathrm{Dec}(n, 2)$. Define

$$A^\perp = \{V^\perp \mid V \in A\}.$$

Then $A^\perp$ is also an antichain in $\mathrm{Mod}(n)$, and the set

$$\sigma = \{(V, V^\perp) \mid V \in A\}$$

is an s-system of order 2. Thus an s-system is in fact a generalization of an antichain. Similarly, the Grassmannian $G(n, k)$ can also be viewed as $\mathrm{Mult}(n; k, n-k)$ through the bijection $V \mapsto (V, V^\perp)$.

In analogy to the construction of the measure ν_k^n on $G(n, k)$, define invariant measures on the sets $\mathrm{Mult}(n; a_1, \ldots, a_r)$ as follows:

For $\delta \in \mathrm{Mult}(n; a_1, \ldots, a_r)$ define a flag $(x_0, x_1, \ldots, x_n) \in \mathrm{Flag}(n)$ to be *compatible* with δ if

1. $x_{a_1} = \delta(1)$ and
2. $x_{a_1+\cdots+a_i}/x_{a_1+\cdots+a_{i-1}} = \delta(i)$ for $i \geq 2$.

Here the quotient $x_{a_1+\cdots+a_i}/x_{a_1+\cdots+a_{i-1}}$ denotes the orthogonal complement of the vector space $x_{a_1+\cdots+a_{i-1}}$ inside the larger space $x_{a_1+\cdots+a_i}$.

For $A \subseteq \mathrm{Mult}(n; a_1, \ldots, a_r)$, let the set of all flags $(x_0, x_1, \ldots, x_n)$ compatible with some $\delta \in A$ be designated by $\mathrm{Flag}(A)$. Define

$$\nu^n_{a_1,a_2,\ldots,a_r}(A) = \frac{1}{[a_1]!\,[a_2]!\cdots[a_r]!}\phi_n(\mathrm{Flag}(A)). \tag{4.2}$$

To justify this normalization combinatorially, note that to choose a flag compatible with $\delta \in A$ one must choose a frame for each of the vector spaces $\delta(i)$, of which there are $[a_i]!$ choices for each i.

The measure $\nu^n_{a_1,\ldots,a_r}$ is evidently invariant under rotations, and we have

$$\nu^n_{a_1,\ldots,a_r}(\mathrm{Mult}(n; a_1, \ldots, a_r)) = \frac{[n]!}{[a_1]!\cdots[a_r]!} = \begin{bmatrix} n \\ a_1, \ldots, a_r \end{bmatrix}.$$

These values, called *multiflag coefficients*, are continuous analogues of the classical multinomial coefficients.

Define a measure $\nu_{n;r}$ on $\mathrm{Dec}(n,r)$ by taking the direct sum of the measures $\nu^n_{a_1,\ldots,a_r}$. That is, for any measurable subset $A \subseteq \mathrm{Dec}(n,r)$, define

$$\nu_{n;r}(A) = \sum_{a_1+\cdots+a_r=n} \nu^n_{a_1,\ldots,a_r}(A \cap \mathrm{Mult}(n;a_1,\ldots,a_r))\,.$$

Just as the continuous Sperner theorem, Theorem 3.5, gives the maximum possible measure for an antichain A in $\mathrm{Mod}(n)$, a generalization of this theorem gives the maximum possible measure for an s-system in $\mathrm{Dec}(n,r)$. En route to such a generalization, we prove a multinomial version of the continuous LYM inequality.

THEOREM 4.1 (Continuous Multinomial LYM Inequality) *Let $\sigma \subseteq \mathrm{Dec}(n,r)$ be an s-system. For $a_1+\cdots+a_r=n$, let*

$$\sigma_{a_1,\ldots,a_r} = \sigma \cap \mathrm{Mult}(n;a_1,\ldots,a_r)$$

so that

$$\sigma = \bigcup_{a_1+\cdots+a_r=n} \sigma_{a_1,\ldots,a_r}$$

is a disjoint union. Then

$$\sum_{a_1+\cdots+a_r=n} \frac{\nu^n_{a_1,\ldots,a_r}(\sigma_{a_1,\ldots,a_r})}{\left[{n \atop a_1,\ldots,a_r}\right]} \leq 1\,. \tag{4.3}$$

PROOF: For $a_1+\cdots+a_r=n$, the measure of flags compatible with $\sigma_{a_1,\ldots,a_r}$ is given by

$$\phi_n(\mathrm{Flag}(\sigma_{a_1,\ldots,a_r})) = \nu^n_{a_1,\ldots,a_r}(\sigma_{a_1,\ldots,a_r})[a_1]!\cdots[a_r]!$$

by the definition (4.2) of $\nu^n_{a_1,\ldots,a_r}$.

Suppose a flag $(x_0,x_1,\ldots,x_n)$ is compatible with both $\gamma,\delta \in \sigma$. Then $\gamma(1)=x_{a_1}$ and $\delta(1)=x_{b_1}$, where $a_1=\dim\gamma(1)$ and $b_1=\dim\delta(1)$. Since $(x_0,x_1,\ldots,x_n)$ is a flag, we have $x_{a_1}\subseteq x_{b_1}$ or vice versa. But σ is an s-system, so that either $\gamma(1)=\delta(1)$ or the two spaces are incomparable. Therefore $\gamma(1)=\delta(1)$ and $a_1=b_1$. Continuing, we have $\gamma(2)=x_{a_1+a_2}/x_{a_1}$ and $\delta(2)=x_{b_1+b_2}/x_{a_1}$ (since $a_1=b_1$). A similar argument then implies that $\gamma(2)=\delta(2)$ and $a_2=b_2$. Continuing in this manner, we conclude that $\gamma(i)=\delta(i)$ for each $1\leq i\leq r$ so that $\gamma=\delta$. In other words, every flag in Flag(n) is compatible with at most one r-decomposition $\delta \in \sigma$. It follows that

$$\sum_{a_1+\cdots+a_r=n} \nu^n_{a_1,\ldots,a_r}(\sigma_{a_1,\ldots,a_r})[a_1]!\cdots[a_r]! = \sum_{a_1+\cdots+a_r=n} \phi_n(\mathrm{Flag}(\sigma_{a_1,\ldots,a_r}))$$
$$= \phi_n(\mathrm{Flag}(\sigma))$$
$$\leq [n]!,$$

so that

$$\sum_{a_1+\cdots+a_r=n} \frac{\nu^n_{a_1,\ldots,a_r}(\sigma_{a_1,\ldots,a_r})}{\begin{bmatrix} n \\ a_1,\ldots,a_r \end{bmatrix}} \leq 1.$$

∎

The multiflag coefficients also satisfy the following property, in analogy to the classical multinomial coefficients:

PROPOSITION 4.2 *Let $r \leq n$ be positive integers, and suppose that $n = rq + b$, where q is a natural number and $0 \leq b \leq r-1$ is the integer remainder. For $a_1 + \cdots + a_r = n$,*

$$\begin{bmatrix} n \\ a_1,\ldots,a_r \end{bmatrix} \leq \begin{bmatrix} n \\ \underbrace{\langle n/r\rangle,\ldots,\langle n/r\rangle}_{r-b}, \underbrace{\langle n/r\rangle+1,\ldots,\langle n/r\rangle+1}_{b} \end{bmatrix}.$$

PROOF: Let $a_1, \ldots, a_r$ be positive integers such that $a_1 + \cdots + a_r = n$. Without loss of generality, suppose that $a_1 < \langle \frac{n}{r} \rangle$. Then $a_i > \langle \frac{n}{r} \rangle$ for some $i > 1$. Again without loss of generality, suppose that

$$a_1 < \left\langle \frac{n}{r} \right\rangle < a_2.$$

Then $a_2 - a_1 \geq 2$, so that

$$a_1 < \left\langle \frac{a_1 + a_2}{2} \right\rangle < a_2.$$

It then follows from Proposition 3.3 that

$$[a_1 + 1]!\,[a_2 - 1]! \leq [a_1]!\,[a_2]!.$$

Replace a_1 with a_1+1 and a_2 with a_2-1. Note that the identity $a_1+\cdots+a_r = n$ is preserved. This process is repeated until $a_i \geq \langle \frac{n}{r} \rangle$ for all $1 \leq i \leq r$, that is, until $a_i = \langle \frac{n}{r} \rangle + 1$ for $1 \leq i \leq b$ and $a_i = \langle \frac{n}{r} \rangle$ for $b+1 \leq i \leq r$, where b is the integer remainder upon division of n by r.

Since each iteration of this procedure decreases the value of the product $[a_1]! \cdots [a_r]!$, it follows that

$$[a_1]! \cdots [a_r]! \geq ([\langle n/r \rangle]!)^{r-b} ([\langle n/r \rangle + 1]!)^b$$

for all $a_1 + \cdots + a_r = n$. Therefore,

$$\frac{[n]!}{[a_1]! \cdots [a_r]!} \leq \frac{[n]!}{([\langle n/r \rangle]!)^{r-b} ([\langle n/r \rangle + 1]!)^b},$$

for all $a_1 + \cdots + a_r = n$. ∎

We are now able to prove a continuous analogue to Meshalkin's theorem [7, 14], a multinomial generalization of the continuous Sperner theorem (Theorem 3.5).

THEOREM 4.3 (Continuous Meshalkin Theorem) *Let σ be an s-system in* $\mathrm{Dec}(n,r)$. *Then*

$$\nu_{n;r}(\sigma) \leq \left[\underbrace{\langle n/r \rangle, \ldots, \langle n/r \rangle}_{r-b}, \underbrace{\langle n/r \rangle + 1, \ldots, \langle n/r \rangle + 1}_{b} \right]^{n},$$

where $n \cong b \mod r$.

PROOF: We reason in analogy to the proof of Meshalkin's theorem due to Hochberg and Hirsch [7]. For $a_1 + \cdots + a_r = n$, let $\sigma_{a_1,\ldots,a_r} = \sigma \cap \mathrm{Mult}(n; a_1, \ldots, a_r)$. Combining (4.3) and Proposition 4.2, we obtain

$$\sum_{a_1+\cdots+a_r=n} \frac{\nu^n_{a_1,\ldots,a_r}(\sigma_{a_1,\ldots,a_r})}{\left[{n \atop \langle n/r \rangle, \ldots, \langle n/r \rangle, \langle n/r \rangle + 1, \ldots, \langle n/r \rangle + 1} \right]} \leq \sum_{a_1+\cdots+a_r=n} \frac{\nu^n_{a_1,\ldots,a_r}(\sigma_{a_1,\ldots,a_r})}{\left[{n \atop a_1,\ldots,a_r} \right]} \leq 1$$

so that

$$\nu_{n;r}(\sigma) = \sum_{a_1+\cdots+a_r=n} \nu^n_{a_1,\ldots,a_r}(\sigma_{a_1,\ldots,a_r}) \leq \left[{n \atop \langle n/r \rangle, \ldots, \langle n/r \rangle, \langle n/r \rangle + 1, \ldots, \langle n/r \rangle + 1} \right].$$

∎

Bibliography

[1] Artin, E., *The Gamma Function*, Holt, Rinehart, and Winston, New York, 1964.

[2] Chen, B., and Rota, G.-C., *Totally invariant set functions of polynomial type*, Comm. Pure Appl. Math. 47, 1994, pp. 187–197.

[3] Erdös, P., *On a lemma of Littlewood and Offord*, Bull. Amer. Math. Soc. 51, 1945, pp. 898–902.

[4] Fisk, S., *Whitney numbers of projective space over* $\mathbb{R}$, $\mathbb{C}$, $\mathbb{H}$, *and the p-adics*, J. Combin. Theory Ser. A 70, No. 1, 1995, pp. 165–169.

[5] Hadwiger, H., *Vorlesungen über Inhalt, Oberfläche, und Isoperimetrie*, Springer-Verlag, Berlin, 1957.

[6] Harper, L. H., and Rota, G.-C., *Matching theory, an introduction*, pp. 169–215 in: *Advances in Probability and Related Topics* 1, Dekker, New York, 1971.

[7] Hochberg, M., and Hirsch, W. M., *Sperner families, s-systems, and a theorem of Meshalkin*, Ann. New York Acad. Sci. 175, 1970, pp. 224–237.

[8] Klain, D., *A short proof of Hadwiger's characterization theorem*, Mathematika 42, 1995, pp. 329–339.

[9] Klain, D., and Rota, G.-C., *Introduction to Geometric Probability*, Cambridge University Press, New York, to appear.

[10] Kung, Joseph P. S., ed., *Gian-Carlo Rota on Combinatorics: Introductory Papers and Commentaries*, Birkhäuser, Boston, 1995.

[11] Lubell, D., *A short proof of Sperner's theorem*, J. Combin. Theory 1, 1966, pp. 209–214.

[12] McMullen, P., *Non-linear angle-sum relations for polyhedral cones and polytopes*, Math. Proc. Cambridge Philos. Soc. 78, 1975, pp. 247–261.

[13] McMullen, P., and Schneider, R., *Valuations on convex bodies*, pp. 170–247 in: *Convexity and Its Applications*, P. M. Gruber and J. M. Wills, eds., Birkhäuser, Boston, 1983.

[14] Meshalkin, L. D., *A generalization of Sperner's theorem on the number of subsets of a finite set*, Theory Probab. Appl. 8, 1963, pp. 203–204.

[15] Nachbin, L., *The Haar Integral*, Van Nostrand, New York, 1965.

[16] Santaló, L. A., *Integral Geometry and Geometric Probability*, Addison-Wesley, Reading, Mass., 1976.

[17] Schneider, R., *Convex Bodies: The Brunn-Minkowski Theory*, Cambridge University Press, New York, 1993.

[18] Sperner, E., *Ein Satz über Untermengen einer endlichen Menge*, Math. Z. 27, 1928, pp. 544–548.

DANIEL A. KLAIN
School of Mathematics
Georgia Institute of Technology
Atlanta, GA 30332-0160
E-mail: `klain@math.gatech.edu`

GIAN-CARLO ROTA
Department of Mathematics
Massachusetts Institute of Technology
Cambridge, MA 02139
E-mail: `rota@math.mit.edu`

Received March 1996.
Revised April 1996.

Proc. Natl. Acad. Sci. USA
Vol. 90, pp. 697–699, January 1993
Mathematics

A stochastic interpretation of the Riemann zeta function

KENNETH S. ALEXANDER*, KENNETH BACLAWSKI†, AND GIAN-CARLO ROTA‡

*Department of Mathematics, University of Southern California, Los Angeles, CA 90089; †College of Computer Science, Northeastern University, Boston, MA 02115; and ‡Department of Mathematics, Massachusetts Institute of Technology, Cambridge, MA 02139

Contributed by Gian-Carlo Rota, August 10, 1992

ABSTRACT We give a stochastic process for which the terms of the Riemann zeta function occur as the probability distributions of the elementary random variables of the process.

Section 1. Introduction

As in ref. 1 we use the term *process* for a probability space Ω together with a parametrized family of random variables. The purpose of this paper is to exhibit a process (Ω, Z_s) parametrized by integers $s > 1$ such that the probability distributions of the Z_s are the terms of the Riemann zeta function—i.e., $\Pr(Z_s = n) = n^{-s}\zeta(s)^{-1}$.

Section 2. The Rédei Zeta Function of an Inverse System

Let $(\{G_\alpha:\alpha \in L\}, \{\phi_{\beta,\alpha}:G_\beta \to G_\alpha\})$ be an inverse system of finite groups over a lattice L. We will write $\hat{0}$ for the minimum element of L, and we will assume that $G_{\hat{0}}$ is a one-element group. The profinite completion $G = \varprojlim G_\alpha$ has a unique normalized Haar measure that we will denote Pr. We may then regard G as a probability space.

If, in addition, we assume that the structure homomorphisms $\phi_{\beta,\alpha}$ are all surjective, then we can define a function ν on pairs of elements (α, β) in L such that $\alpha \le \beta$, by the formula $\nu(\alpha, \beta) = |\mathrm{Coker}(\phi_{\beta,\alpha})|$, which then satisfies the condition

$$\text{for any triple } \alpha \le \gamma \le \beta, \qquad \nu(\alpha, \beta) = \nu(\alpha, \gamma)\nu(\gamma, \beta).$$

A lattice with such a function is called a *lattice of Dirichlet type*, and ν is called its *order function*. We abbreviate $\nu(\hat{0}, \alpha)$ to $\nu(\alpha) = |G_\alpha|$.

Lattices of Dirichlet type were introduced in ref. 2, and for such lattices a formal Dirichlet series was introduced called the *Rédei zeta function*. This function is defined by

$$\rho(s; L) = \sum_{\alpha\in L} \mu(\hat{0}, \alpha)\nu(\alpha)^{-s},$$

where μ is the Möbius function of L (see ref. 3).

It is easy to check that inverse limits preserve products. In particular, from the inverse system above one can form the *kth power inverse system*

$$(\{G_\alpha^k:\alpha \in L\}, \{\phi_{\beta,\alpha}^{(k)}:G_\beta^k \to G_\alpha^k\}),$$

whose limit will be denoted G^k. The lattices of the kth power inverse systems are all isomorphic as lattices, but they differ from one another as lattices of Dirichlet type. Let $L^{(k)}$ be the lattice of Dirichlet type associated with the kth power inverse system. It is easy to check that

$$\rho(s; L^{(k)}) = \rho(sk; L), \qquad \text{for every } k \ge 1.$$

The publication costs of this article were defrayed in part by page charge payment. This article must therefore be hereby marked "*advertisement*" in accordance with 18 U.S.C. §1734 solely to indicate this fact.

A lattice L with minimum element $\hat{0}$ is said to be *homogeneous* if

$$\text{for every } \alpha \in L, \{\beta \in L:\beta \ge \alpha\} \cong L.$$

The image of $\gamma \in \{\beta:\beta \ge \alpha\}$ in L via this isomorphism will be written γ/α.

A lattice L of Dirichlet type is said to be *homogeneous* if L is homogeneous as a lattice and if in addition

$$\text{for every } \alpha \le \beta \text{ in L}, \qquad \nu(\beta/\alpha) = \nu(\beta)/\nu(\alpha).$$

An inverse system $(\{G_\alpha\}, \{\phi_{\beta,\alpha}\})$ is said to be *homogeneous* if the following conditions apply:

(*i*) The lattice of the inverse system is homogeneous.

(*ii*) For every $\alpha \le \beta$ in L, $\mathrm{Ker}(\phi_{\beta,\alpha}) \cong G_{\beta/\alpha}$.

(*iii*) The isomorphisms above commute with the structure homomorphisms of the inverse system.

It is easy to check that the lattice of a homogeneous inverse system is also homogeneous as a Dirichlet lattice.

Section 3. Möbius Inversion on an Infinite Lattice

We will use Möbius inversion on L, so it is useful to formalize the basic convergence result we need.

THEOREM 1. *Let* P *be a locally finite poset. Let* $g:P \to C$ *be a function such that* $\sum_{y\ge x} g(y)$ *converges absolutely for every* $x \in P$. *Write* $f(x)$ *for the sum* $\sum_{y\ge x} g(y)$. *If the double sum* $\sum_{w\ge x}\sum_{y\ge w}\mu(x, w)g(y)$ *converges absolutely for every* $x \in P$, *then* $\sum_{y\ge x}\mu(x, y)f(y)$ *converges absolutely to* $g(x)$ *for every* $x \in P$.

Proof: Absolutely convergent series are arbitrarily rearrangeable. Therefore

$$\sum_{w\ge x} \mu(x, w) \sum_{y\ge w} g(y) = \sum_{y\ge x} g(y) \sum_{y\ge w\ge x} \mu(x, w),$$

$$\text{for every } x \in P.$$

Now the left-hand side of the equation above is $\sum_{w\ge x}\mu(x, w)f(w)$, while the right-hand side is $\sum_{y\ge x} g(y)\delta(x, y) = g(x)$, by definition of the Möbius function on a locally finite poset. Absolute convergence clearly also holds. ∎

We now give an example of the use of Möbius inversion in the setting of *Section 2*.

THEOREM 2. *Let* L *be a homogeneous lattice of Dirichlet type with order function* ν. *If*

(i) $|\mu(\hat{0}, \alpha)|$ *is bounded by a polynomial in* $\nu(\alpha)$ *of degree* k ≥ 0, *and*

(ii) $\sum_{\alpha\in L}\nu(\alpha)^{-s}$ *converges absolutely for* $Re(s) > s_0$; *then*

(i) $\rho(s; L)$ *converges absolutely for* $Re(s) > s_0 + k$, *and*

(ii) $\rho(s; L)\,(\sum_{\alpha\in L}\nu(\alpha)^{-s}) = 1$, *for* $Re(s) > s_0 + k$.

Proof: We first check conclusion *i*. By condition *i*, we have that, for some constant $C > 0$,

$$\sum_{\alpha\in L} |\mu(\hat{0}, \alpha)||\nu(\alpha)^{-s}| \le C \sum_{\alpha\in L} \nu(\alpha)^k|\nu(\alpha)^{-s}|$$

$$= \sum_{\alpha\in L} \nu(\alpha)^{k-Re(s)}.$$

By condition *ii*, the last expression above is finite for $\mathrm{Re}(s) > s_0 + k$. This gives conclusion *i* by the definition of ρ.

To show conclusion *ii*, one proceeds as in the proof of *Theorem 1*. Form the following double sum:

$$\sum_{\alpha,\beta\in L} \mu(\hat{0}, \beta)\nu(\beta)^{-s}\nu(\alpha)^{-s}.$$

We first show that this double sum converges absolutely for $\mathrm{Re}(s) > s_0 + k$ by using the following bounds:

$$\begin{aligned}\sum_{\alpha,\beta\in L} |\mu(\hat{0}, \beta)||\nu(\beta)^{-s}||\nu(\alpha)^{-s}| &\le C\sum_{\alpha,\beta} \nu(\beta)^{k}\nu(\beta)^{-\mathrm{Re}(s)}\nu(\alpha)^{-\mathrm{Re}(s)}\\ &\le C\sum_{\alpha} \nu(\alpha)^{-\mathrm{Re}(s)}\sum_{\beta} \nu(\beta)^{k-\mathrm{Re}(s)}\\ &< \infty, \quad \text{for } \mathrm{Re}(s) > s_0 + k.\end{aligned}$$

The last bound above follows from the fact that both of the series converge in the specified region by condition *ii* and conclusion *i*. It follows that the double sum above may be rearranged in any order. In particular, it follows that:

$$\begin{aligned}\sum_{\alpha,\beta\in L} \mu(\hat{0}, \beta)\nu(\beta)^{-s}\nu(\alpha)^{-s} &= \sum_{\beta} \mu(\hat{0}, \beta)\nu(\beta)^{-s}\sum_{\alpha} \nu(\alpha)^{-s}\\ &= \rho(s; L)\left(\sum_{\alpha} \nu(\alpha)^{-s}\right),\\ &\quad \text{for } \mathrm{Re}(s) > s_0 + k.\end{aligned}$$

On the other hand, by homogeneity of L, for each pair $\alpha,\beta \in L$, there is a unique $\gamma \in L$ such that $\gamma/\beta = \alpha$. Moreover, for this γ we have $\nu(\alpha)\nu(\beta) = \nu(\gamma)$. For a fixed β, a sum over all $\alpha \in L$ is then equivalent to a sum over $\gamma \ge \beta$. Thus the double sum can also be rearranged as follows:

$$\begin{aligned}\sum_{\alpha,\beta\in L} \mu(\hat{0}, \beta)\nu(\beta)^{-s}\nu(\alpha)^{-s} &= \sum_{\beta} \mu(\hat{0}, \beta)\sum_{\gamma\ge\beta} \nu(\gamma)^{-s}\\ &= \sum_{\gamma}\left(\sum_{\hat{0}\le\beta\le\gamma} \mu(\hat{0}, \beta)\right)\nu(\gamma)^{-s}\\ &= \delta(\hat{0}, \gamma)\nu(\gamma)^{-s}\\ &= \nu(\hat{0})^{-s}\\ &= 1, \quad \text{for } \mathrm{Re}(s) > s_0 + k.\end{aligned}$$

The theorem now follows. ∎

Section 4. A Stochastic Interpretation of the Rédei Zeta Function

Let $(\{G_\alpha\}, \{\phi_{\beta,\alpha}\})$ be an inverse system as in *Section 2* with profinite completion G. Let Y be the random variable on G with values in $L \cup \{\infty\}$ given by

$$Y(x) = \begin{cases}\sup\{\alpha \in L: x_\alpha = e\},\\ \quad \text{if the supremum exists in L, and}\\ \infty, \text{ otherwise,}\end{cases}$$

where e denotes the identity element of the group G_α. Similarly, let $Y^{(t)}$ for $t \ge 1$ be the corresponding random variable on the tth power inverse system of $(\{G_\alpha\}, \{\phi_{\beta,\alpha}\})$. This sequence of random variables gives the following stochastic interpretation of the Rédei zeta function:

THEOREM 3. *Let* G *be the profinite limit of a homogeneous inverse system over a lattice* L, *having associated random variables* $Y^{(t)}$, *where* $t > 0$ *is an integer. If* L *satisfies the hypotheses of* Theorem 2 *and if* $Pr(Y^{(t)} = \hat{0}) > 0$, *for* $t > t_0$, *then*

$$Pr(Y^{(t)} = \alpha|Y^{(t)} \ne \infty) = \nu(\alpha)^{-t}\rho(t; L), \textit{ for } t > \max(t_0, s_0+k),$$

where k *and* s_0 *are specified by condition* ii *of* Theorem 2.

Proof: By homogeneity, we know that

$$\Pr(Y^{(t)} = \alpha) = \nu(\alpha)^{-t}\Pr(Y^{(t)} = \hat{0}).$$

Hence

$$\begin{aligned}\Pr(Y^{(t)} \ne \infty) &= \sum_{\alpha\in L} \Pr(Y^{(t)} = \alpha)\\ &= \Pr(Y^{(t)} = \hat{0})\sum_{\alpha\in L} \nu(\alpha)^{-t}.\end{aligned}$$

If $t > t_0$, then $\Pr(Y^{(t)} = \hat{0}) > 0$ and hence also $\Pr(Y^{(t)} \ne \infty) > 0$. So in this case we have

$$\Pr(Y^{(t)} = \hat{0}|Y^{(t)} \ne \infty)\sum_{\alpha\in L} \nu(\alpha)^{-t} = 1, \quad \text{for } t > t_0.$$

By *Theorem 2* we may then conclude that

$$\Pr(Y^{(t)} = \hat{0}|Y^{(t)} \ne \infty) = \rho(t; L), \quad \text{for } t > \max(t_0, s_0 + k).$$

Finally, using homogeneity once more, we obtain the result in general. ∎

The random variables $Y^{(t)}$ have the desired probability distributions, but they are on different probability spaces. We would like a single probability space that supports all of these random variables in a natural way.

For an inverse system $(\{G_\alpha\}, \{\phi_{\beta,\alpha}\})$ with profinite completion G, let G^∞ denote the product of countably many copies of G, i.e., $\Pi_{i=1}^{\infty}G$. For every positive integer t, let $\pi_{[1,t]}:G^\infty \to G^t$ denote the projection onto the first t components of G^∞. Finally, let $Z_t:G^\infty \to C$ be the composition $Y^{(t)} \circ \pi_{[1,t]}$. Because the probability measure on G^∞ is Haar measure, and hence the product measure, it follows that $Y^{(t)}$ and Z_t have the same distribution.

We summarize these considerations in the following:

COROLLARY 1. *Given the hypotheses of* Theorem 3, *there is a probability space* G^∞ *and sequence of random variables* Z_t, *for integers* $t > t_0$, *with values in* L *such that*

$$Pr(Z_t = \alpha|Z_t \ne \infty) = \nu(\alpha)^{-t}\rho(t; L), \quad \textit{for } t > \max(t_0, s_0 + k).$$

Section 5. The Riemann Zeta Function

We now specialize to the following classical situation. Let L = N be the lattice of positive integers ordered by divisibility. The groups G_n are the cyclic groups Z/nZ, and the order function is given by $\nu(n) = n$. The profinite completion is known to be $\hat{Z} \cong \Pi_{p \text{ prime}} Z_p$, where Z_p is the ring of p-adic integers. Note that the minimum element of N is $\hat{0} = 1$. The random variable $Y^{(t)}$ on $\hat{Z}^t$ takes values in $N \cup \{\infty\}$ and is defined by

$$Y^{(t)}(x) = \begin{cases}\sup\{n: x_n^{(1)} = x_n^{(2)} = \cdots = x_n^{(t)} = e\},\\ \quad \text{if this supremum exists, and}\\ \infty, \text{ otherwise.}\end{cases}$$

We then have the following stochastic interpretation of the Riemann zeta function:

THEOREM 4. *For every integer* $s > 1$, *we have*

$$Pr(Y^{(s)} = n) = n^{-s}\zeta(s)^{-1}.$$

Proof: First apply *Theorem 3*. Since the inverse system defined above is obviously homogeneous, we must check that the hypotheses of *Theorem 2* hold and also that $\Pr(Y^{(s)} = \hat{0}) > 0$ for $s > 1$. Now the Möbius function of N takes values ±1 and 0, so the first hypothesis holds with $k = 0$. The sum

$\sum_{a \in L} \nu(a)^{-s}$ is, in this case, the sum defining the Riemann zeta function $\zeta(s) = \sum_{n=1}^{\infty} n^{-s}$. This is known to converge for $\mathrm{Re}(s) > s_0 = 1$.

To check that $\Pr(Y^{(s)} = 1) > 0$, we introduce the random variables $X_p^{(s)}$ on $\hat{\mathbb{Z}}_p^s$ by $X_p^{(s)}(x) = x_p \in (\mathbb{Z}/p\mathbb{Z})^s$, for every prime p. Since $\hat{\mathbb{Z}}^s \cong \Pi_{p \text{ prime}} \mathbb{Z}_p^s$, and since Pr is a Haar measure, the X_p are independent random variables. Now $Y^{(s)}$ has value $\hat{0} = 1$ if and only if every $X_p^{(s)}$ is not the identity element of $(\mathbb{Z}/p\mathbb{Z})^s$. Hence

$$\begin{aligned} \Pr(Y^{(s)} = 1) &= \prod_p \Pr(X_p^{(s)} \neq e) \\ &= \prod_p (1 - p^{-s}) \\ &= \zeta(s)^{-1}, \qquad \text{for } s > 1, \end{aligned}$$

by a well-known product expansion for $\zeta(s)$ valid for $\mathrm{Re}(s) > 1$.

Therefore, by *Theorem 3*, we have that

$$\Pr(Y^{(s)} = n | Y^{(s)} \neq \infty) = n^{-s}\zeta(s)^{-1}, \qquad \text{for } s > 1.$$

On the other hand, setting $n = 1$ in the formula above and comparing it with the earlier unconditional probability computed above yields the fact that $\Pr(Y^{(s)} \neq \infty) = 1$. The result then follows. ∎

We remark that one can show

$$\Pr(Y^{(s)} \neq \infty) = \begin{cases} 0, & \text{if } s = 1, \text{ and} \\ 1, & \text{if } s > 1. \end{cases}$$

This is consistent with the fact that $(Y^{(s)} = \infty)$ is a "tail event."

As in *Corollary 1* of *Section 4*, we can restate *Theorem 4* in terms of a process on a probability space. This gives the result stated in *Section 1:*

Corollary 2. *There is a probability measure on $\hat{\mathbb{Z}}^\infty$, and a sequence of random variables $\mathbb{Z}_s$ on $\hat{\mathbb{Z}}^\infty$, for $s > 1$, such that*

$$Pr(\mathbb{Z}_s = n) = n^{-s}\zeta(s)^{-1}, \qquad \textit{for every } s > 1.$$

This research was supported by National Science Foundation Grants DMS-90006395 to K.S.A., MCS-7903029 to K.B., and MCS-7308445 to G.-C.R.

1. Baclawski, K., Cerasoli, M. & Rota, G.-C. (1984) *Introduction to Probability and Random Processes* (Unione Matematica Italiana, Bologna, Italy) (published in Italy under the title *Introduzione alla Probabilità*).
2. Kung, J., Murty, M. & Rota, G.-C. (1980) *J. Number Theory* 12, 421–436.
3. Rota, G.-C. (1964) *Z. Wahrsch. Verw. Gebiete* 2, 340–368.

An Example of Profinite Combinatorics

GIAN-CARLO ROTA

Let us begin with a piece of history-fiction and fantasize how Riemann might have discovered the Riemann zeta function. Professor Riemann was aware that arithmetic density is of fundamental importance in number theory. If A is a subset of the set of positive integers N, then the *arithmetic density* of the set A is defined by the formula

$$\operatorname{dens}(A) = \lim_{n\to\infty} \frac{|A \cap \{1, 2, \ldots, n\}|}{n}$$

whenever the limit exists. For example, $\operatorname{dens}(N) = 1$. If A_p is the set of multiples of the prime p, then

$$\operatorname{dens}(A_p) = \frac{1}{p};$$

what is more appealing, one can easily compute that

$$\operatorname{dens}(A_p \cap A_q) = \frac{1}{pq}$$

for any two primes p and q. If density were a probability, we would infer that the events that a randomly chosen number is divisible by either of two primes are independent. Unfortunately, arithmetic density shares some but not all properties of a probability.

After a period of soul-searching, Riemann was able to find a remedy to all the deficiencies of arithmetic density by a brilliant leap of imagination. He chose a real number s greater than 1, and defined the measure of a positive integer n to equal $1/n^s$; in this way, the measure of the set N turned out to equal

$$\zeta(s) = \sum_{n=1}^{\infty} \frac{1}{n^s}.$$

Therefore, he could define a probability P_s on the set N of positive integers by setting

$$P_s(A) = \frac{1}{\zeta(s)} \sum_{n\in A} \frac{1}{n^s}.$$

Riemann then proceeded to verify what he had sensed all along, namely, the fundamental property

$$P_s(A_p \cap A_q) = P_s(A_p)P_s(A_q) = \frac{1}{pq},$$

This essay is extracted from Rota's notes for the Third Colloquium Lecture "Combinatorial Snapshots" delivered at the annual meeting of the American Mathematical Society, Baltimore, January 9, 1998. This version was edited by Joseph P. S. Kung.

which implies that the events that a randomly chosen number n is divisible by either of two primes p and q are independent relative to the probability P_s. The zeta function was good for something after all.

I will now use a rhetorical device that was effectively employed by one of my undergraduate teachers, Professor Salomon Bochner. In the classroom, Bochner would prefix the statement of a theorem by the words, "Subject to technical assumptions, the following is true", without, of course, ever disclosing what his technical assumptions were. I have never had the *chutzpah* to imitate Bochner until now.

Riemann then proceeded to show that, subject to technical assumptions on the set A,

$$\lim_{s \to 1} P_s(A) = \text{dens}(A).$$

Even though arithmetic density is not a probability, it is, under suitable conditions, the limit of probabilities.

Long after Riemann was gone, it was shown, again subject to technical assumptions, that the probabilities P_s are the only probabilities defined on the set N for which the events of divisibility by different primes are independent. This fact seems to lend support to the program of proving results of number theory by probabilistic methods.

Why didn't Riemann ever publish this wonderful idea? The answer is not hard to find. True, some theorems of number theory, for example, Dirichlet's theorem on primes in an arithmetic progression, can be proved probabilistically by this limiting process. However, deeper number theoretic results have to this day eluded this approach. For example, no one has succeeded in proving the prime number theorem by this method. Riemann, aware of this deficiency, threw his notes into the wastebasket and proceeded to link his zeta function to the distribution of primes in an altogether different way, by stating the hypothesis that bears his name.

Why am I telling you this bit of history-fiction? I want to propose another probabilistic interpretation of the Riemann zeta function that is quite different from the interpretation just outlined.

Let us consider a problem in combinatorial enumeration. Suppose that a cyclic group C_r of order r is given. Every character χ of the group C_r has a kernel which is a subgroup of C_r. More generally, every sequence $\chi_1, \chi_2, \ldots, \chi_s$ of characters of C_r has a joint kernel which is also a subgroup of C_r; here, the *joint kernel* of a sequence of characters is simply the intersection of the kernels of each of the characters in the sequence. If a sequence $\chi_1, \chi_2, \ldots, \chi_s$ of characters is chosen independently and at random, what is the probability that its joint kernel equals a given subgroup C_n of C_r?

The probability of the event that the kernel of a randomly chosen character will contain the subgroup C_n equals $1/n$, since there are r characters of the group C_r and r/n such characters will vanish on C_n. Therefore, the probability that the joint kernel of a randomly chosen sequence $\chi_1, \chi_2, \ldots, \chi_s$ of s characters will contain the subgroup C_n equals $(1/n)^s$. Denoting by P_{C_n} the probability that the joint kernel of the characters $\chi_1, \chi_2, \ldots, \chi_s$ will contain the subgroup C_n, we have the identity

$$\frac{1}{n^s} = \sum_{n|d|r} P_{C_d}.$$

Here, we use the fact that the partially ordered set of subgroups of a cyclic group C_r is isomorphic to the partially ordered set of divisors of the integer r, ordered by divisibility.

We now use the Möbius inversion formula of number theory, thereby obtaining

$$P_{C_n} = \sum_{n|d|r} \mu(d/n)\frac{1}{d^s}.$$

Here, $\mu(j)$ is the Möbius function of number theory.

After the change of variable $d = nj$, we can recast the right hand side as follows:

$$P_{C_n} = \frac{1}{n^s}\sum_{j} \mu(j)\frac{1}{j^s},$$

where the variable j on the right ranges over some subset of divisors of the integer r, which we need not know exactly. Now the sum

$$\sum_{j} \mu(j)\frac{1}{j^s},$$

with j ranging over all positive integers, equals $1/\zeta(s)$, the inverse of the zeta function. If we could change our combinatorial problem to get an unrestricted sum on the right hand side, then we would have a probabilistic interpretation of the zeta function.

This is done by replacing the finite cyclic group C_n by a "profinite" group. The group C_∞ of rational numbers modulo 1 (under addition) is such a group. Every finitely generated subgroup of C_∞ is a finite cyclic group C_n. The character group C_∞^* of C_∞ is a compact group. It has a Haar measure which is a probability measure P.

The set of all characters of the group C_∞ which vanish on a subgroup C_n has a Haar measure equal to $1/n$. Thus, if we choose a sequence of s characters of C_∞ independently and at random, the probability that their joint kernel will contain the group C_n equals $(1/n)^s$. If we again denote by P_{C_n} the probability that the joint kernel of a sequence of s characters equals the group C_n, then we have the identity

$$\frac{1}{n^s} = \sum_{n|d} P_{C_d},$$

where the sum on the right is now infinite. Again by the Möbius inversion formula, we obtain

$$P_{C_n} = \sum_{n|d} \mu(d/n)\frac{1}{d^s} = \frac{1}{n^s\zeta(s)}.$$

This is the promised probabilistic interpretation of the Riemann zeta function. Some properties of the Riemann zeta function can be derived probabilistically using this interpretation. An example is the product formula. It remains to be seen what other properties of the Riemann zeta function can be proved in this way.

The preceding argument is an instance of a generalization of an enumeration problem on a finite set to an enumeration problem on a profinite "set". Such a replacement of a finite set by a profinite set works in other combinatorial problems. Will we ever have a profinite combinatorics on profinite sets side by side with combinatorics on a finite set?

Editorial note. In the essay, Rota invoked the character group of C_∞ and a Haar measure on it. This can be replaced by an elementary argument.

The group C_∞ of rational numbers modulo one under addition is not profinite in the usual sense of the word. Rather, since

$$C_\infty = \bigcup_{n=1}^{\infty} \{0, \frac{1}{n}, \frac{2}{n}, \dots, \frac{n-1}{n}\},$$

the group is *in*finite, or perhaps, *in*cyclic, the injective limit of a directed system of finite cyclic groups. One might view it as a profinite group in the opposite category if one wishes.

It is easy to see that for every non-negative integer n, there is exactly one finite subgroup of order n, namely, the subgroup generated by $1/n$. This subgroup is denoted C_n. We do not need to know exactly what the character group of C_∞ is. What we need is the fact that when r is an integer, the functions $\chi_r : C_\infty \to \mathbb{C}$, defined by

$$\chi_r(x) = \exp(2\pi r x i)$$

are characters. These characters form a subgroup of the character group isomorphic to the integers under addition. Such a character χ_r has kernel containing the subgroup C_n if and only if n divides r. It is not difficult to make the following statement precise: the probability that a randomly chosen integer r is divisible by a fixed integer n is $1/n$. Indeed, this "follows" from the fact that the integers are divided naturally into n congruence classes modulo n, all of the same size. Hence, the probability that a random character χ_r vanishes on C_n is $1/n$. With this model, Rota's argument can now be carried through.

A different version of this argument is used in the paper "A stochastic interpretation of the Riemann zeta function" reprinted in this chapter. The profinite limit is taken there, resulting in a product of p-adic integers, and Rota's argument with the character group and Haar measure works rigorously.

Continuous and Profinite Combinatorics

Daniel A. Klain and Joseph P. S. Kung

1. Geometric probability

Gian-Carlo Rota believed that mathematics is a unity, in the deep sense that the same themes recur – as analogies – in its many branches. Thus, it comes as no surprise that Rota perceived combinatorial themes in continuous mathematics (as well as the other way around). In the preface to his book *Introduction to Geometric Probability* [10], written with the first author, Rota suggested only partly in jest that the field of geometric probability be renamed "continuous combinatorics". Two of the papers reprinted in this chapter paper belong to the research program described in [10].

Perhaps the central idea behind Rota's "continuous combinatorics" is the analogy between counting and measure, especially measures that are invariant with respect to some symmetry or group action. Here, the word "measure" is used in the broadest sense to include finitely additive measures which may admit no countably additive extension. These finitely additive measures, also called *valuations*, provide the intermediate hues in a spectrum of set functionals that extends from the purely discrete (such as lattice point enumerators and the Euler characteristic) to the analytic measures of Lebesgue theory.

Although far more attention has been given in the last century to the two extreme cases (combinatorics and real analysis), constructions in convex and integral geometry going back as far as Minkowski offer a panoply of invariant valuations that are neither wholly analytic nor combinatorial in nature. Functionals on polytopes and convex sets such as the mean width, projection functions onto flats, and more general families of intrinsic volumes [10, 12], mixed volumes [18], and dual mixed volumes [11], provide examples whose fundamental properties are still poorly understood as compared to Lebesgue measure and simple counting. It was Rota's contention that the best way to develop a comprehensive theory of these intermediate functionals is to determine how they connect analogous structures observed in combinatorics and real analysis, structures that are most evident in the contexts of combinatorial and analytic convex geometry.

The analogies between the intrinsic volumes or *Quermassintegrals* (characterized by Hadwiger [6] as the fundamental valuations invariant under rigid motions), the Ehrhart coefficients of lattice polytopes (which are affine unimodular invariants later characterized by Betke and Kneser [1]), and fundamental families of enumerative functionals on simplicial complexes and finite vector spaces (for example, face and subspace enumerators [5, 8, 9, 10]), provide further evidence that a comprehensive theory of invariant set functionals is waiting in the wings.

The motivation for the paper "Totally invariant set functions of polynomial type", written with Beifang Chen, can be found in Problem Five in [14]. This problem, "Set functions on convex bodies", is to prove the "correct" statement of the conjecture:

> *Subject to technical assumptions,* every invariant set function defined on convex bodies is associated with a symmetric function.

This conjecture is motivated by the observation that the intrinsic volumes of a rectangular parallelotope are elementary symmetric functions of the lengths of its sides (see [10] and the expository papers [15, 16]). In the reprinted paper, Chen and Rota proved a version of this conjecture, but, as Rota himself put it in [14], the technical assumptions are "preposterous." Rota's fifth problem awaits the development of a truly combinatorial approach to defining intrinsic volumes and other set functions on convex bodies invariant under Euclidean motions.

Sperner's theorem, that the maximum size of an antichain in the Boolean algebra of subsets of an n-element set equals the maximum binomial coefficient $\binom{n}{k}$, where k is $\lceil n/2 \rceil$ or $\lfloor n/2 \rfloor$, began extremal set theory. An analogue holds for the lattice of subspaces of vector space over a finite field. In the paper "A continuous analogue of Sperner's theorem", Rota and Klain proved an analogue for the lattice of subspaces of a finite-dimensional real vector space. The tools for doing this are a continuous analogue of the LYM-inequality and careful choices of normalizations for measures on Grassmannians, inspired in turn by normalizations that transform the Quermassintegrals into the intrinsic volumes on convex bodies. Variations on this theme can be found in [10].

2. Profinite limits

Perhaps the most striking example of a profinite limit is due to von Neumann [13] (see also [2], p. 237). Von Neumann observed that there is a natural embedding of projective geometries over a finite field $\mathrm{GF}(q)$ of order q which doubles the rank. Thinking of $\mathrm{PG}(n,q)$ as the lattice of subspaces of an n-dimensional subspace over $\mathrm{GF}(q)$, this embedding

$$\mathrm{PG}(2^m,q) \to \mathrm{PG}(2^m,q) \times \mathrm{PG}(2^m,q) \to \mathrm{PG}(2^{m+1},q)$$

is given by

$$a \mapsto (a,a) \mapsto (a,0) \vee (0,a).$$

Taking the completion of the profinite limit of the directed system

$$\mathrm{PG}(1,q) \to \mathrm{PG}(2,q) \to \mathrm{PG}(4,q) \to \dots \to \mathrm{PG}(2^m,q) \to \mathrm{PG}(2^{m+1},q) \to \dots,$$

von Neumann obtains the "pointless" *continuous geometry* $\mathrm{CG}(q)$ *over the finite field of order* q in which there are subspaces of "dimension" r for every real number r in the unit interval $[0,1]$.

This example must have fascinated Rota, and in his paper [5] with Jay Goldman, he proposed using it to study the theory of infinite q-series. As Rota has shown (see [5] or [14]), one can prove finite q-identities by finding bijections between objects (such as subspaces, bases, and linear transformations) defined on finite vector or projective spaces. However, identities between infinite q-series (such as the Rogers-Ramanujan identities) cannot be directly proved in this way. Rota suggested using the continuous geometries $\mathrm{CG}(q)$ for this purpose. He also suggested that there is a q-analog of the Poisson process on $\mathrm{CG}(q)$. Another natural problem is to find an analogue of Sperner's theorem for $\mathrm{CG}(q)$.

Rota suggested that von Neumann's construction can also be done with lattices of partition of finite sets. This is done in [3]. See also [7]. However, the next step ([17], Section 4), that of using such continuous partition lattice to study entropy, has yet to be taken.

Von Neumann's construction is one example of a general method for constructing an infinite object from a directed system of finite objects. Rota was fascinated by such "profinite" objects; they seem to offer another bridge between the finite and the infinite. Rota wrote only one paper devoted specifically to his ideas on profinite combinatorics. This is the paper "A stochastic interpretation of the Riemann zeta function", written with Kenneth Alexander and Kenneth Baclawski. In this paper, a profinite limit of cyclic groups is used to give an interpretation of zeta functions defined by critical problems. Rota was optimistic – perhaps too optimistic – about this probabilistic approach. His ultimate aim is to obtain a bijective proof of the functional equation. However, among a host of technical problems, this would require a compatible stochastic interpretation of the gamma function. The paper "An example of profinite combinatorics" is extracted from his notes for a lecture to a general audience. It gives an elementary account of the probabilistic interpretation of the Riemann zeta function.

References

[1] U. Betke and M. Kneser, Zerlegungen und Bewertungen von Gitterpolytopen. *J. Reine Angew. Math.* **358** (1985), 202–208.

[2] G. Birkhoff, *Lattice Theory*. 3rd edition, Amer. Math. Soc., Providence R.I., 1967.

[3] A. Björner, Continuous partition lattices. *Proc. Nat. Acad. Sci. U. S. A.* **84**(1987), 6327–6329.

[4] H. Crapo and D. Senato, eds., Algebraic Combinatorics and Computer Science, a Tribute to Gian-Carlo Rota. Springer-Verlag Italia, Milano, 2001, pp. 27–39.

[5] J. R. Goldman and G.-C. Rota, On the foundations of combinatorial theory. IV. Finite vector spaces and Eulerian generating functions. *Stud. Appl. Math.* **49**(1970), 239–258. Reprinted in J. P. S. Kung, ed., *Gian-Carlo Rota on Combinatorics,* Birkhäuser, Boston and Basel, 1995, pp. 226–260.

[6] H. Hadwiger, *Vorlesungen über Inhalt, Oberfläche, und Isoperimetrie.* Springer, Berlin, 1957.

[7] M. Haiman, On realization of Björner's continuous partition lattices by measurable partitions. *Trans. Amer. Math. Soc.* **343**(1994), 695–711.

[8] D. A. Klain, Kinematic formulas for finite lattices, *Ann. Comb.*, **1**(1997), 353-366.

[9] D. A. Klain, Kinematic formulas for finite vector spaces, *Discrete Math.*, **179**(1998), 121-132.

[10] D. A. Klain and G.-C. Rota, *Introduction to Geometric Probability.* Cambridge Univ. Press, Cambridge, 1997.

[11] E. Lutwak, Intersection bodies and dual mixed volumes. *Adv. Math.* **71**, No. 2 (1988), 232–261.

[12] P. McMullen, Valuations and dissections. In *Handbook of Convex Geometry,* P. M. Gruber and J. M. Wills, eds., North-Holland, Amsterdam, 1993, pp. 933–988.

[13] J. von Neumann, Examples of continuous geometries. *Proc. Nat. Acad. Sci. U. S. A.* **22**(1936), 101–108.

[14] G.-C. Rota, Ten mathematics problems I will never solve. *Mitt. Dtsch. Math.-Ver.,* **2**(1998), 45–52.

[15] G.-C. Rota, Geometric probability. *Math. Intelligencer,* **20**(1998), 11–16.

[16] G.-C. Rota, The adventures of measure theory. In [4], pp. 27–39.

[17] G.-C. Rota, Twelve problems in probability no one likes to bring up. In [4], pp. 57–93.

[18] R. Schneider, *Convex Bodies: The Brunn-Minkowski Theory.* Cambridge Univ. Press, Cambridge and New York, 1993.

CHAPTER 7

Probability Theory

Rend. Sem. Mat. Univ. Padova, Vol. 59 (1978)

A Measure Theoretic Approach to Logical Quantification.

David P. Ellerman (*) - Gian-Carlo Rota (**)

1. Introduction.

It is an old philosophical thesis that logic is, in some sense, a limiting case of probability theory. However, this thesis has not given rise to a mathematical treatment of quantification theory (i.e., first-order logic) by the methods of probability theory because of certain measure-theoretic difficulties. The principal difficulty is that it is impossible to put a *positive* finitely-additive real-valved probability measure on an uncountable set X (i.e., on the power-set Boolean algebra of X). If there was such a measure μ on a set X, then the existential quantification of a unary predicate on X could be obtained as the support of the expectation of the indicator or characteristic function of the predicate and the quantification of n-ary relations could be similarly treated by means of conditional expectation operators. However, if X is uncountable, then consider the sets $X_n = \{x \in X | \mu(\{x\}) > 1/n\}$ where n is a positive integer. Since $X = \bigcup_n X_n$ and since a countable union of countable sets is countable, at least one of the X_n is uncountable. But that contradicts $\mu(X) = 1$, so such a positive measure μ is impossible. A non-negative finitely-additive measure (e.g., an ultrafilter) can always be defined on any set X, but if the measure is not positive, then it ignores certain non-empty subsets

(*) Indirizzo dell'A.: University of Massachusetts, Boston, Mass. U.S.A.
(**) Indirizzo dell'A.: Massachusetts Institute of Technology, Cambridge, Mass. U.S.A. Work sponsored by NSF Grant No. MCS 78-02743.

and thus it does not permit a faithful treatment of logical quantification.

This difficulty can be circumvented by associating with each Boolean algebra B, a ring $G(B)$, called the *generator ring* of B, and a universal measure $|\ |: B \to G(B)$. The measure is universal in the sense that any finitely-additive measure $\mu: B \to R$, with values in any ring R (all rings are taken to be commutative with unity), factors through the universal measure by a unique ring homomorphism $\hat{\mu}: G(B) \to R$. In addition to allowing a measure-theoretic approach to logic, generator rings and the associated universal positive measures should be of use in probability theory itself.

It has long been known (e.g., Wright [12]) that there is an analogy between averaging operators such as the conditional expectation operators of probability theory and the algebraic (existential) quantifiers in Halmos' theory of polyadic algebras (Halmos [5]) or the cylindrifications used by Tarski and his co-workers in the theory of cylindric algebras (Henkin, Monk, and Tarski [7]). All these operators satisfy an averaging condition which has the general form: $A(f)\cdot A(g) = A\cdot \cdot(f\cdot A(g))$. We will provide some theoretical underpinning for this « analogy » by constructing the logical quantifiers using an abstract rendition of conditional expectation operators on a ring of simple random variables.

The propositional operations will be treated ring-theoretically using the *valuation rings* defined by Rota ([10] and [11]). The valuation rings of Boolean algebras might be viewed by logicians as a generalization of Boolean rings which occur as a trivial special case and they might be viewed by probability theorists as an intrinsic algebraic construction which generalizes the rings of simple random variables found in probability theory.

2. Valuation rings.

Let L be a distributive lattice with maximal element u and minimal element z, and let A be a ring (always commutative with unity). Let $F(L, A)$ be the free A-module on the elements of L which consists of all the finite formal sums $\sum_i a_i x_i$ for $a_i \in A$ and $x_i \in L$. A ring structure is put upon $F(L, A)$ by defining multiplication as $x\cdot y = = x\wedge y$ for lattice elements $x, y \in L$ and then extending by linearity to all the elements of $F(L, A)$. Let J be the submodule generated by all the elements of the form $x\vee y + x\wedge y - x - y$ for $x, y \in L$.

LEMMA. J is an ideal.

PROOF. Since J is a submodule, it suffices to check that J is closed under multiplication by lattice elements w. Now

$$w(x\vee y + x\wedge y - x - y) = w\wedge(x\vee y) + w\wedge(x\wedge y) - w\wedge x - w\wedge y\,.$$

Since L is a distributive lattice, we have

$$w\wedge(x\vee y) = (w\wedge x)\vee(w\wedge y) \quad \text{and} \quad w\wedge(x\wedge y) = (w\wedge x)\wedge(w\wedge y)\,.$$

Hence

$$w(x\vee y + x\wedge y - x - y) =$$
$$= (w\wedge x)\vee(w\wedge y) + (w\wedge x)\wedge(w\wedge y) - w\wedge x - w\wedge y$$

which is a generator of J.

The *valuation ring of L with values in A* is then defined as

$$V(L, A) = F(L, A)/J\,.$$

A *valuation* on L with values in an abelian group G is a function $\nu\colon L\to G$ such that $\nu(x\vee y) + \nu(x\wedge y) = \nu(x) + \nu(y)$ for all x and y in L. The injection $L\to(L, A)$ is a valuation and it is universal for valuations on L with values in an A-module.

THEOREM 1. Let M be an A-module and let $\nu\colon L\to M$ be a valuation. Then there exists a unique linear transformation (i.e., A-module homomorphism) $\hat{\nu}\colon V(L, A)\to M$ such that the following diagram commutes.

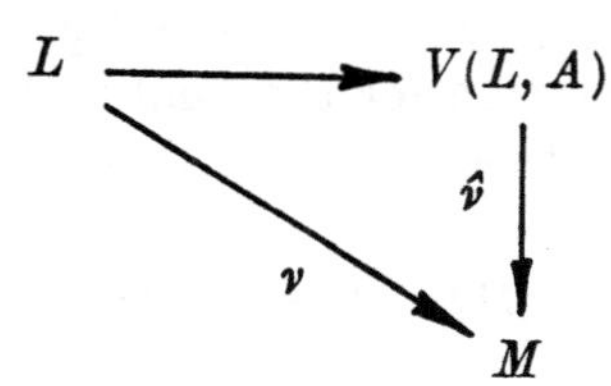

PROOF. By the universality property of the free module $F(L, A)$, there is a unique linear transformation $\bar{\nu}\colon F(L, A)\to M$ such that the left-hand triangle in the following diagram commutes.

$$\begin{array}{ccccc} L & \rightarrow F(L, A) & \rightarrow & V(L, A) = F(L, A)/J \\ & \searrow^{\nu} & \searrow^{\bar{\nu}} & \downarrow \hat{\nu} \\ & & & M \end{array}$$

Since ν is a valuation, the kernal of $\bar{\nu}$ contains J, so $\bar{\nu}$ extends to the unique linear transformation $\hat{\nu}$: $V(L, A) \rightarrow M$ as desired.

Corollary. The valuations on L with values in A are in one-to-one correspondence with the linear functionals on $V(L, A)$. The construction $L \rightarrow V(L, A)$ defines a functor from the category of distributive lattices (with u and z, and (u, z)-preserving lattice homomorphisms) to the category of A-algebras.

For any x in L, $u \cdot x = u \wedge x = x$ so the maximal element u of L serves as the unity of the valuation ring $V(L, A)$ (although the minimal element z is not the zero of the ring). In the situation of Theorem 1, if the A-module M is also an A-algebra and if $\nu(x \wedge y) = \nu(x) \cdot \cdot \nu(y)$ for all x and y in L, then the factor map $\hat{\nu}$ is an A-algebra homomorphism. The map $L \rightarrow A$ which carries each lattice element to 1 is a valuation satisfying that multiplicative condition, so there is an A-algebra homomorphism ε: $V(L, A) \rightarrow A$ which takes $\sum_i a_i x_i$ to $\sum_i a_i$. Hence $V(L, A)$ is an augmented algebra with the augmentation ε. Since $z \cdot \sum a_i x_i = \sum_i a_i (z \wedge x_i) = z \sum_i a_i$, the minimal element z of the lattice functions in the valuation ring as an integral which « computes » the augmentation $\varepsilon(\sum a_i x_i) = \sum a_i$ of any ring element.

As in any augmented algebra, there is another natural multiplication that can be put on the A-module structure of $V(L, A)$ in order to obtain a ring, i.e., $f \vee g = \varepsilon(g) \cdot f + \varepsilon(f) g - f \cdot g$ for all f and g in $V(L, A)$. This join notation for the dual multiplication is appropriate since if f and g are lattice elements, then

$$\varepsilon(g) f + \varepsilon(f) g - fg = f + g - f \wedge g = f \vee g$$

is the join of f and g. In the dual valuation ring, endowed with the V-multiplication, the roles of u and z are reversed, i.e., z is the unit and u is the integral.

The dual role of the « meet » and « join » multiplications is not the only aspect of Boolean duality which extends to arbitrary valuation rings. Even if the distributive lattice L is not complemented, a *complementation* endomorphism $\tau\colon V(L, A) \to V(L, A)$ can be defined on the valuation ring by $\tau(f) = \varepsilon(f)(u+z) - f$ for any f in $V(L, A)$. If x is a lattice element, then $\tau(x) = u + z - x$ so $\tau(u) = z$ and $\tau(z) = u$. Moreover if x does have a complement x' in L, then $x \vee x' + x \wedge x' - - x - x' = u + z - x - x' =$ in $V(L, A)$ so $\tau(x) = x'$. Complementation is also idempotent in the sense that $\tau^2(f) = f$ for any f in $V(L, A)$. Let us denote the valuation ring with the usual « meet » multiplication by $(V(L, A), \wedge)$ and let $(V(L, A), \vee)$ denote the ring with the « join » multiplication. Then complementation

$$\tau\colon (V(L, A), \wedge) \to (V(L, A), \vee)$$

is an isomorphism of augmented algebras and, since τ^2 is the identity mapping,

$$\tau\colon (V(L, A), \vee) \to (V(L, A), \wedge)$$

is also an isomorphism.

A valuation v is *normalized* if $v(z) = 0$. Given a valuation ring $V(L, A)$, the ring obtained by setting z equal to zero, i.e., $V(L, A)/(z)$, will be called the *normalized valuation ring*. If « normalized valuation » is substituted throughout for « valuation », then Theorem 1 will describe the universality property enjoyed by normalized valuation rings. Boolean algebras, when constructed as Boolean rings, occur as a rather special case of normalized valuation rings. When L is a Boolean algebra B and $A = 2$ $(= Z_2)$, then $(V(B, 2), \wedge)/(z)$ is simply B constructed as a Boolean ring with « meet » multiplication and $(V(B, 2), \vee)/(u)$ is B as a Boolean ring with « join » multiplication. For a history of the two interpretations of a Boolean algebra as a Boolean ring, with either the « meet » or the « join » multiplication, consult Church [11], pp. 103-104. In this special case of $L = B$ and $A = 2$, the complementation isomorphisms given above reduce to the familiar De Morgan's laws, i.e., $(x \wedge y)' = x' \vee y'$ and $(x \vee y)' = x' \wedge y'$. The generalization of Boolean duality to arbitrary valuation rings is due to Ladnor Geissinger, whose papers ([2], [3] and [4]) should be consulted for further analysis and applications of valuation rings.

In view of the above isomorphism, we can ignore the « join » multiplication, and always consider $V(L, A)$ as being endowed with the usual « meet » multiplication. We have heretofore refrained from normalizing the valuation ring because it is only the unnormalized valuation ring which enjoys the extensive duality theory given above. However, for a variety of reasons we will henceforth consider all valuation rings as being normalized unless otherwise specified. Hence $V(L, A)$ will now denote the *normalized* valuation ring on L with values in A. The normalization eliminates the augmentation, the « join » multiplication and the idempotent endofunction of complementation. However when a lattice element x has a complement x' in L, then $u - x = x'$.

3. Propositional calculus.

The conventional algebraic treatment of the propositional calculus utilizes the free Boolean algebra $B = B(P)$ which is free on the set P of propositional variables. We will formulate a treatment in the general setting of the (normalized) valuation ring $V(B, A)$, whose A is any commutative ring—instead of the special case of $V(B, 2) \cong B$ (where $2 = Z_2$). Instead of constructing the free Boolean algebra $B = B(P)$ and then $V(B, A)$, we will give a direct characterization of $V(B, A)$ for a free Boolean algebra B.

For any (commutative) ring A and any set P, let $A[P]$ be the polynomial ring over A with the elements of P as indeterminates. Let I be the ideal generated by the polynomials $p^2 - p$ for all p in P. Then $A[P]/I$ will be called an *idempotent polynomial ring* since it behaves like a polynomial ring except that the indeterminates are all idempotent.

CHARACTERIZATION THEOREM. The normalized valuation rings on free Boolean algebras are precisely the idempotent polynomial rings, *i.e.* for any commutative ring A (with unity) and any set P,

$$V(B(P), A) \cong A[P]/I\,.$$

PROOF. Let B_0 be the set of idempotents in $A[P]/I$. They form a Boolean algebra with the Boolean operations of $f \vee g = f + g - fg$, $f \wedge g = fg$, and $f' = 1 - f$ for f, g in B_0. Since $P \subseteq B_0$, by the universality property of the free Boolean algebra $B = B(P)$, there exists a unique Boolean algebra homomorphism $\nu\colon B \to B_0 \subseteq A[P]/I$ which commutes with the insertion of P. Since ν is a Boolean algebra homomorphism, we have

$$\nu(z) = 0\,, \qquad \nu(x \wedge y) = \nu(x) \wedge \nu(y) = \nu(x) \cdot \nu(y)\,,$$

and

$$\nu(x \vee y) = \nu(x) \vee \nu(y) = \nu(x) + \nu(y) - \nu(x) \cdot \nu(y) = \nu(x) + \nu(y) - \nu(x \wedge y)$$

for any x and y in B. Hence ν is a normalized multiplicative valuation on B with values in the A-algebra $A[P]/I$, so by Theorem 1 (for normalized multiplicative valuations), there exists a unique A-algebra homomorphism $\hat{\nu}$ such that the following diagram commutes;

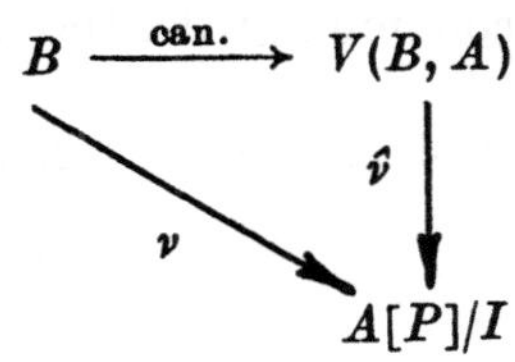

Let $w: P \to V(B, A)$ be the insertion of the elements of P into the A-algebra $V(B, A)$. Then by the universality property of $A[P]$ (e.g., Lang [8], p. 113), there exists a unique A-algebra homomorphism $\overline{w}$ such that the left-hand triangle in the following diagram commutes.

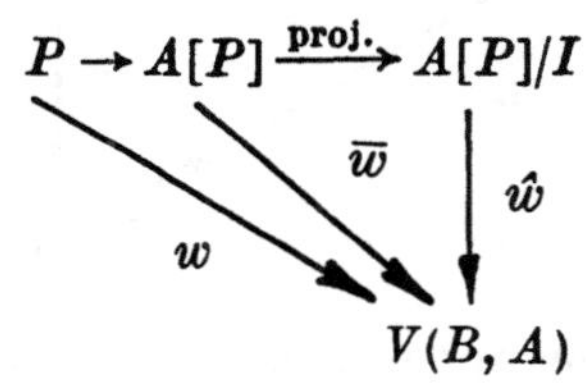

Then for any p in P, $\overline{w}(p^2 - p) = \overline{w}(p)^2 - \overline{w}(p) = p^2 - p = p - p = 0$ so $I \subseteq \ker(\overline{w})$. Hence there exists a unique A-algebra homomorphism $\hat{w}$ such that $\overline{w} = \hat{w} \cdot \text{proj}$, i.e., such that the right-hand triangle commutes.

Now consider the following diagram.

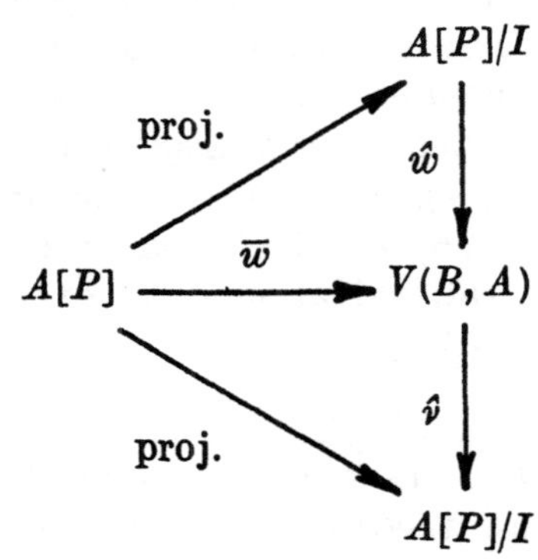

16

We have just seen that the upper triangle commutes. To see that the lower triangle commutes, note that $\hat{v}\cdot\overline{w}$ and proj are both A-algebra homomorphisms $A[P]\to A[P]/I$ which commute with the insertion of P. By the universality property of $A[P]$, there is only one such map so $\hat{v}\cdot\overline{w} = \text{proj}$. Since $\overline{w} = \hat{w}\cdot\text{proj}$, we have that $\text{proj} = \hat{v}\cdot\overline{w} = \hat{v}\cdot\hat{w}\cdot\text{proj}$, i.e., the outer triangle commutes. By the universality property of quotient rings, the identity is the unique A-algebra homomorphism $A[P]/I\to A[P]/I$ which commutes with the projections, so $\hat{v}\cdot\hat{w}$ is the identity on $A[P]/I$.

It remains to consider the following diagram.

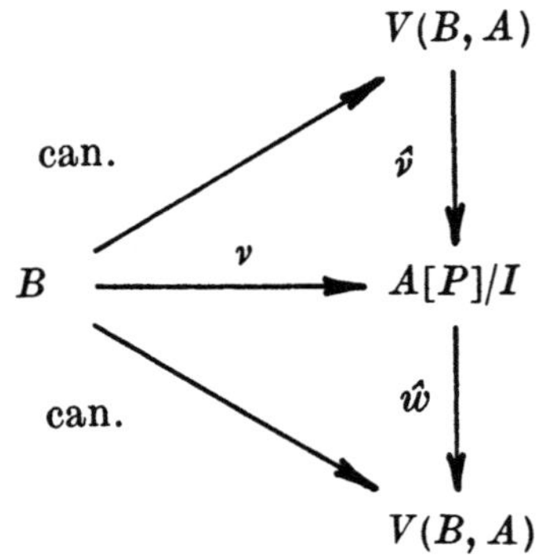

We have seen that the upper triangle commutes. To see that the lower triangle commutes, note that when the A-algebra homomorphism $\hat{w}$ is restricted to B_0, the Boolean algebra of idempotents in $A[P]/I$, then it constitutes a Boolean algebra homomorphism into the Boolean algebra of idempotents of $V(B, A)$. Hence $\hat{w}\cdot v$ and can. are both Boolean algebra homomorphisms $B\to V(B, A)$ which agree on the insertion of P. By the universality property of the free Boolean algebra $B = B(P)$, there is only one such map so $\hat{w}\cdot v = \text{can.}$, i.e. the lower triangle commutes. Since $v = \hat{v}\cdot\text{can.}$, we have that can. $=$ $= \hat{w}\cdot v = \hat{w}\cdot\hat{v}$ can., i.e., the outer triangle commutes. By the universality property of $V(B, A)$, the identity is the unique A-algebra homomorphism $V(B, A)\to V(B, A)$ which commutes with the normalized multiplicative (canonical) valuation can., so $\hat{w}\cdot\hat{v}$ is the identity on $V(B, A)$. Hence $V(B, A)\cong A[P]/I$.

Corollary. $B(P)\cong Z_2[P]/I$.

This characterization theorem reduces much of propositional logic (e.g., the completeness theorem) to elementary polynomial algebra.

The elements of P will be construed as propositional variables, and the elements of the idempotent polynomial ring $A[P]/I \cong V(B(P), A)$ will be called *Boolean polynomials*. The Boolean operations are rendered as ring operations in the following manner: for $f, g \in V(B(P), A)$, $f \vee g = f + g - fg$, $f \wedge g = fg$, $f \supset g = u - f + fg$, and $-f = u - f$, where u is the unity in $V(B(P), A) \cong A[P]/I$. A Boolean polynomial f is said to be a *theorem* if $f = u$ and is said to be *refutable* if $f = 0$ (see Halmos [6], pp. 45-46 for a justification of these definitions in the case of $A = Z_2 = 2$). In any axiomatization of the propositional calculus, each axiom is a Boolean polynomial equal to unity, and if $f = u$ and $f \supset g = u$, then $g = u$.

A ***truth-table*** or ***t-t valuation*** is a function $v\colon P \to 2$. A t-t valuation induces a Boolean algebra homomorphism $Z_2[P]/I \cong B(P) \to 2$. In general, for any (non-zero) commutative ring A with unity, we can view a t-t valuation v as taking values in $2 \subseteq A$. Hence it induces an A-algebra homomorphism $\bar{v}\colon A[P] \to A$ which vanishes on the generators $p^2 - p$ of the ideal I. Thus we have the induced A-algebra homomorphism $\hat{v}\colon A[P]/I \cong V(B(P), A) \to A$. A Boolean polynomial f is said to be a *tautology* if $\hat{v}(f) = 1$, or a *contradiction* if $\hat{v}(f) = 0$, for any t-t valuation $v\colon P \to 2$. Since any induced $\hat{v}$ is an A-algebra homomorphism, all theorems are tautologies and all refutable polynomials are contradictions.

Completeness theorem for propositional logic. For any $f \in V(B(P), A) \cong A[P]/I$, if f is a tautology, then $f = u$.

Proof. Since f is a tautology if and only if $u - f$ is a contradiction, we will prove the equivalent proposition: if f is a contradiction, then $f = 0$. Since f is actually an equivalence class of polynomials in $A[P]/I$, we consider any representative $f \in A[P]$. The proof is by induction over n, the number of propositional variables in f, which we assume for the sake of notational convenience to be $p_1, \ldots, p_n$.

Basis step: $n = 1$ so $f = f(p_1)$. For any t-t valuation $v\colon P \to 2$, $\bar{v}(f) = f(\bar{v}(p_1))$. Since f is a contradiction, $f(0) = 0$ so the constant term in the polynomial $f(p_1)$ is zero. Then p_1 may be factored out to obtain $f(p_1) = p_1 \cdot g(p_1)$. Now $f(1) = 1 \cdot g(1) = 0$ so $p_1 - u$ divides $g(p_1)$, i.e., for some $h(p_1)$, $g(p_1) = (p_1 - u)h(p_1)$. Hence

$$f(p_1) = p_1(p_1 - u)h(p_1) = (p_1^2 - p_1)h(p_1) \qquad \text{so } f = 0 \text{ in } A[P]/I\,.$$

Induction step: $f = f(p_1, \ldots, p_n)$ where we assume the theorem for polynomials of less than n variables. We note first that there is a polynomial in the equivalence class of f where p_n occurs with degree 1. For example, any monomial in the form $g(p_1, \ldots, p_{n-1})p_n^m$ for $m \geq 2$ is equivalent modulo $p_n^2 - p_n$ to $g(p_1, \ldots, p_{n-1})p_n^{m-1}$ since $gp_n^m - gp_n^{m-1} = gp_n^{m-2}(p_n^2 - p_n)$. In this manner, each occurrence of p_n in f can be reduced to an occurrence of degree 1 without changing the equivalence class. Hence we may assume that f has the form

$$f(p_n, \ldots, p_n) = f_0(p_1, \ldots, p_{n-1}) + f_1(p_1, \ldots, p_{n-1})p_n$$

where p_n does not occur in f_0 or f_1. For any t-t valuation ν such that $\nu(p_n) = 0$, we have $\bar{\nu}(f) = \bar{\nu}(f_0) = 0$. But p_n does not occur in f_0, so $\bar{\nu}(f_0) = 0$ for all t-t valuations ν. Since f_0 is a contradiction with less than n variables, we have, by the induction hypothesis, $f_0 = 0$ in $A[P]/I$. For any ν such that $\nu(p_n) = 1$, $\bar{\nu}(f) = \bar{\nu}(f_0) + \bar{\nu}(f_1) = \bar{\nu}(f_1) = 0$. But p_n does not occur in f_1 either so it is a contradiction. Hence f_1 and thus $f = f_0 + f_1 p_n$ is zero in $A[P]/I \cong V(B(P), A)$.

In view of this approach to propositional logic, it would seem quite possible and appropriate to continue in the same spirit to generalize the algebraitization of first-order logic which has been developed by Tarski and by Halmos. The Boolean algebras used by Tarski and Halmos would be replaced by valuation rings, and their quantifiers or cylindrification operators on Boolean algebras would be generalized to similar operators on valuation rings. However, such an abstract axiomatic approach to quantification would not fulfill our goal in this paper of providing a measure-theoretic treatment of the logical quantifiers. Hence we shall herein follow the alternative course of defining universal positive measures so that the logical quantifiers can be constructed, by measure-theoretic methods, in their natural habitat (i.e. when they quantify over a set).

4. Generator rings.

Let B_0 be a Boolean algebra and let A and R be commutative rings with unity. A *measure* on B_0 is a map $\mu\colon B_0 \to R$ such that for all b, b_1, and b_2 in B_0;

(1) $$\mu(u) = 1 .$$

(2) $$\mu(b) + \mu(b') = 1 ,$$

and

(3) $$\mu(b_1 \vee b_2) + \mu(b_1 \wedge b_2) = \mu(b_1) + \mu(b_2)\,.$$

Let $A[B_0]$ be the polynomial ring generated over the ring A by taking all the elements of B_0 as indeterminates (denoted $|b|$ for $b \in B_0$), and let K be the ideal generated by the polynomials

$$\{1-|u|,\ 1-|b|-|b'|,\ |b_1 \vee b_2| + |b_1 \wedge b_2| - |b_1| - |b_2| \text{: for } b, b_1, b_2 \in B_0\}\,.$$

Then $G(B_0, A) = A[B_0]/K$ is called the *generator ring on* B_0 *over* A. The insertion of the generators $B_0 \xrightarrow{||} G(B_0, A)$ is a measure on B_0 and it is universal for all measures on B_0 with values in an A-algebra.

THEOREM 2. Let R be an A-algebra and let $\mu\colon B_0 \to R$ be a measure. Then there is a unique A-algebra homomorphism $\hat{\mu}\colon G(B_0, A) \to R$ such that the following diagram commutes;

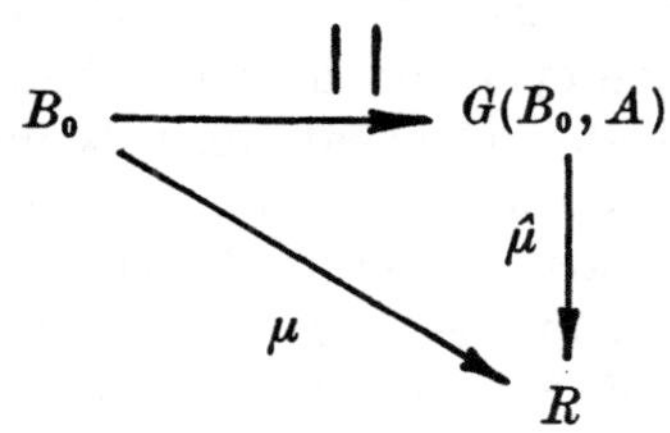

PROOF. By the universality property of polynomial rings, there is a unique A-algebra homomorphism $\bar{\mu}\colon A[B_0] \to R$ such that the left-hand triangle in the following diagram commutes.

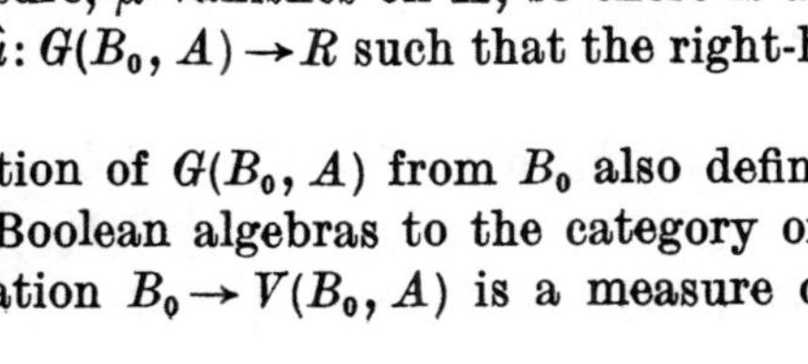

Since μ is a measure, $\bar{\mu}$ vanishes on K, so there is a unique A-algebra homomorphism $\hat{\mu}\colon G(B_0, A) \to R$ such that the right-hand triangle commutes.

The construction of $G(B_0, A)$ from B_0 also defines a functor from the category of Boolean algebras to the category of A-algebras. The normalized valuation $B_0 \to V(B_0, A)$ is a measure on B_0 with values in an A-algebra, so there is a canonical map $\varphi_{B_0}\colon G(B_0, A) \to V(B_0, A)$ which is a surjective A-algebra homomorphism. The quotient map φ_{B_0} « interpretes » the multiplication of generators $|b_1|\cdot|b_2|$ in $G(B_0, A)$ as being their intersection $b_1 \wedge b_2$ in $V(B_0, A)$. The maps $\{\varphi_{B_0}\colon B_0$ a Boolean algebra$\}$ constitute a natural transformation $\varphi\colon G(\cdot, A) \to \to V(\cdot, A)$ from the generator ring functor to the (normalized) valuation ring functor.

If there is a positive cone (semi-ring) C_A defined in A so that A is partially ordered by the relation $a_1 \leqslant a_2$ if $a_2 - a_1 \in C_A$, then we may extend the partial ordering to $G(B_0, A)$ by defining the positive cone C as the semi-ring generated over C_A by the « absolute values » $|b|$ for b in B_0 (i.e., close $C_A \cup \{|b|\colon b \in B_0\}$ under sums and products). Then $G(B_0, A)$ is a partially ordered ring with the ordering $x_1 \leqslant x_2$ if $x_2 - - x_1 \in C$. Let $C^+ = C - \{0\}$ so that $x_1 < x_2$ iff $x_2 - x_1 \in C^+$. Hence, if A is equipped with a positive cone, then there is an induced ordering on $G(B_0, A)$ so that the universal measure $B_0 \to G(B_0, A)$ is positive.

As an example of the use of generator rings in probability theory, we consider expectation operators on rings of simple random variables. If B_0 is a Boolean algebra of events on a (sample) space X and if R is the reals, then the (normalized) valuation ring $V(B_0, R)$ is the ring of simple real random variables on X. A random variable X in $V(B_0, R)$ can always be written in the form $x = \sum r_i p_i$ where $r_i \in R$ and $\{p_i\}$ is a finite partition of B_0. Given a real-valued measure $\mu\colon B_0 \to R$ (i.e., a finitely additive probability measure), the *expectation operator with respect to* μ, $E_\mu\colon V(B_0, R) \to V(B_0, R)$, is defined by

$$E_\mu\left(\sum r_i p_i\right) = \sum r_i \mu(p_i) u\,.$$

There is a different expectation operator E_μ for every different measure μ. However, by using the generator ring $G(B_0, R)$, we may define a *universal expectation operator*

$$E\colon V(B_0, G(B_0, R)) \to V(B_0, G(B_0, R)) \qquad \text{by } E\left(\sum \alpha_i p_i\right) = \sum \alpha_i |p_i| u\,.$$

For any measure $\mu\colon B_0 \to R$, the particular expectation operator E_μ can then be obtained by « specializing » the universal operator in the following manner:

$$\begin{array}{ccc} V(B_0, G(B_0, R)) & \xrightarrow{E} & V(B_0, G(B_0, R)) \\ \cup\!\shortmid & & \downarrow V(B_0, \hat{\mu}) \\ V(B_0, R) & \overset{E_\mu}{\dashrightarrow} & V(B_0, R)\,. \end{array}$$

measures on B_0, then $\mu_1 \times \mu_2$ is such a measure on $B_0 \otimes B_0$. Given a product measure $\mu_1 \times \mu_2$, a conditional expectation operator on the valuation ring is determined by a ***conditioning subalgebra*** B of $B_0 \otimes B_0$. We will only consider the conditioning subalgebras which are « independent » of specified coordinates. Thus $u \otimes B_0$ is the subalgebra of elements of the form $u \otimes q$ for q in B_0 and $B_0 \otimes u$ is the subalgebra of elements $p \otimes u$ for p in B_0. We will consider the subalgebra $u \otimes B_0$ which ignores the first coordinate since the other coordinates would be treated similarly. The conditional expectation operator

$$E_{\mu_1 \times \mu_1}(\cdot \,|u \otimes B_0): V(B_0 \otimes B_0, R) \to V(B_0 \otimes B_0, R)$$

is then defined by

$$E_{\mu_1 \times \mu_2}: \Big(\sum_{i,j} r_{ij} p_i \otimes q_j | u \otimes B_0\Big) = \sum_j \Big[\sum_i r_{ij} \mu_1(p_i)\Big] u \otimes q_j$$

since each value of this linear operator is independent of the first coordinate and since it does not involve μ_2, we may simplify the notation to $E_{1,\mu_1}(\cdot)$. Hence if $f = \sum_{i,j} r_{ij} p_i \otimes q_j$, then $E_{1,\mu_1}(f)$ is the random variable whose value on $u \otimes q_j$ is the average $\sum_i r_{ij} \mu_1(p_i)$ of the j-th column in the $m \times n$ matrix of coefficients $[r_{ij}]$.

By using the generator ring $G(B_0, R)$, we may define the ***universal conditional expectation operator***

$$E(\cdot \,|u \otimes B_0) = E_1(\cdot): V(B_0 \otimes B_0, G(B_0, R)) \to V(B_0 \otimes B_0, G(B_0, R))$$

by

$$E_1\Big(\sum_{i,j} \alpha_{ij} p_i \otimes q_j\Big) = \sum_j \Big[\sum_i \alpha_{ij} |p_i|\Big] u \otimes q_j \,.$$

The particular operator E_{1,μ_1} is then obtained as;

$$\begin{array}{ccc} V(B_0 \otimes B_0, G(B_0, R)) & \xrightarrow{E_1} & V(B_0 \otimes B_0, G(B_0, R)) \\ \cup\!\!| & & \downarrow V(B_0 \otimes B_0, \hat{\mu}_1) \\ V(B_0 \otimes B_0, R) & \xrightarrow{E_{1,\mu_1}} & V(B_0 \otimes B_0, R)\,. \end{array}$$

Universal expectation operators can be used to treat the quantification of unary predicates. Since certain universal conditional expectation operators will be used to treat the quantification of n-ary relations, we will first consider such operators in probability theory. If B_1 and B_2 are Boolean algebras, then their *tensor product* $B_1 \otimes B_2$ is the Boolean algebra obtained by taking the ring-theoretic tensor product (over Z_2) of B_1 and B_2 as Z_2-algebras. The tensor product $B_1 \otimes B_2$ is generated by the elements $p \otimes q$ for $p \in B_1$ and $q \in B_2$, and each element of $B_1 \otimes B_2$ can be expressed as a sum of disjoint generators. The (Boolean) sum obeys the rules $(p_1 + p_2) \otimes q = p_1 \otimes q + p_2 \otimes q$ and $p \otimes (q_1 + q_2) = p \otimes q_1 + p \otimes q_2$. Generators are multiplied « componentwise », i.e., $p_1 \otimes q_1 \cdot p_2 \otimes q_2 = p_1 p_2 \otimes q_1 q_2$, so $p_1 \otimes q_1 \leqslant p_2 \otimes q_2$ if $p_1 \leqslant p_2$ and $q_1 \leqslant q_2$. Also if z is the minimal element in each algebra, then $z(p \otimes q) = (zp) \otimes q = p \otimes (zq) =$ minimal element of $B_1 \otimes B_2$. If $\mu_1 \colon B_1 \to R$ and $\mu_2 \colon B_2 \to R$ are measures with values in a ring R, then the map $\mu_1 \times \mu_2 \colon B_1 \otimes B_2 \to R$ defined by $\mu_1 \times \mu_2 (p \otimes q) = \mu_1(p) \cdot$ $\cdot \mu_2(q)$ (and extended additively to all of $B_1 \otimes B_2$) is a measure on $B_1 \times B_2$ called the *product measure*. If B_1 and B_2 are Boolean algebras of subsets of the sets X_1 and X_2 respectively, then $B_1 \otimes B_2$ is (isomorphic to) the Boolean algebra of subsets of $X_1 \times X_2$ that is generated by the rectangular sets $p \times q$ for p in B_1 and q in B_2. In this instance, we would identify $p \otimes q$ with the subset $p \times q$ so that $B_1 \otimes B_2$ would be a Boolean subalgebra of the power set Boolean algebra $\mathcal{P}(X_1 \times X_2)$.

We now consider the tensor power $B_0 \otimes B_0$ and the (normalized) valuation ring $V(B_0 \otimes B_0, R)$. If $\pi_1 = \{p_1, \ldots, p_m\}$ and $\pi_2 = \{q_1, \ldots, q_n\}$ are two partitions of B_0, then their *tensor product* $\pi_1 \otimes \pi_2 = \{p_i \otimes q_j\}_{i,j}$ is a partition of $B_0 \otimes B_0$. Every element in the valuation ring $V(B_0 \otimes B_0, R)$ can be taken as being defined on a partition of $B_0 \otimes B_0$, and each partition of $B_0 \otimes B_0$ is refined by a partition of the form $\pi_1 \otimes \pi_2$. Hence we will assume that each element in $V(B_0 \otimes B_0, R)$ is presented in the form

$$f = \sum_{i,j} r_{ij} p_i \otimes q_j$$

for some partitions $\pi_1 = \{p_1, \ldots, p_m\}$ and $\pi_2 = \{q_1, \ldots, q_n\}$ of B_0—so that the coefficients r_{ij} will form an $m \times n$ matrix of reals.

If B_0 is a Boolean algebra of subsets (events) of a space X, then the elements of $V(B_0 \otimes B_0, R)$ are simple random variables on the product space $X \times X$. If μ_1 and μ_2 are finitely additive probability

5. The logical quantifiers.

We will consider the standard case where there is a countably infinite set of variables indexed by the positive integers Z^+. The tensor power of a Boolean algebra B_0 over the index set Z^+ can be constructed as a direct union. Identify $\bigotimes_{i=1}^{n} B_0$ as a subalgebra of $\bigotimes_{i=1}^{n+1} B$ by identifying $p_1 \otimes \ldots \otimes p_n$ with $p_1 \otimes \ldots \otimes p_n \otimes u$ for any $p_1, \ldots, p_n$ in B_0. Then the infinite tensor power B_0^∞ is constructed as the direct union

$$B_0^\infty = \bigcup_{n \in Z^+} \bigotimes_{i=1}^{n} B_0 .$$

It is generated by the elements $\bigotimes_{n \in Z^+} p_n$ where all but a finite number of the p_n are equal to u, and each element in B_0^∞ can be expressed as a sum of disjoint generators.

Quantification over certain « rectangular » relations can be treated using universal conditional expectation generators. In treating logic, there is no need to restrict the ring of coefficients in the generator ring to the reals. Hence we will use the generator ring $G(B_0, A)$ where A is any commutative ring with unity. It is convenient for calculations if each element in the (normalized) valuation ring $V(B_0^\infty, G(B_0, A))$ is taken as being defined on some partition of B_0^∞ of the form $\bigotimes_{n \in Z^+} \pi_n$ where each π_n is a finite partition of B_0 and all but a finite number of the partitions π_n are equal to the maximal partition $\{u\}$ (so there are only a finite number of blocks in the partition $\bigotimes_n \pi_n$). For the sake of expository simplicity we will restrict attention to two or three coordinates. A typical element f in the valuation ring $V(B_0^\infty, G(B_0, A))$ has the form

$$f = \sum_{i,j,\ldots,k} \alpha_{ij\ldots k} p_i \otimes q_j \otimes \ldots \otimes r_k \otimes u \otimes \ldots .$$

The universal conditional expectation operator

$$E_1(\cdot) = E(\cdot \,|u \otimes B_0 \otimes B_0 \ldots) \colon V(B_0^\infty, G(B_0, A)) \to V(B_0^\infty, G(B_0, A))$$

is then defined by

$$E_1(f) = \sum_{j,\ldots,k}\Big[\sum_i a_{ij\ldots k}|p_i|\Big]\, u\otimes q_i\otimes\ldots\otimes r_k\otimes u\otimes\ldots .$$

The other operators $E_n(\cdot)$ would be defined in the same manner. With these operators, as with those previously considered, it must be verified that they are well-defined, but we will skip the details.

In any valuation ring $V(B, R)$, where B is a Boolean algebra and R is a ring, if $\{b_i\}$ is a partition of B, then the *support* of an element $f = \sum_i r_i b_i$ is defined as $\mathrm{Supp}(f) = \sum_{r_i\neq 0} b_i$. The *co-support* of an element f is defined as $\text{Co-Supp}(f) = \sum_{r_i=1} b_i$. The support and the co-support of any element $f = \sum r_i b_i$ are both Boolean algebra elements $\Big($i.e., $\bigvee_{r_i\neq 0} b_i$ and $\bigvee_{r_i=1} b_i$ respectively$\Big)$ in the image of the insertion $B \to V(B, R)$. The co-support operator could also be defined in terms of the support operator, i.e., $\text{Co-Supp}(f) = u - \mathrm{Supp}(u - f)$.

The universal conditional expectation operators E_n are all *linear* operators which are *idempotent* in the sense that $E_n(E_n(f)) = E_n(f)$. Moreover, the operators *commute* in the sense that $E_m(E_n(f)) = E_n\cdot$ $\cdot(E_m(f))$. The existential and universal quantification operators are defined by taking the support and co-support respectively of the conditional expectation operators. For any positive integer n, the operator $\exists_n$: $V(B_0^\infty, G(B_0, A)) \to V(B_0^\infty, G(B_0, A))$ for *existential quantification over the n-th variable* is defined by $\exists_n(f) = \mathrm{Supp}(E_n(f))$ for any f in $V(B_0^\infty, G(B_0, A))$. The operator $\forall_n$ for *universal quantification over the n-th variable* is defined by $\forall_n(f) = \text{Co-Supp}(E_n(f))$. Since $\forall_n(f)$ is equal to $1 - \mathrm{Supp}(1 - E_n(f)) = 1 - \mathrm{Supp}(E_n(1 - f)) = 1 -$ $- \exists_n(1 - f)$, where $1 = u\otimes u\otimes\ldots$, we will henceforth deal only with the existential quantifier.

We now consider a simple concrete example. Let B_0 be a Boolean algebra of subsets of X and let $p_1, q_1 \neq u, z$ be elements of B_0. Let $p_2 = p_1'$ and $q_2 = q_1'$ so that $\pi_1 = \{p_1, p_2\}$ and $\pi_2 = \{q_1, q_2\}$ are two non-trivial partitions of B_0. Let $R = (p_1\times q_1)\cup(p_2\times q_2)\cup(p_1\times q_2)$ so that R is a binary relation on X that is the union of three disjoint rectangular relations. The «formula» $R(v_1, v_2)$, where v_1 and v_2 are the first two variables, would then be represented by the element

$$R(v_1, v_2) = p_1\otimes q_1\otimes u\otimes\ldots + p_2\otimes q_1\otimes u\otimes\ldots + p_1\otimes q_2\otimes u\otimes\ldots$$

in the valuation ring $V(B_0^\infty, G(B_0, A))$. Then we have

$$E_1(R(v_1, v_2)) = [|p_1| + |p_2|]u \otimes q_1 \otimes u \otimes \ldots + |p_1| u \otimes q_2 \otimes u \otimes \ldots .$$

Since $|p_1| + |p_2| = 1$ and $0 \neq |p_1| \neq 1$,

$$\exists_1 R(v_1, v_2) = \sum_{j=1}^{2} u \otimes q_j \otimes u \otimes \ldots \qquad \text{and} \qquad \forall_1 R(v_1, v_2) = u \otimes q_1 \otimes u \otimes \ldots .$$

The primary limitation of the machinery developed so far is that it can only accomodate n-ary relations which (like the binary relation R) can be « paved » or « tiled » with disjoint (generalized) rectangles. The analogous problem arises in measure theory where a notion of area in R^2 or « content » in R^n can be easily defined for rectangles in R^2 or generalized rectangles in R^n. The content of certain irregularity shaped sets must then be defined as the limit of the contents of the approximating paved sets (e.g., Loomis and Sternberg [9], p. 331). We will utilize an algebraic version of this old Archimedian technique of measuring irregularity shaped objects by approximating with certain regularity shaped objects.

Our purpose is to give an abstract measure theoretic treatment of the logical quantifiers in their natural setting, i.e., when they quantify over some universe of discourse X. While the constructions will be abstract and algebraic, it will be convenient to henceforth give a set theoretical interpretation to the Boolean algebras used. Let B_0 be the power set Boolean algebra on some set X. We now wish to define the Boolean algebra B which is the Boolean algebra component of the cylindric set algebra associated with the full relational structure of all finitary relations on X (e.g., Henkin, Monk, and Tarski [7], pp. 9-10). Let X^{Z^+} be the set of sequences in X, let $R \subseteq X^n$, and let $v_{i_1}, \ldots, v_{i_n}$ be any n variables. Then the graph of the formula $R(v_{i_1}, \ldots, v_{i_n})$ is the set of sequences $x = (x_1, x_2, \ldots)$ such that the n-tuple $(x_{i_1}, \ldots, x_{i_n})$ is in R. The x_j for $j \neq i_1, \ldots, i_n$ are arbitrary. Let B be the Boolean algebra of all subsets of X^{Z^+} which, for some n, are the graphs of formulas $R(v_{i_1}, \ldots, v_{i_n})$ for some n-ary relation $R \subseteq X^n$ and some set of f variables $v_{i_1}, \ldots, v_{i_n}$. We will identify a B_0^∞ element $p \otimes q \otimes \ldots \otimes r \otimes u \otimes \ldots$ with the set of sequences $p \times q \times \ldots \times r \times u \times \ldots$ so that B_0^∞ is a subalgebra of B and $V(B_0^\infty, G(B_0, A))$ is a subring of $V(B, G(B_0, A))$. B is not a complete Boolean algebra, but each element of B is the supremum of the B_0^∞ elements below it.

Let $\{b_i\}$ be a finite partition of B and let $f = \sum \alpha_i b_i$ be an element in $V(B) = V(B, G(B_0, A))$ (where we may assume that the non-zero α_i are distinct). Then an element in $V(B_0^\infty) = V(B_0^\infty, G(B_0, A))$ of the form $\alpha \cdot p \otimes q \otimes \ldots \otimes r \otimes u \otimes \ldots$ will be called a *rectangular approximation* to f if for some i, $p \otimes q \otimes \ldots \otimes r \otimes u \otimes \ldots \leqslant b_i$ and $\alpha = \alpha_i$. For any element f in $V(B)$, let $RA(f)$ be the set of rectangulai approximations to f. If we apply the expectation operator E_n to all the rectangular approximations in $RA(f)$ and then apply the support operator, the result will be a set of Boolean algebra elements from $B_0^\infty \subseteq V(B_0^\infty)$ whose supremum exists in $B \subseteq V(B)$. Hence we define the *existential quantifier* (on the n-th variable)

$$\exists_n \colon V(B) \to V(B) \quad \text{by} \quad \exists_n(f) = \operatorname{Sup}\Big(\operatorname{Supp}\big(E_n(RA(f))\big)\Big).$$

The *universal quantifier* (on the n-th variable) $\forall_n \colon V(B) \to V(B)$ is defined by

$$\forall_n(f) = \operatorname{Sup}\Big(\text{Co-Supp}\big(E_n(RA(f))\big)\Big).$$

If $f = R(v_{i_1}, \ldots, v_{i_n})$, then it is easily seen that these algebraically defined operators give the correct results from the set theoretic viewpoint.

We have utilized the Boolean algebra B which includes the graphs of formulas $R(v_{i_1}, \ldots, v_{i_n})$ for *all* finitary relations R on X. In general, a cylindric set algebra would not treat all relations on the universe X. The machinery can be easily adapted to the general case where B^* is the Boolean algebra component of any locally finite (diagonal-free) cylindric set algebra of dimension ω over the universe X (see Henkin, Monk, and Tarski [7], pp. 164-166). The only difference is that B_0^∞ is not necessarily a subalgebra of B^* so the order condition $p \otimes q \otimes \ldots \otimes r \otimes u \otimes \ldots \leqslant b_i$ in the definition of rectangular approximation would have to be formulated in B (or the subalgebra of B generated by B_0^∞ and B^*). The quantifiers $\exists_n, \forall_n \colon V(B^*) \to V(B^*)$ could then be defined as before since the existence of the required suprema in B^* is guaranteed by B^* being the Boolean algebra component of a cylinderic set algebra.

It remains to show how any finite transformation of variables can be accomplished. In the theory of polyadic algebras, substitution operators are included in the structure of a polyadic algebra. In the theory of cylindric algebras, the equality (or diagonal) relations are included in the structure of a cylindric algebra and then the result of substituting the j-th variable for the i-th variable in f is defined to be $\exists_i(f \wedge v_i = v_j)$. In a locally-finite algebra of infinite dimension, any finite transformation of variables can be obtained by an appropriate product of substitutions. Hence we only need to define $S(i/j)$: $V(B) \to V(B)$ such that $S(i/j)(f)$ is the result of substituting the j-th variable for the i-th variable in any free occurrence in f. A rectangular approximation has the general form $\alpha \cdot p \otimes q \otimes \ldots \otimes r \otimes u \otimes \ldots$ where p is the first component, q is the second component, and so forth. We will say that a rectangular approximation is *square in the i-th and j-th components* if the i-th and j-th component are the same element of B_0. Let $RA_{(i,j)}(f)$ be the set of rectangular approximations to the $V(B)$ element f which are square in the i-th and j-th components. The *substitution operator* $S(i/j)$: $V(B) \to V(B)$, for any distinct positive integers i and j, is defined by

$$S(i/j)(f) = \mathrm{Sup}\left(\mathrm{Supp}\left(E_i(RA_{(i,j)}(f))\right)\right).$$

For example, if $f = R(v_1, v_2, \ldots, v_n)$, then $S(1/2)(f)$ would be the B-element which is the set of sequences $x = (x_1, x_2, \ldots)$ such that the n-tuple $(x_2, x_2, x_3, \ldots, x_n)$ was in R, i.e., the graph of the formula $R(v_2, v_2, v_3, \ldots, v_n)$.

BIBLIOGRAPHY

[1] A. Church, *Introduction to Mathematical Logic*, Princeton University Press, Princeton, 1956.

[2] L. Geissinger, *Valuations on Distributive Lattices I*, Arch. Math., **24** (1973), pp. 230-239.

[3] L. Geissinger, *Valuations on Distributive Lattices II*, Arch. Math., **24** (1973), pp. 337-345.

[4] L. Geissinger, *Valuations on Distributive Lattices III*, Arch. Math., **24** (1973), pp. 475-481.

[5] P. Halmos, *Algebraic Logic*, Chelsea Publishing C., New York, 1962.

[6] P. Halmos, *Lectures on Boolean Algebras*, Van Nostrand Co., Princeton, 1963.

[7] L. Henkin - D. Monk - A. Tarski, *Cylindric Algebras Part I*, North-Holland Publishing Co., Amesterdam, 1971.

[8] S. Lang, *Algebra*, Addison-Wesley Publishing C., Reading, Mass., 1965.

[9] L. Loomis - S. Sternberg, *Advanced Calculus*, Addison-Wesley Publishing Co., Reading, Mass., 1968.
[10] G.-C. Rota, *On the Combinatorics of the Euler Characteristic*, Studies in Pure Mathematics, edited by L. Mirsky, Academic Press, London, (1971), pp. 221-233.
[11] G.-C. Rota, *The Valuation Ring of a Distributive Lattice*, Proc. Univ. of Houston Lattice Theory Conf., Houston (1973), pp. 575-628.
[12] F. B. Wright, *Convergence of Quantifiers and Martingales*, Illinois J. of Math., **6** (1962), pp. 296-307.

Manoscritto pervenuto in redazione il 1° dicembre 1978.

The Annals of Probability
1997, Vol. 25, No. 3, 1257–1283

STOCHASTIC INTEGRALS: A COMBINATORIAL APPROACH

BY GIAN-CARLO ROTA AND TIMOTHY C. WALLSTROM

Massachusetts Institute of Technology, Los Alamos National Laboratory and Catholic University of America and Los Alamos National Laboratory

Dedicated to W. T. Martin

A combinatorial definition of multiple stochastic integrals is given in the setting of random measures. It is shown that some properties of such stochastic integrals, formerly known to hold in special cases, are instances of combinatorial identities on the lattice of partitions of a set. The notion of stochastic sequences of binomial type is introduced as a generalization of special polynomial sequences occuring in stochastic integration, such as Hermite, Poisson–Charlier and Kravchuk polynomials. It is shown that identities for such polynomial sets have a common origin.

1. Introduction. Few subjects in modern probability have undergone as many disparate presentations and have been rediscovered in as many different guises as the theory of stochastic integrals. Wiener's homogeneous chaos [38], Wiener and Wintner's discrete chaos [39], the Fock spaces of quantum field theory [2], Itô's stochastic integrals [13, 14], integration over semimartingales [24, 28, 7, 4], Segal's tensor algebras over Hilbert spaces [34], Kakutani's maximal Gaussian subspaces [15, 16], are only some of the theories that have evolved in the last fifty years around one fundamental idea [23, 20, 21, 22, 25, 12, 18].

The variety of notations, ranging from Cameron and Martin's products of Hermite polynomials [3] to Wick's "dots" [37], has obscured the basic simplicity of the underlying concept. What is more, the lack of communication among various schools, notably between physicists aiming at the development of nonlinear quantum field theories [36, 40] and probabilists in search of new point processes that would not turn out to be Poisson distributions in disguise [17, 6], have delayed and in some cases prevented a full understanding of the possibilities of stochastic integration.

The purpose of the present work is modest. It is to present the basic idea of stochastic integration in what we believe to be its simplest and most natural form, stripped of all analytic superstructure and of all physical irrelevancies. Not surprisingly, we have found that the basic infrastructure of stochastic integration is a combinatorial technique which had heretofore remained hidden, namely, Möbius inversion over the lattice of partitions of a set [30]. Insofar as the underlying analytic machinery goes, we rely heavily on the pioneering

Received March 1996.

AMS 1991 *subject classifications*. Primary 60H05, 05A18; secondary 05E05, 05E35, 11B65, 60G57, 81T18.

Key words and phrases. Multiple stochastic integrals, partitions of sets, discrete and homogeneous chaos, orthogonal polynomials, symmetric functions, Kailath–Segall formula.

work of Engel [9], whose proof of countable additivity of product measures allows us to proceed directly to the combinatorial and algorithmic aspects of stochastic integration.

Systematic use of Möbius inversion not only simplifies definition of multiple stochastic integrals and their computation, but displays the unity of the ad hoc formulas that seemed to vary capriciously from one example to the next. In the presentation given below, identities holding among the special functions that had arisen in various contexts of stochastic integration are shown to stem from one and the same set theoretic identity. Thus, Hermite, Poisson–Charlier, Kravchuk and other sets of polynomials [5] are seen as special cases of one and the same sequence of set functions, which we have decided to name stochastic sequences of binomial type. Similarly, formerly mysterious computations with Feynman diagrams [1] that went with the integration of products of such polynomials are now seen to reduce to a single identity on the partition lattice.

Our purpose is not only to present a unified theory of known results. It is, rather, to clear the air for the development of further stochastic processes and, we hazard to guess, some rigorous instances of nonlinear quantum field theories. To keep the length of the present paper within bounds, we have left for a later publication the systematic development of chaos theory, the study of new special functions that arise in varying the stochastic measure, as well as the systematization of other notions that are scattered in theories of second quantization, and whose intimate relationship had remained hidden because of the lack of a unifying principle.

We are indebted to D. W. Stroock, who read an earlier draft of the present paper thoroughly and suggested substantial improvements. We also wish to thank P.-A. Meyer, whose recent monograph [25] provided the initial spur to writing this paper.

2. Synopsis. The basic difficulty of stochastic integration is the following. We are given a measure ϕ on a set S, and we wish to extend such a measure to the product set S^n. There is a well known and established way of carrying out such an extension, namely, taking the product measure.

While the product measure is adequate in most instances dealing with a scalar valued measure, it turns out to be woefully inadequate when the measure is vector-valued, or, in the case dealt with presently, random-valued. The product measure of a nonatomic scalar measure will vanish on sets supported by lower-dimensional linear subspaces of S^n. This is not the case, however, for random measures.

The oldest and best known example of this phenomenon is the Gaussian measure. If a measure σ is given on the set S, then the Gaussian measure ϕ is the measure such that, for $\sigma(A) < \infty$, the value $\phi(A)$ is a Gaussian (normal) random variable of mean zero and variance equal to $\sigma(A)$.

The product measure of the Gaussian measure on the set S^2 does not vanish on the diagonal, that is, on the subset of S^2 consisting of all points (s, s). In fact, the restriction of such a product measure to the diagonal is none other than the measure σ, as one may verify by an application of the law of large numbers.

The problem therefore arises of modifying the definition of product measure of a random measure in such a way that the resulting measure will vanish on lower-dimensional subsets of S^n, or *diagonal sets*, as we call them. This is the problem that is solved by the definition of the stochastic measure. Diagonal sets, that is, subsets of the cube S^n consisting of points that have two or more coordinates equal, play a game of musical chairs, which previous workers have managed to avoid by various clever devices. A notable exception is the thesis of Engel [9], whose pioneering work has been the starting point of the present work.

We choose to face directly the combinatorics of the intersection structure of the diagonal sets. This is not hard, once one realizes that the family of diagonal sets is isomorphic to the lattice of partitions of the set $\{1, 2, \ldots, n\}$. The sieving out of the overlaps among the various diagonal sets is none other than the Möbius inversion formula on such a lattice of partitions.

After facing up to the need of sieving out overlaps among diagonal sets by the Möbius inversion formula on the lattice of partitions, the definition of a stochastic measure becomes all but trivial, and its properties, classical and new, become a combinatorial game played with partitions.

One may ask for the purpose of defining a stochastic measure, or more precisely, for a natural, or "functorial" definition. The present combinatorial approach leads to the discovery of such a natural definition. In fact, after sieving out all diagonal sets from the set S^n, all one needs to do is restrict the product measure of ϕ to the remaining set, thereby obtaining the stochastic measure. In this way, one sees that the stochastic measure is symmetric under permutations of coordinates, and that furthermore it is supported on the subset of S^n consisting of points which are n-tuples of *distinct* coordinates. In other words, the stochastic measure is seen to be the natural measure on n-subsets of the set S induced by a given random measure ϕ.

The bulk of the present work is concerned with the generalization to arbitrary stochastic measures of facts that were previously known in special cases. The celebrated Kailath–Segall [35] formula is given a combinatorial setting that not only substantially generalizes it, but renders it all but obvious.

Perhaps a notable new result in the present work is a formula for the simplified computation of stochastic measures which relies on taking meets of partitions. This universal formula contains as special cases the orthogonality properties of Hermite, Poisson–Charlier and Kravchuk polynomials, as well as all Feynman diagram computations of integrals of products of Hermite polynomials. In fact, it can be viewed as a rigorous version of the method of Feynman diagrams. What is more, the meet formula (Section 5) leads to an extension of the method of Feynman diagrams to the discrete chaos of Wiener and Wintner. We have not developed this line, but we give an example that shows how such an extension might be developed.

Various special polynomial sequences that arose in stochastic integration are here unified under the concept of a stochastic sequence of binomial type.

Such sequences, giving a very broad generalization of the theory of polynomial sequences of binomial type [32], are ripe objects for further investigation.

A closing word about notation. The symbol S^n usually denotes the n-dimensional cube of side S, that is, the set of all sequences of n entries out of S. In other contexts, the symbol S^n has been used to denote the set of all functions from the set $[n] = \{1, 2, \ldots, n\}$ to the set S. In the present work, this ambiguity is avoided. The symbol S^b denotes the set of all functions from a subset b of the integers to the set S. In particular, the symbol $S^{[n]}$ denotes the set of all functions from the set $[n]$ to the set S. One may visualize such functions as sequences, but in the proofs it is preferable to keep in mind that one is dealing with functions.

3. Partitions. We review some basic facts about the lattice of partitions of a finite set b, which is denoted by $\Pi(b)$ [29, 30]. A partition σ of b is a family of nonempty disjoint subsets of b, called blocks, whose union is the set b. The set $\Pi(b)$ is given a partial order by setting $\sigma \leq \pi$ if and only if every block of the partition σ is contained in some block of the partition π. In fact, the partially ordered set $\Pi(b)$ is a lattice, in which the meet of two partitions $\sigma \wedge \pi$ is the partition whose blocks are all the nonempty pairwise intersections of some block of π and some block of σ. Similarly, the join $\sigma \vee \pi$, namely, the smallest partition containing both σ and π is easily characterized set theoretically.

A *segment* $[\sigma, \pi]$ of the lattice $\Pi(b)$ is defined as the set of all partitions ρ such that $\sigma \leq \rho \leq \pi$. The zero partition of $\Pi(b)$ (namely, the minimal element of the partially ordered set) is the partition consisting of a separate block for each element; it is denoted by $\hat{0}$. Similarly, the one partition (namely, the maximum element of the partially ordered set) is the partition consisting of a single block containing all elements; it is denoted by $\hat{1}$.

We will require the Möbius inversion formula on the lattice $\Pi(b)$, and we recall some of the fundamental definitions. The incidence algebra of $\Pi(b)$ is the algebra spanned by all functions $f(\sigma, \pi)$ with values in a field K, such that $f(\sigma, \pi) = 0$ unless $\sigma \leq \pi$. The sum of any elements of the incidence algebra is formally defined, and the product is defined by convolution, namely,

$$f * g(\sigma, \pi) = \sum_{\sigma \leq \rho \leq \pi} f(\sigma, \rho) g(\rho, \pi).$$

The identity of the incidence algebra is the Kronecker delta: $\delta(\sigma, \pi) = 0$ unless $\sigma = \pi$, in which case it equals 1. The zeta function of the incidence algebra of $\Pi(b)$, in symbols $\zeta(\sigma, \pi)$, is defined to equal 1 if $\sigma \leq \pi$ and 0 otherwise. The Möbius function $\mu(\sigma, \pi)$ is defined to be the inverse of the zeta function in the incidence algebra of $\Pi(b)$, namely, the inverse relative to the convolution product.

If G is a function on the lattice of partitions and

$$F(\pi) = \sum_{\sigma \geq \pi} G(\sigma),$$

Let S^b denote the set of all functions from a finite subset b of positive integers to the set S. When $b = [n]$, we sometimes represent such a function from $[n]$ to S as a sequence $(s_1, \ldots, s_n)$.

If b is a subset of the positive integers and if A_i, $i \in b$, are measurable sets, a *rectangle set* is the set of functions f from b to S such that $f(i) \in A_i$ for all $i \in b$. We indicate this set by the notation $\mathrm{X}_{i\in b} A_i$. When $b = [n]$, we also indicate rectangle sets by the notation

$$A \times B \times \cdots \times C;$$

that is, the set of all functions such that $f(1) \in A$, $f(2) \in B, \ldots, f(n) \in C$.

A subset of S^b of the form $A^b = A \times \cdots \times A$ ($|b|$ times), where A is a measurable set in Σ, is said to be a *cubical set*. A subset of S^b of the form $\mathrm{X}_{i\in b} A_i$, where each A_i is measurable and where the A_i are pairwise disjoint, is said to be a *triangular* set.

A collection of subsets $\mathscr{E}$ of a set S is a *(Boolean) ring* if it contains the empty set and is closed under unions and differences. It is a *σ-ring* if it is also closed under countable unions. If a ring (σ-ring) is closed under complementation, so that in particular it contains S, then it is called a *(Boolean) algebra* (*σ-algebra*).

We denote by Σ^b the set of all rectangle sets in S^b, by $\Sigma_{\text{fin}}^{\otimes b}$ the Boolean algebra of subsets of S^b generated by all rectangle sets, and by $\Sigma^{\otimes b}$ the Boolean σ-algebra generated by Σ^b. Elements of $\Sigma^{\otimes b}$ will also be called measurable sets.

Every function $f\colon b \to S$ has a kernel which is the partition of b whose blocks are defined as follows: i and j of b belong to the same block of the kernel of f whenever $f(i) = f(j)$.

Given a partition π of the set b and a set $\mathscr{A}$ in $\Sigma^{\otimes b}$, we denote by $\mathscr{A}_\pi$ the set of all functions in $\mathscr{A}$ whose kernel is the partition π, and we denote by $\mathscr{A}_{(\geq\pi)}$ the set of all functions whose kernel is some partition $\sigma \geq \pi$. A set of the form $\mathscr{A}_\pi$ is called a *diagonal* set. Among all diagonal sets $\mathscr{A}_\pi$, the set of the form $\mathscr{A}_{\hat{0}}$ consists of all one-to-one functions, and the set $\mathscr{A}_{\hat{1}}$ consists of all functions which take only one value.

A set of the form $\mathscr{A}_{\hat{1}}$, is called a *pure diagonal* set. If $\mathscr{A} = A^b$, we write $\mathscr{D}^b(A)$ in place of $\mathscr{A}_{\hat{1}}$. The set $(A^b)_\pi$ is often written A^b_π.

The group of permutations of $[n]$ acts on $S^{[n]}$. If w is a permutation of $[n]$ and if $(s_1, \ldots, s_n)$ is a point of $S^{[n]}$, we set $w(s_1, \ldots, s_n) = (s_{w(1)}, \ldots, s_{w(n)})$. If $\mathscr{A} \subset S^{[n]}$, we set

$$w(\mathscr{A}) = \{w(s_1, \ldots, s_n)\colon (s_1, \ldots, s_n) \in \mathscr{A}\}.$$

In particular, $w(A_1 \times \cdots \times A_n) = A_{w(1)} \times \cdots \times A_{w(n)}$.

A subset $\mathscr{A}$ of $\Sigma^{\otimes b}$ is said to be *symmetric* when $w(\mathscr{A}) = \mathscr{A}$ for all permutations w of b. The family of all symmetric sets in $\Sigma_{\text{fin}}^{\otimes b}$ is a Boolean subalgebra of $\Sigma_{\text{fin}}^{\otimes b}$ which will be denoted by $\Sigma_{\text{fin, sym}}^{\otimes b}$. Similarly, the family of all symmetric sets in $\Sigma^{\otimes b}$ is a Boolean σ-algebra, which will be denoted by $\Sigma_{\text{sym}}^{\otimes b}$.

Let ϕ be a set function on some ring of sets $\mathscr{R}$, taking values in a Banach space X. We say that ϕ is *finitely additive* if $\phi(\varnothing) = 0$ and if, whenever

then the Möbius inversion formula states that

$$G(\pi) = \sum_{\sigma \geq \pi} \mu(\pi, \sigma) F(\sigma).$$

We also have use for partitions λ of an integer n, which is a distinct concept from partitions of a set, though closely related. We say $\lambda = (\lambda_1, \dots, \lambda_r)$ is a partition of n, in symbols $\lambda \vdash n$, if $\lambda_i \geq \lambda_{i+1}$ and $\lambda_1 + \lambda_2 + \cdots + \lambda_r = n$. We also write $\lambda = (1^{r_1} 2^{r_2} \cdots n^{r_n})$ to indicate a partition of n in which r_i of the λ_j's are equal to i, for $i = 1, \dots, n$.

The *class* of a segment $[\sigma, \pi]$, where σ and π are partitions of a set b, is the partition $(\lambda_1, \dots, \lambda_{|\pi|})$ of $|\sigma|$ such that λ_i is the number of blocks of σ contained in the ith block of π. (We may assume the blocks are numbered so that the sequence $\{\lambda_i\}$ is nonincreasing.) Equivalently, it is the partition $(1^{r_1} 2^{r_2} \cdots)$ of $|\sigma|$ such that r_i blocks of π contain exactly i blocks of σ. We write $r = |\pi|$ and $n = |\sigma|$. Note that

$$n = \lambda_1 + \cdots + \lambda_r = r_1 + 2r_2 + \cdots + nr_n$$

and

$$r = r_1 + \cdots + r_n.$$

We sometimes write $\lambda(\sigma, \pi)$ for the class of the segment $[\sigma, \pi]$, and $\lambda(\pi)$ for the class of the segment $[\hat{0}, \pi]$.

It is shown [31] that the Möbius function $\mu(\sigma, \pi)$ depends only on the class of the segment $[\sigma, \pi]$, and that it is given by the following formula:

$$(*) \qquad \mu(\sigma, \pi) = (-1)^{n-r} (2!)^{r_3} (3!)^{r_4} \cdots ((n-1)!)^{r_n},$$

where $n = |\sigma|$, $r = |\pi|$ and $\lambda(\sigma, \pi) = (1^{r_1} 2^{r_2} \cdots n^{r_n})$.

We denote by $[n]$ the set $\{1, 2, \dots, n\}$. We write μ_n for $\mu(\hat{0}, [n])$. Thus, $\mu_n = (-1)^{n-1}(n-1)!$. The number of partitions of class λ of the set $[n]$ is given by

$$\begin{bmatrix} n \\ \lambda \end{bmatrix} = \frac{n!}{1^{r_1} r_1! 2^{r_2} r_2! \cdots n^{r_n} r_n!}.$$

Let b_i, $i = 1, \dots, r$, be the blocks of π, and let λ_i be the number of blocks of σ contained in the ith block of π. Then the segment $[\sigma, \pi]$ is isomorphic, as a lattice, to the direct product of the lattices $[\hat{0}, [\lambda_i]]$, $i = 1, \dots, r$. The Möbius function of a direct product lattice is the product of the Möbius function on factors in the product. One easily checks that $(*)$ is consistent with this requirement. (For a treatment of lattices, including a definition of their direct products and Möbius functions, see [30, 31].)

4. Sets and measures. Let S be a set, and Σ a σ-field of subsets of S. Elements of Σ will be called measurable sets.

$A = \bigcup_{i=1}^m A_i$, where the A_i are disjoint sets in $\mathscr{R}$,

$$\phi(A) = \sum_{i=1}^{m} \phi(A_i).$$

We say that ϕ is (norm) *countably additive* if $\phi(\varnothing) = 0$ and if, whenever $A = \bigcup_{i=1}^\infty A_i$, where the A_i are disjoint elements of $\mathscr{R}$,

$$\lim_{m\to\infty} \phi\left(\bigcup_{i=1}^{m} A_i\right) = \phi(A),$$

where the limit is taken in X.

A *vector-valued measure* on $\mathscr{R}$ is a norm countably additive set function on $\mathscr{R}$ taking values in a Banach space X. Finitely additive vector-valued measures are defined analogously.

Let $(\Omega, \mathscr{P}, P)$ be a probability space, and let $\mathscr{M}$ be the space of real-valued measurable functions on $(\Omega, \mathscr{P}, P)$, that is, real random variables. If f and g are in $\mathscr{M}$, we define their product fg by

$$(fg)(\omega) = f(\omega)g(\omega), \qquad \omega \in \Omega.$$

The expectation of a random variable f is denoted $\langle f \rangle$.

A *random measure* is a vector-valued measure taking values in a Banach space X of random variables; for example, $X = L^2(\Omega, \mathscr{P}, P)$. The space X, as a subspace of $\mathscr{M}$, inherits its multiplicative structure. Note, however, that fg need not belong to X, and a fortiori, multiplication need not be continuous. Thus, X is not a Banach algebra, but rather a Banach space equipped with a not-everywhere-defined multiplicative structure.

If ϕ is a random-valued measure, we write $\langle \phi \rangle$ for the real measure given by $\langle \phi \rangle(A) = \langle \phi(A) \rangle$.

We next consider products of measures. A set function $\nu(A_1, A_2, \ldots, A_n)$, where the n variables $A_1, A_2, \ldots, A_n$ range over Σ, is called a *multimeasure* if it is a measure in each variable when each of the other variables is held fixed. Given an X-valued multimeasure ν on $\Sigma^{[n]}$, let $\mathscr{R}_\nu$ be the ring generated by the rectangles

$$\{\mathscr{A} \in \Sigma^{[n]}: \nu(A_1, \ldots, A_n) \in X\}.$$

Then there exists a unique finitely additive measure m_n on $\mathscr{R}_\nu$ such that

$$m_n(A_1 \times \cdots \times A_n) = \nu(A_1, \ldots, A_n).$$

The measure m_n will be said to be *associated* to the multimeasure ν.

The ordinary notion of product of two measures η and ψ can be defined by means of a multimeasure. Let $b = \{i_1, \ldots, i_k\}$ be a subset of $[n]$, with $i_1 < \cdots < i_k$ and let $b^c = \{j_1, \ldots, j_{n-k}\}$ with $j_1 < \cdots < j_{n-k}$ be the complement of b. A multimeasure ν is defined on rectangular sets by setting

$$\nu(A_1, \ldots, A_n) = \eta(A_{i_1} \times \cdots \times A_{i_k})\psi(A_{j_1} \times \cdots \times A_{j_{n-k}}).$$

The finitely additive measure on $\mathscr{R}_\nu$ associated with ν will be denoted $\eta^b \otimes \psi^{b^c}$. A similar construction may and will be used to define the product of more than two measures.

We will assume from now on that all diagonal sets belong to $\Sigma^{\otimes[n]}$. We will also assume that ϕ is a random measure, taking values in a Banach space of random variables X. Finally, we assume that all finitely additive product measures, defined in terms of an associated multimeasure as above, have unique countably additive extensions to some ring containing $\mathscr{R}_\nu$, where $\mathscr{R}_\nu$ is defined above. We will say that the random measure ϕ is *good* (for lack of a better term) when the space (S, Σ) and the process ϕ satisfy these assumptions.

The assumption that the diagonals are contained in the product σ-algebra is satisfied when S is a complete metrizable separable space, that is, a Polish space, with Σ the corresponding Borel algebra.

The proof that certain finitely additive product measures have unique countably additive extensions is given by Engel [9]. Engel's results can be used to establish the existence of a product measure for measures defined from multimeasures as above. Our examples satisfy Engel's assumptions, and thus deal with countably additive measures. In particular, we have benefited from Engel's ideas in the proof of Proposition 12.

5. Stochastic measures. We repeat the definitons of $\mathscr{A}_{(\geq\pi)}$ and $\mathscr{A}_\pi$ in slightly different language, when $b = [n]$. Write $i \sim_\pi j$ when i and j are in the same block of π. Then

$$\mathscr{A}_{(\geq\pi)} = \{(s_1, s_2, \ldots, s_n) \in \mathscr{A}\colon s_i = s_j \text{ if } i \sim_\pi j\}$$

and

$$\mathscr{A}_\pi = \{(s_1, s_2, \ldots, s_n) \in \mathscr{A}\colon s_i = s_j \text{ if and only if } i \sim_\pi j\}.$$

We stress the facts that, in both cases, the points s_i and s_j coincide whenever i and j are in the same block b of π, but in the definition of $\mathscr{A}_\pi$ we require in addition that s_i and s_j shall *not* coincide if i and j are in different blocks of π.

The following trivial partition of $\mathscr{A}_{(\geq\pi)}$ is fundamental:

$$(*) \qquad \mathscr{A}_{(\geq\pi)} = \bigcup_{\sigma\geq\pi} \mathscr{A}_\sigma \quad \text{(disjoint union).}$$

DEFINITION 1. Let ϕ be a good random measure on (S, Σ). Define, for $\mathscr{A} \in \Sigma^{\otimes b}$, and for π a partition of the set b,

$$\phi_\pi^b(\mathscr{A}) = \left(\bigotimes_{i\in b} \phi\right)(\mathscr{A}_{(\geq\pi)})$$

and

$$\mathrm{St}_\pi^b(\mathscr{A}) = \left(\bigotimes_{i\in b} \phi\right)(\mathscr{A}_\pi).$$

Neither ϕ_π^b nor St_π^b is defined unless π is a partition of the set b. Since $\mathscr{A}_{(\geq \hat{0})} = \mathscr{A}$, we have

$$\phi_{\hat{0}}^b = \bigotimes_{i \in b} \phi.$$

The product measure is $\phi_{\hat{0}}^b$. We will consistently use this notation for product measures.

The following proposition is fundamental.

PROPOSITION 1. *The measures ϕ_π^b and St_π^b are related as follows*:

$$(**)\qquad \begin{aligned} \phi_\pi^b &= \sum_{\sigma \geq \pi} \mathrm{St}_\sigma^b, \\ \mathrm{St}_\pi^b &= \sum_{\sigma \geq \pi} \mu(\pi, \sigma) \phi_\sigma^b. \end{aligned}$$

PROOF. The first equation follows directly from $(*)$. The second follows from Möbius inversion. □

The following special case of the preceding proposition is one of our main results.

THEOREM 1. *For $\pi = \hat{0}$ and $b = [n]$, we have the identity*

$$\mathrm{St}_{\hat{0}}^{[n]} = \sum_{\sigma \geq \hat{0}} \mu(\hat{0}, \sigma) \phi_\sigma^{[n]}.$$

The measure $\mathrm{St}_{\hat{0}}^{[n]}$ will be called the *stochastic measure of degree n*, or, when no confusion is possible, the *stochastic measure* associated to the measure ϕ.

An integral

$$\int f(s_1, \ldots, s_n) \mathrm{St}_{\hat{0}}^{[n]}(ds_1 \cdots ds_n)$$

is said to be an *nth multiple stochastic integral* or simply, when no confusion may arise, the *stochastic integral* of the function f. It is evident from its definition that $\mathrm{St}_{\hat{0}}^{[n]}$ is symmetric. That is,

$$\mathrm{St}_{\hat{0}}^{[n]}(w(\mathscr{A})) = \mathrm{St}_{\hat{0}}^{[n]}(\mathscr{A}),$$

where w is a permutation of the set $[n]$.

As already remarked, the measure $\phi_{\hat{0}}^{[n]}$ is a product measure. We will now see that, more generally, the measure $\phi_\pi^{[n]}$ is expressible as the product over the blocks b of π of the measures $\phi_{\hat{1}}^b$.

PROPOSITION 2. *The measure $\phi_\pi^{[n]}$ is a product measure*:

$$\phi_\pi^{[n]} = \bigotimes_{b \in \pi} \phi_{\hat{1}}^b.$$

PROOF. Both measures are the restriction of the product measure $\phi_{\hat{0}}^{[n]}$ to the set of elements of $S^{[n]}$ whose kernel σ is greater than or equal to π. □

The measures $\phi_{\hat{1}}^{b}$ are concentrated on the pure diagonal sets $\mathscr{A}_{\hat{1}}$:

$$\phi_{\hat{1}}^{b}(\mathscr{A}) = \phi_{\hat{0}}^{b}(\mathscr{A}_{\hat{1}}).$$

If the set b has a single element, then $\phi_{\hat{1}}^{b} = \phi$. Another immediate consequence of the definitions is the identity $\mathrm{St}_{\hat{1}}^{b} = \phi_{\hat{1}}^{b}$.

We have shown that $\mathrm{St}_{\pi}^{[n]}$ could be expressed as the Möbius sum of measures $\phi_{\sigma}^{[n]}$. We now show that $\mathrm{St}_{\pi}^{[n]}$ can also be computed by a simple recursion over products of measures $\phi_{\hat{1}}^{b}$ and lower order stochastic measures. The next theorem is a generalization of a formula first given by Kailath and Segall [35]. We shall see that it can be applied to derive recursion relations for orthogonal polynomials.

THEOREM 2. *The following identity holds on* $\Sigma^{\otimes[n]}$:

$$\mathrm{St}_{\hat{0}}^{[n]} = \sum \mu_{|b|}\, \phi_{\hat{1}}^{b} \otimes \mathrm{St}_{\hat{0}}^{b^c},$$

where the sum ranges over all subsets b of $[n]$ containing the element 1.

PROOF. By (**),

$$\mathrm{St}_{\hat{0}}^{[n]} = \sum_{\pi} \mu(\hat{0}, \pi)\phi_{\pi}^{[n]}.$$

Given a partition π of $[n]$, let b be the block of π containing the element 1, and let π' be the partition of b^c consisting of the remaining blocks of π. Now $\phi_{\pi}^{[n]} = \phi_{\hat{1}}^{b} \otimes \phi_{\pi'}^{b^c}$, and $\mu(\hat{0}, \pi) = \mu_{|b|}\mu(\hat{0}, \pi')$. Therefore,

$$\sum_{\pi} \mu(\hat{0}, \pi)\phi_{\pi}^{[n]} = \sum_{b} \sum_{\pi' \in \Pi(b^c)} \mu_{|b|}\mu(\hat{0}, \pi')\phi_{\hat{1}}^{b} \otimes \phi_{\pi'}^{b^c}$$
$$= \sum_{b} \mu_{|b|}\phi_{\hat{1}}^{b} \otimes \mathrm{St}_{\hat{0}}^{b^c}. \qquad \square$$

Proposition 1 represents $\mathrm{St}_{\hat{0}}^{b}$ as a linear combination of the measures ϕ_{π}^{b}, where the sum extends over all partitions π. We next show that the product of stochastic measures can also be expressed as a linear combination of the measures ϕ_{π}^{b}, where the sum extends over all partitions in an appropriate segment.

THEOREM 3. *Let τ be a partition of $[n]$. Then*

$$\bigotimes_{b\in\tau} \mathrm{St}_{\hat{0}}^{b} = \sum_{\hat{0}\leq\pi\leq\tau} \mu(\hat{0}, \pi)\phi_{\pi}^{[n]}.$$

PROOF. If b is a block of τ, and τ' is the partition on b^c consisting of remaining blocks of τ, then

$$\sum_{\hat{0}\leq\pi\leq\tau} \mu(\hat{0},\pi)\phi_\pi^b = \Big(\sum_{\rho\in\Pi(b)} \mu(\hat{0},\rho)\phi_\rho^b\Big)\bigotimes\Big(\sum_{\hat{0}\leq\pi'\leq\tau'} \mu(\hat{0},\pi')\phi_{\pi'}^{b^c}\Big)$$
$$= \mathrm{St}_{\hat{0}}^b\bigotimes\Big(\sum_{\hat{0}\leq\pi'\leq\tau'} \mu(\hat{0},\pi')\phi_{\pi'}^{b^c}\Big).$$

The conclusion follows by induction. □

The product of stochastic measures may also be expressed as a sum over measures $\mathrm{St}_\sigma^{[n]}$. The following result is fundamental in the combinatorial interpretation of orthogonal polynomials.

THEOREM 4. *Let π be a partition of $[n]$. Then*

$$\bigotimes_{b\in\pi} \mathrm{St}_{\hat{0}}^b = \sum_{\{\sigma:\ \sigma\wedge\pi=\hat{0}\}} \mathrm{St}_\sigma^{[n]}.$$

PROOF. By Theorem 3,

$$\bigotimes_{b\in\pi} \mathrm{St}_{\hat{0}}^b = \sum_{\hat{0}\leq\sigma\leq\pi} \mu(\hat{0},\sigma)\phi_\sigma^{[n]}.$$

But

$$\phi_\sigma^{[n]} = \sum_{\tau\geq\sigma} \mathrm{St}_\sigma^{[n]}.$$

Therefore,

$$\bigotimes_{b\in\pi} \mathrm{St}_{\hat{0}}^b = \sum_{\hat{0}\leq\sigma\leq\pi} \mu(\hat{0},\sigma)\sum_{\tau\geq\sigma} \mathrm{St}_\tau^{[n]}$$
$$= \sum_{\sigma\geq\hat{0}} \mu(\hat{0},\sigma)\zeta(\sigma,\pi)\sum_{\tau\geq\hat{0}} \zeta(\sigma,\tau)\mathrm{St}_\tau^{[n]}$$
$$= \sum_{\tau\geq\hat{0}} \mathrm{St}_\tau^{[n]}\sum_{\sigma\geq\hat{0}} \mu(\hat{0},\sigma)\zeta(\sigma,\pi\wedge\tau)$$
$$= \sum_{\tau\geq\hat{0}} \mathrm{St}_\tau^{[n]}\sum_{\hat{0}\leq\rho\leq(\pi\wedge\tau)} \mu(\hat{0},\rho)$$
$$= \sum_{\{\tau:\ \pi\wedge\tau=\hat{0}\}} \mathrm{St}_\tau^{[n]}.$$

The final equality follows from the fact that the sum of the Möbius function over a segment $[\sigma,\pi]$ is zero unless $\sigma=\pi$, in which case it equals one. □

We shall now state a property which characterizes a stochastic measure associated with a given measure ϕ. We begin with a preliminary definition.

Recall that a measurable set $A_1 \times \cdots \times A_n \in \Sigma^{[n]}$ is *triangular* when the measurable sets $A_1, \ldots, A_n$ in Σ are pairwise disjoint. If $\mathscr{A} = A_1 \times \cdots \times A_n$ is a triangular set, then as w ranges over all permutations of $[n]$, the family of $n!$ sets $A_{w(1)} \times \cdots \times A_{w(n)}$ consists of pairwise disjoint sets. Hence, the set $\bigcup_w w(\mathscr{A})$ belongs to $\Sigma_{\text{fin}}^{\otimes[n]}$. We denote this set by $\text{perm}(A_1 \times \cdots \times A_n)$. Note that $\text{perm}(A_1 \times \cdots \times A_n)$ is not defined unless the set $A_1 \times \cdots \times A_n$ is triangular.

The family of all sets of the form $\text{perm}(A_1 \times \cdots \times A_n)$ is closed under unions and intersections (but not under complements). In other words, it is a distributive lattice of sets.

PROPOSITION 3. *The σ-field generated by sets of the form* $\text{perm}(A_1 \times \cdots \times A_n)$*, where $A_1 \times \cdots \times A_n$ is triangular coincides with the σ-field of all symmetric sets contained in $S_{\hat{0}}^{[n]}$.*

The proof is omitted.

We are now ready to give an intrinsic characterization of the stochastic measure $\text{St}_{\hat{0}}^{[n]}$ associated to a random measure ϕ.

THEOREM 5. *For a given countably additive random-valued good measure ϕ on (S, Σ), there exists one and only one measure $\text{St}_{\hat{0}}^{[n]}$ on symmetric sets contained in $S_{\hat{0}}^{[n]}$, such that*

$$\text{St}_{\hat{0}}^{[n]}(\text{perm}(A_1 \times \cdots \times A_n)) = n!\phi(A_1) \cdots \phi(A_n).$$

PROOF. In view of the preceding proposition, such a measure is unique if it exists. Its existence is guaranteed by the Möbius inversion formula

$$\text{St}_{\hat{0}}^{[n]} = \sum_{\pi} \mu(\hat{0}, \pi) \phi_{\pi}^{[n]}.$$

Indeed, since $A_1 \times \cdots \times A_n$ is triangular, $(A_1 \times \cdots \times A_n)_\pi = \varnothing$ unless $\pi = \hat{0}$. Thus $\phi_\pi^{[n]}(A_1 \times \cdots \times A_n) = 0$ unless $\pi = \hat{0}$. When $\pi = \hat{0}$,

$$\phi_{\hat{0}}^{[n]}(\text{perm}(A_1, \ldots, A_n)) = n!\, \phi(A_1) \times \cdots \times \phi(A_n). \qquad \square$$

6. Stochastic sequences of binomial type. A sequence of measures ψ_n on (S, Σ) may be defined by setting $\psi_n(A)$ to equal the stochastic measure $\text{St}_{\hat{0}}^{[n]}$ of cubical set $A^{[n]}$. The stochastic measure of any set can then be approximated by sums of products of the ψ_n. The $\psi_n(A)$, furthermore, are of binomial type (v. below, Theorem 6), and special cases of these sequences are sequences of orthogonal polynomials (Section 7).

DEFINITION 2. Given a good random measure ϕ on (S, Σ) we define the *stochastic sequence of binomial type* $\psi_n(A)$, $n = 1, 2, \ldots$, to be the sequence of random-valued set functions

$$\psi_n(A) = \mathrm{St}_{\hat{0}}^{[n]}(A^{[n]})$$

for all $A \in \Sigma$ of finite measure.

PROPOSITION 4. *If $\mathscr{A} \in \Sigma_{\mathrm{fin}}^{\otimes[n]}$, then $\mathrm{St}_{\hat{0}}^{[n]}(\mathscr{A})$ is the sum of products of the form*

$$\psi_{i_1}(A_1)\cdots\psi_{i_r}(A_r),$$

where $i_1 + \cdots + i_r = n$, and the sets $A_1, \ldots, A_r$ are disjoint.

PROOF. Any set $\mathscr{A} \in \Sigma_{\mathrm{fin}}^{\otimes[n]}$ is the finite union of disjoint rectangles. It can easily be shown that any rectangle can be expressed as the disjoint union of rectangles whose sides are pairwise equal or disjoint. Let $\mathscr{A} = A_1 \times \cdots \times A_n$ be such a rectangle, considered as a function from $[n]$ to Σ, and let τ be the kernel of $\mathscr{A}$. That is, if b is a block of τ, then $A_i = A_j$ for any two elements i and j of b; we denote this common value by the symbol A_b. Conversely, if i and j are not in the same block of τ, then A_i and A_j are disjoint. If $\sigma > \tau$, then $\mathscr{A}_\sigma = \varnothing$, and $\phi_\sigma^{[n]}(\mathscr{A}) = \phi_{\hat{0}}^{[n]}(\mathscr{A}_{(\geq\sigma)}) = 0$. Thus, by Theorem 3,

$$\begin{aligned}\mathrm{St}_{\hat{0}}^{[n]}(\mathscr{A}) &= \sum_{\pi\geq\hat{0}} \mu(\hat{0}, \pi)\phi_{\hat{0}}^{[n]}(\mathscr{A}) \\ &= \sum_{\hat{0}\leq\pi\leq\tau} \mu(\hat{0}, \pi)\phi_{\hat{0}}^{[n]}(\mathscr{A}) \\ &= \bigotimes_{b\in\tau} \mathrm{St}_{\hat{0}}^{b}(\mathscr{A}) \\ &= \prod_{b\in\tau} \psi_{|b|}(A_b).\end{aligned}$$

□

The following theorem is fundamental and justifies our terminology.

THEOREM 6. *If $\psi_n(A)$ is the stochastic sequence of binomial type associated with a measure ϕ on a good measure space (S, Σ, σ), and if A and B are disjoint sets in Σ of finite measure, we have*

$$\psi_n(A \cup B) = \sum_{k=0}^{n} \binom{n}{k} \psi_k(A)\psi_{n-k}(B).$$

PROOF.

$$\mathrm{St}_{\hat{0}}^{[n]}((A \cup B)^{[n]}) = \sum_{b\subset[n]} \mathrm{St}_{\hat{0}}^{[n]}(A^b \times B^{b^c}).$$

A typical summand on the right-hand side is

$$\mathrm{St}_{\hat{0}}^{[n]}(A^b \times B^{b^c}) = \sum_{\pi\geq\hat{0}} \mu(\hat{0}, \pi)\phi_{\pi}^{[n]}(A^b \times B^{b^c}).$$

As in the previous proof, $\phi_\pi^{[n]}(A^b \times B^{b^c}) = 0$ if $\pi > \tau$, where τ is the partition $\{b, b^c\}$. The Möbius sum thus terminates at $\pi = \tau$, and by Theorem 3,

$$\mathrm{St}_{\hat{0}}^{[n]}(A^b \times B^{b^c}) = \mathrm{St}_{\hat{0}}^{b}(A^b)\mathrm{St}_{\hat{0}}^{b^c}(B^{b^c}).$$

But if $|b| = k$,

$$\mathrm{St}_{\hat{0}}^{b}(A^b) = \mathrm{St}_{\hat{0}}^{[k]}(A^{[k]}),$$

and similarly for $\mathrm{St}_{\hat{0}}^{b^c}$. There are $\binom{n}{k}$ partitions of $[n]$ into blocks $\{b, b^c\}$ with $|b| = k$. □

In a similar way, one may prove a multinomial expansion.

PROPOSITION 5. *If $A = \bigcup_{i=1}^r A_i$ and the A_i are pairwise disjoint, then*

$$\psi_n(A) = \sum_{i_1+\cdots+i_r=n} \binom{n}{i_1 \cdots i_r} \psi_{i_1}(A_1) \cdots \psi_{i_r}(A_r).$$

Useful as the measures ψ_n are, we require some further measures, namely the diagonal measures Δ_n. These are defined in terms of the measures $\phi_{\hat{1}}^{[n]}$.

DEFINITION 3. For every positive integer n, the *nth diagonal measure* Δ_n of ϕ is the measure on (S, Σ) given by

$$\Delta_n(A) = \phi_{\hat{1}}^{[n]}(A^{[n]}) = \mathrm{St}_{\hat{1}}^{[n]}(A^{[n]})$$

We omit the verification that Δ_n is a measure.

Theorem 2 implies the following recursion relation for the measures ψ_n.

THEOREM 7. *If $\psi_n(A)$ is the stochastic sequence of binomial type associated with the random measure ϕ, then*

$$\psi_n(A) = \sum_{k=1}^{n} (-1)^{k-1}(n-1)_{k-1}\Delta_k(A)\psi_{n-k}(A).$$

Theorem 1 implies the following expression for ψ_n in terms of the diagonal measures:

THEOREM 8.

$$\psi_n(A) = \sum_{\lambda \vdash n} \binom{n}{r_1 \cdots r_n} \frac{(-1)^{r_1+\cdots+r_n-n}}{\prod_{i=1}^n i^{r_i}} \Delta_1^{r_1}(A) \cdots \Delta_n^{r_n}(A),$$

where $\lambda = (1^{r_1}2^{r_2} \cdots n^{r_n})$.

The proof is omitted.

Given any partition $\lambda \vdash n$, one can define, in analogy to ψ_n, the measure

$$\Psi_\lambda^n(A) = \mathrm{St}_\pi^{[n]}(A^{[n]}),$$

where π is any partition of class λ. The measure Ψ^n_λ is well defined, by virtue of the symmetry of the argument $A^{[n]}$. The measures Ψ^n_λ are of independent interest, and hint at a rich theory which extends the theory of orthogonal polynomials.

7. Examples. In the following examples, all random measures are assumed to be good.

EXAMPLE D (Discrete). Let S be a set, and let Σ be the σ-algebra generated by all singletons $\{s\}$. Suppose that the measure ϕ is so defined that $\phi(s)$ equals either 0 or 1 for every element $s \in S$, and suppose that $\phi(S) < \infty$. Then for every measurable function $f(s_1, \dots, s_n)$ we have

$$\int f(s_1, \dots, s_n) \mathrm{St}_{\hat{0}}^{[n]}(ds_1, \dots, ds_n) = \sum f(s_1, \dots, s_n),$$

where the sum ranges over all sequences $(s_1, \dots, s_n)$ of elements of S all of whose entries are distinct. The sequence of binomial type associated with the discrete measure is

$$\psi_n(A) = (|A|)_n,$$

where $(x)_n = x(x-1)\cdots(x-n+1)$ is the falling factorial.

The diagonal measures are given by

$$\Delta_n(A) = |A|, \qquad n \geq 1.$$

EXAMPLE SP (Symmetric polynomials) [8]. Let Σ be the Boolean algebra of all subsets of a set S. Let $Z[S]$ be the polynomial ring with integer coefficients in the set of variables S. Define a measure ϕ as follows: for every finite subset A of S,

$$\phi(A) = \sum_{a \in A} a.$$

Set $p_n(A) = \sum_{a\in A} a^n$, and for every partition π of $[n]$, set $p_\pi(A) = \prod_{b\in\pi} p_{|b|}$. The polynomials $p_n(A)$ are known as the power sum symmetric polynomials.

One verifies that $\phi_\pi^{[n]}(A^n) = p_\pi(A)$, and $\psi_n(A) = n! e_n(A)$, where e_n is the nth elementary symmetric polynomial:

$$e_n(A) = \sum_{i_1 < i_2 < \cdots < i_n} a_{i_1} a_{i_2} \cdots a_{i_n},$$

where $(a_1, a_2, \dots)$ is some linear ordering of the elements of A.

The symmetric polynomials $\mathrm{St}_\pi^{[n]}(A^n)$ are known as the monomial symmetric polynomials, and are denoted by $m_\pi(A) = m_\pi(a_1, a_2, \dots)$.

The diagonal measures are given by

$$\Delta_n(A) = \sum_{a\in A} a^n.$$

EXAMPLE G (Gaussian measure). Let σ be a nonatomic real measure on (S, Σ), and let ϕ be an L^2 random-valued measure on (S, Σ) such that for any set $A \in \Sigma$ with $\sigma(A)$ finite, $\phi(A)$ is a Gaussian random variable with mean zero and variance $\sigma(A)$. Furthermore, assume that if $\{A_i\}$ is a set of pairwise disjoint elements of Σ with $\sigma(A)$ finite, then the random variables $\{\phi(A_i)\}$ are independent.

The stochastic sequence of binomial type associated with the Gaussian measure is the sequence

$$\psi_n(A) = H_n^{(v)}(\phi(A)),$$

where $v = \sigma(A)$. The polynomials $H_n^{(v)}(x)$ are the *Hermite polynomials of variance* v in the variable x. We take the preceding equation as the definition of the Hermite polynomials; all properties of the Hermite polynomials are consequences of this definition. For example, we have the identity ([15], [32], page 722)

$$H_n^{(v+w)}(x+y) = \sum_{k\geq 0} \binom{n}{k} H_k^{(v)}(x) H_k^{(w)}(y).$$

We next compute the diagonal measures for the Gaussian process, using Proposition 12 of Section 9.

PROPOSITION 6. *The diagonal measures of the Gaussian measure are given by*

$$\Delta_2(A) = \sigma(A),$$
$$\Delta_n(A) = 0, \qquad n > 2.$$

PROOF. For every positive integer N, let $\{A_{Ni},\ i = 1, \ldots, N\}$ be a collection of disjoint sets such that $\bigcup_{i=1}^N A_{Ni} = A$ and $\sigma(A_{Ni}) = \sigma(A)/N$. Proposition 12 gives conditions under which

$$\Delta_n(A) = \lim_{N\to\infty} \sum_{i=1}^{N} \phi(A_{Ni})^n$$

in L^2. However,

$$\sum_{i=1}^{N} [\phi(A_{Ni})]^n = \frac{1}{N^{n/2-1}} \left(\frac{1}{N} \sum_{i=1}^{N} (\sqrt{N}\phi(A_{Ni}))^n \right).$$

The Gaussian random variables $\sqrt{N}\phi(A_{Ni})$ are independent and each one has variance $\sigma(A)$. The term in parentheses is finite in L^2 for any $n > 1$. For $n > 2$, therefore, the right-hand side converges to zero as $m \to \infty$. When $n = 2$, it converges to $\sigma(A)$ by the law of large numbers. The hypotheses of Proposition 12 are easily verified. □

Theorem 7 now gives the classical recursion formula for the Hermite polynomials:

$$H_n^{(v)}(x) = xH_{n-1}^{(v)}(x) - (n-1)vH_{n-2}^{(v)}(x).$$

Theorem 8 implies the following known expression for $H_n^{(v)}(x)$ [19]:

$$H_n^{(v)}(x) = \sum_{j=0}^{\lfloor n/2 \rfloor} x^{n-2j}(-v)^j c(n, j),$$

where

$$c(n, j) = \frac{n!}{(n-2j)!(2j)!!},$$

and $\lfloor n/2 \rfloor$ is the largest integer less than $n/2$.

EXAMPLE P (Poisson). Let σ be a nonatomic measure on (S, Σ). Let ϕ be the random process such that, for every measurable set A, the random variable $\phi(A)$ has a Poisson distribution with parameter $\sigma(A)$, and such that if $\{A_i\}$ is a set of pairwise disjoint sets, then $\{\phi(A_i)\}$ is a set of independent random variables.

PROPOSITION 7. *The diagonal measures of the Poisson process are*

$$\Delta_n(A) = \phi(A), \qquad n = 1, 2, \ldots .$$

PROOF. Let $\{A_{Ni}\}$ be as above, with σ now the (nonatomic) measure defining the Poisson process. We can thus use Proposition 12 to calculate $\Delta_n(A)$ as $\sum_{i=1}^{N} \phi(A_{Ni})^n$. But

$$\left\| \sum_{i=1}^{N} \phi(A_{Ni})^n - \phi(A) \right\|^2 = \sum_{i=1}^{N} \|\phi(A_{Ni})^n - \phi(A_{Ni})\|^2 + \sum_{i \neq j} \langle (\phi(A_{Ni})^n - \phi(A_{Ni}))(\phi(A_{Nj})^n - \phi(A_{Nj})) \rangle$$

To estimate these terms, recall that for the Poisson process,

$$P(\{\phi(A) = j\}) = \frac{\sigma(A)^j e^{-\sigma(A)}}{j!}.$$

Thus,

$$\begin{aligned} \|\phi(A_{Ni})^n - \phi(A_{Ni})\|^2 &= \sum_{j=0}^{\infty} \frac{(j^n - j)^2}{j!} \left(\frac{\sigma(A)}{N} \right)^j e^{-\sigma(A)/N} \\ &= \frac{1}{N^2} \sum_{j=2}^{\infty} \frac{(j^n - j)^2}{j!} \frac{\sigma(A)^j}{N^{j-2}} e^{-\sigma(A)/N} \\ &\leq \frac{C_1}{N^2}. \end{aligned}$$

Similarly,

$$\langle(\phi(A_{Ni_1})^n - \phi(A_{Ni_1}))(\phi(A_{Ni_2})^n - \phi(A_{Ni_2}))\rangle$$
$$= \sum_{j,\,k=0}^{\infty} \frac{(j^n - j)(k^n - k)}{j!k!}\left(\frac{\sigma(A)}{N}\right)^{j+k} e^{-2\sigma(A)/N} = \frac{C_2}{N^4}.$$

Thus,

$$\left\|\sum_{i=1}^{N} \phi(A_{Ni})^n - \phi(A)\right\|^2 \leq \frac{C_1}{N} + \frac{C_2}{N^2},$$

which goes to zero as $N \to \infty$. The hypotheses of Proposition 12 are easily verified. □

Let $\psi_n(A)$ be the stochastic sequence of binomial type associated with the Poisson measure ϕ. Then by Theorem 7,

$$\psi_n(A) = \sum_{k=1}^{n}(-1)^{k-1}(n-1)_{k-1}\phi(A)\psi_{n-k}(A)$$
$$= \phi(A)\psi_{n-1}(A) + \sum_{j=1}^{n-1}(-1)^j(n-1)_j\phi(A)\psi_{n-1-j}(A)$$
$$= (\phi(A) - n + 1)\psi_{n-1}(A).$$

Thus, we have shown that for the Poisson measure we have

$$\psi_n(A) = (\phi(A))_n.$$

EXAMPLE PC (Poisson–Charlier). With the same assumptions as in Example P, set

$$\phi_C(A) = \phi(A) - \sigma(A).$$

One finds that the stochastic sequence associated with the Poisson–Charlier measure is

$$\psi_n(A) = \mathscr{C}_n^{(\alpha)}(\phi_C(A)),$$

where $\alpha = \sigma(A)$. The preceding equation can be taken as a definition of the *Poisson-Charlier polynomials* $\mathscr{C}_n^{(\alpha)}(x)$ [5, 27, 10]. All properties of Poisson–Charlier polynomials can be derived from this definition.

For the Poisson–Charlier measure, we have

$$\Delta_1(A) = \phi_C(A) = \phi(A) - \sigma(A),$$
$$\Delta_n(A) = \phi(A), \qquad n > 2,$$

where ϕ is the Poisson measure.

EXAMPLE K (Kravchuk) [26, 5]. Let S be a finite set and let $X_1, X_2, \ldots$ be independent identically distributed Bernoulli random variables such that $\langle X \rangle = p$. Define

$$\phi(A) = \sum_{i \in A} X_i - p|A|.$$

We find that the stochastic sequence of binomial type associated with the Kravchuk measure is

$$\psi_n(A) = k_n^{(p)}(\phi(A)),$$

where $k_n^{(p)}(x)$ are the Kravchuk polynomials.

For the Kravchuk measure, we have

$$\Delta_n(A) = \sum_{i \in A} (X_i - p)^n, \qquad n \geq 1.$$

EXAMPLE CP (Compound Poisson) [11]. Let ϕ be a Poisson process as in Example P, and let $X_1, X_2, \ldots$ be independent identically distributed positive integer-valued random variables. We stress the assumption that $P(X_i = 0) = 0$. Let α be the common probability distribution of the random variables X_i, so that $\alpha_j = P(X_i = j)$.

The *compound Poisson process* with distribution α is a measure ϕ_α defined on (S, Σ) as

$$\phi_\alpha(A) = X_1 + X_2 + \cdots + X_{\phi(A)}$$

for every $A \in \Sigma$ such that $\sigma(A) < \infty$.

Here we find that

$$\mathrm{St}_\pi^{[n]}(A \times \cdots \times A) = m_\pi(X_1, X_2, \ldots, X_{\phi(A)}),$$

and the stochastic sequence of binomial type associated with the compound Poisson process is

$$\psi_n(A) = n! e_n(X_1, X_2, \ldots, X_{\phi(A)}).$$

The diagonal measures for the compound Poisson process are

$$\Delta_n(A) = X_1^n + X_2^n + \cdots + X_{\phi(A)}^n.$$

EXAMPLE U (Umbral). This example requires some acquaintance with the notation introduced in [33]. Let ϕ and σ be as in Example P, and let $\alpha_1, \alpha_2, \ldots$ be exchangeable umbrae. Define

$$\phi_\alpha(A) = \alpha_1 + \alpha_2 + \cdots + \alpha_{\phi(A)}.$$

Again, one finds that

$$\psi_n(A) = n!\, e_n(\alpha_1, \ldots, \alpha_{\phi(A)}).$$

For the umbral process, we have

$$\Delta_n(A) = \alpha_1^n + \alpha_2^n + \cdots + \alpha_{\phi(A)}^n.$$

8. Combinatorial orthogonality. A measure $\phi(A)$ defined for $A \in \Sigma$, whose values are random variables, is said to be a *completely random* measure when the following hold.

1. For every set $A \in \Sigma$ such that $\sigma(A) < \infty$, $\phi(A)$ is a random variable with finite expectation.
2. If $A_1, A_2, \ldots, A_n$ are pairwise disjoint measurable sets of finite measure, then the random variables $\phi(A_1), \phi(A_2), \ldots, \phi(A_n)$ are independent.

A completely random measure ϕ is said to be *multiplicative* if for every positive integer n and for every partition π of $[n]$,

$$\langle \mathrm{St}_\pi^{[n]} \rangle = \bigotimes_{b \in \pi} \langle \mathrm{St}_{\hat{1}}^{b} \rangle.$$

All stochastic measures in the examples of the previous section are multiplicative, except for the measures in Examples D and SP. The following proposition provides a necessary and sufficient condition for ϕ to be multiplicative.

PROPOSITION 8. *The measure ϕ is multiplicative if and only if the real-valued measures $\langle \Delta_n \rangle$ are nonatomic for all positive integers n.*

PROOF. For ease of computation, we assume that $\mathscr{A}$ is a rectangle set: $\mathscr{A} = A_1 \times \cdots \times A_n$. We then have

$$\begin{aligned}
\langle \mathrm{St}_\pi^{[n]} \rangle(\mathscr{A}) &= \langle \mathrm{St}_\pi^{[n]}(\mathscr{A}_\pi) \rangle \\
&= \left\langle \prod_{b \in \pi} \mathrm{St}_{\hat{1}}^{b}\left(\left(\mathop{\mathsf{X}}_{i \in b} A_i\right)_{\hat{1}}\right) \right\rangle \\
&= \left\langle \prod_{b \in \pi} \Delta_{|b|}\left(\bigcap_{i \in b} A_i\right) \right\rangle \\
&= \prod_{b \in \pi} \left\langle \Delta_{|b|}\left(\bigcap_{i \in b} A_i\right) \right\rangle \\
&= \bigotimes_{b \in \pi} \langle \mathrm{St}_{\hat{1}}^{b} \rangle(\mathscr{A}_\pi).
\end{aligned}$$

But

$$\bigotimes_{b \in \pi} \langle \mathrm{St}_{\hat{1}}^{b} \rangle(\mathscr{A}) = \bigotimes_{b \in \pi} \langle \mathrm{St}_{\hat{1}}^{b} \rangle(\mathscr{A}_{(\geq \pi)}).$$

We need only show, therefore, that

$$\bigotimes_{b \in \pi} \langle \mathrm{St}_{\hat{1}}^{b} \rangle(\mathscr{A}_{(> \pi)}) = 0,$$

where $\mathscr{A}_{(> \pi)}$ is defined as $\mathscr{A}_{(\geq \pi)} \backslash \mathscr{A}_\pi$.

If $\sigma > \pi$, then

$$\bigotimes_{b \in \pi} \langle \mathrm{St}_{\hat{1}}^{b} \rangle(\mathscr{A}_\sigma) = \bigotimes_{b \in \pi} \langle \Delta_{|b|} \rangle(\tilde{\mathscr{A}}_{\sigma'}),$$

where $\tilde{\mathscr{A}} \in S^{[r]}$ and $\sigma' > \hat{0}$ is a partition of $[r]$, where $r = |\pi|$. This is zero if and only if the diagonal measures are nonatomic for all n. □

Theorem 4 expresses the product of two or more stochastic measures as the sum over stochastic measures. When ϕ is multiplicative, the expectation of these stochastic measures can be expressed as the product of expectations of diagonal measures. We state these results first for the general case, and then specialize to results concerning $\psi_n(A)$ and $\Delta_n(A)$.

PROPOSITION 9. *Let ϕ be a multiplicative completely random measure. Let π be a partition of $[n]$. Then*

$$\left\langle \bigotimes_{b\in\pi} \mathrm{St}_{\hat{0}}^{b} \right\rangle = \sum_{\{\sigma:\ \sigma\wedge\pi=\hat{0}\}} \bigotimes_{b\in\sigma} \langle \mathrm{St}_{\hat{1}}^{b} \rangle$$

COROLLARY 1 (Feynman). *If $\langle\phi\rangle = 0$, and if b is a subset of $[n]$, then*

$$\langle \mathrm{St}_{\hat{0}}^{b} \otimes \mathrm{St}_{\hat{0}}^{b^c} \rangle = \begin{cases} \sum_{f\in \mathrm{bi}(b,b^c)} \bigotimes_{i\in b} \langle \mathrm{St}_{\hat{1}}^{\{i,f(i)\}} \rangle, & \text{if } |b| = |b^c|, \\ 0, & \text{otherwise}, \end{cases}$$

where $\mathrm{bi}(b, b^c)$ *is the set of bijections from b to b^c.*

PROOF. In the notation of the theorem, σ is the partition of $[n]$ consisting of the block $\{b, b^c\}$, and π is the zero partition. If $\tau \wedge \sigma = \pi$, then each block of τ must contain no more than one element of each set. On the other hand, if a partition τ contains any blocks with only one element, then $\langle \mathrm{St}_\tau^{[n]} \rangle = 0$, since $\mathrm{St}_\tau^{[n]}$ is multiplicative and $\langle\phi\rangle = 0$. Therefore, the only contribution to $\langle \mathrm{St}_\tau^{[n]} \rangle$ is from partitions whose blocks each contain exactly one element of each set. If $|b| \neq |b^c|$, no such partitions exist. If $|b| = |b^c|$, there is one such partition for every bijection between b and b^c. □

COROLLARY 2. *If $|b| > 1$, $|b^c| \geq 1$ and $\langle\phi\rangle = 0$, then*

$$\langle \mathrm{St}_{\hat{0}}^{b} \otimes \phi_{\hat{1}}^{b^c} \rangle = 0.$$

PROOF. Again, $\sigma = \{b, b^c\}$, but $\pi = \{\{i\colon i \in b\}, b^c\}$. If $|b| > 1$, each τ with $\sigma \wedge \tau = \pi$ contains a block with exactly one element, so $\langle \mathrm{St}_\tau^{[n]} \rangle = 0$. If $\sigma \wedge \tau = \pi$, one block of τ will contain b^c and at most one element of b, and the other blocks will each consist of one of the remaining elements of b. If $|b| > 1$, therefore, each τ will contain at least one block which contains only one element, and therefore $\langle \mathrm{St}_\tau^{[n]} \rangle = 0$. □

We now restate these results in terms of stochastic sequences of binomial type.

THEOREM 9. *Let $\psi_n(A)$ be the stochastic sequence of binomial type associated with a multiplicative completely random measure ϕ. Let π be a partition of the set $[n]$, whose blocks are of size $n_1, n_2, \ldots, n_k$. Then*

$$\langle \psi_{n_1}(A) \cdots \psi_{n_k}(A) \rangle = \sum_{\{\sigma:\ \sigma \wedge \pi = \hat{0}\}} \prod_{b \in \sigma} \langle \Delta_{|b|} \rangle (A).$$

PROPOSITION 10 (Orthogonality). *Let $\psi_n(A)$ be the stochastic sequence of binomial type associated with a multiplicative completely random measure ϕ. If $\langle \phi \rangle = 0$, then*

$$\langle \psi_n(A) \psi_m(A) \rangle = \begin{cases} n! \langle \Delta_2(A) \rangle^n, & \text{if } m = n, \\ 0, & \text{if } m \neq n. \end{cases}$$

PROPOSITION 11. *Under the same assumptions as Proposition 10, if $n > 1$ and $m \geq 1$, then*

$$\langle \psi_n(A) \Delta_m(A) \rangle = 0.$$

EXAMPLE G (Continued). We illustrate Theorem 9 by computing the expectation of a finite product of Hermite polynomials of arbitrary degrees. Recall that

$$\psi_n(A) = H_n^{(\sigma(A))}(\phi(A)).$$

It will be convenient to use the stochastic sequence of binomial type notation $\psi_n(A)$ instead of the Hermite notation. Our objective is to compute

$$\langle \psi_{n_1}(A) \psi_{n_2}(A) \cdots \psi_{n_k}(A) \rangle.$$

Theorem 9 can be applied in the following way. Choose any partition π of the set $[n_1 + n_2 + \cdots + n_k]$ having blocks b_i of size n_i. Define $n = n_1 + n_2 + \cdots + n_k$. Then Theorem 9 states that

$$\langle \psi_{n_1}(A) \psi_{n_2}(A) \cdots \psi_{n_k}(A) \rangle = \sum_{\{\sigma:\ \sigma \wedge \pi = \hat{0}\}} \bigotimes_{b \in \sigma} \langle \Delta_{|b|} \rangle (A).$$

However, $\langle \Delta_{|b|} \rangle = 0$ unless $|b| = 2$, in which case $\Delta_2 = \sigma$. The problem therefore boils down to listing all partitions σ, all of whose blocks have two elements such that $\sigma \wedge \pi = \hat{0}$. If C is equal to the number of such partitions, then we have that

$$\langle \psi_{n_1}(A) \psi_{n_2}(A) \cdots \psi_{n_k}(A) \rangle = C \sigma(A)^n.$$

For example, if n is odd, then $C = 0$.

The preceding computation is usually carried out using Feynman diagrams. We hope to have finally pinpointed the source of such computations.

EXAMPLE P (Continued). We compute

$$\langle \psi_n(A)\psi_k(A)\rangle,$$

where $\psi_n(A)$ is the sequence of binomial type associated with the Poisson measure.

Let π be a partition of the set $[n+k]$ having a block b_1 with k elements and a block b_2 with n elements. Say $k \le n$. Then Theorem 9 tells us that

$$\langle \psi_n(A)\psi_k(A)\rangle = \sum_{\sigma\wedge\pi=\hat{0}} \bigotimes_{b\in\sigma} \langle \Delta_{|b|}\rangle(A).$$

Any partition σ of the set $[n+k]$ such that $\sigma\wedge\pi=\hat{0}$ has blocks containing either one element or two elements. Every such partition can be obtained by choosing subsets c_1 and c_2 of b_1 and b_2 such that $|c_1| = |c_2|$, and matching the elements of c_1 with the elements of c_2 in an arbitrary way. In this way we obtain

$$\langle \psi_n(A)\psi_k(A)\rangle = \sum_{i\ge 0} \binom{n}{i}\binom{k}{i} i!\,\sigma(A)^{n+k-i},$$

where we have used the fact that $\Delta_1(A) = \Delta_2(A) = \phi(A)$ and $\langle \phi(A)\rangle = \sigma(A)$.

9. Diagonal measures. Under reasonable assumptions, the diagonal measures Δ_n of a completely random L^2 measure are given by

$$\Delta_n(A) = \lim_{N\to\infty} \sum_{i=1}^{N} \phi(A_{iN})^n,$$

where the limit is taken in the L^2 sense, and where, for each N, $\{A_{iN}: i = 1,\ldots,N\}$ is a disjoint partition of A into sets of equal measure.

PROPOSITION 12. *Let ϕ be good completely random measure, taking values in $L^2(\Omega)$. Let σ be a nonatomic measure and let the sets $\{A_{iN}: i = 1,\ldots,N\}$ be defined as above. Let a positive integer n be given, and suppose that for each positive integer $m < n$ there exist positive constants c_m and d_m such that*

$$\langle(\Delta_m - \langle\Delta_m\rangle)^2\rangle \le c_m\sigma \quad \textit{and} \quad |\langle\Delta_m\rangle| \le d_m\sigma.$$

Then

$$\Delta_n(A) = \lim_{N\to\infty} \sum_{i=1}^{N} \phi(A_{iN})^n,$$

where the limit is taken in L^2.

PROOF. We have

$$\Delta_n(A) = \Delta_n\left(\bigcup_{i=1}^{N} A_{iN}\right)$$
$$= \sum_{i=1}^{N}[\phi(A_{iN})^n - \mathrm{St}^{[n]}_{<\hat{1}}(A^{[n]}_{iN})],$$

where

$$\mathrm{St}^{[n]}_{<\hat{1}}(A^{[n]}_{iN}) = \sum_{\hat{0}\leq\pi<\hat{1}} \mathrm{St}^{[n]}_{\pi}(A^{[n]}_{iN}).$$

Therefore,

$$(*)\qquad \begin{aligned}\left\|\Delta_n(A) - \sum_{i=1}^{N}\phi(A_{iN})^n\right\|^2 &= \left\|\sum_{i=1}^{N}\mathrm{St}^{[n]}_{<\hat{1}}(A^{[n]}_{iN})\right\|^2 \\ &= \sum_{i=1}^{N}\left\|\mathrm{St}^{[n]}_{<\hat{1}}(A^{[n]}_{iN})\right\|^2 + \sum_{\substack{i,\,j=1\\ i\neq j}}^{N}\langle \mathrm{St}^{[n]}_{<\hat{1}}(A^{[n]}_{iN})\mathrm{St}^{[n]}_{<\hat{1}}(A^{[n]}_{jN})\rangle.\end{aligned}$$

We will show that the right-hand side goes to zero as $N\to\infty$.

We first show that if $\sigma(A) < 1$, then

$$\left\|\mathrm{St}^{[n]}_{\pi}(A^{[n]})\right\|^2 \leq K_n\sigma(A)^{|\pi|},$$

where K_n is a constant depending only on n. To this end, set $\tilde{\Delta}_m = \Delta_m - \langle\Delta_m\rangle$. Suppose $|\pi| = r$, and say that the blocks of π have sizes $m_1, \ldots, m_r$. Then

$$\mathrm{St}^{[n]}_{\pi}(A^{[n]}) = (\Delta_{m_1}\otimes\cdots\otimes\Delta_{m_r})(A^{[r]}_{\hat{0}})$$
$$= \sum_{b\subset[r]}\Phi^{[r]}_b(A^{[r]}_{\hat{0}}),$$

where

$$\Phi^{[r]}_b = \bigotimes_{j\in b}\langle\Delta_{m_j}\rangle \otimes \bigotimes_{j\in b^c}\tilde{\Delta}_{m_j} \equiv \bar{\Psi}^b\otimes\tilde{\Psi}^{b^c}.$$

Let $C_n = \max_{m<n}\{1, c_m, d_m\}$. Then

$$\|\bar{\Psi}^b(A^b)\| \leq C_n^{|b|}\sigma(A)^{|b|}.$$

We will show that

$$(**)\qquad \|\tilde{\Psi}^b(A^b_{\hat{0}})\| \leq \sqrt{|b|!C_n^{|b|}\sigma(A)^{|b|}}.$$

Recalling that $C_n \geq 1$ and $\sigma(A) < 1$, this inequality implies that

$$\|\Phi^{[r]}_b(A^{[r]}_{\hat{0}})\| \leq C_n^r\sqrt{r!}\,\sigma(A)^{r/2}$$

and thus that

$$\|\mathrm{St}_{\pi}^{[n]}(A^{[n]})\| \le 2^{|r|}\sqrt{r!}\, C_n^r \sigma(A)^{r/2}.$$

The assertion follows by setting $K_n = 2^n\sqrt{n!}\, C_n^n$.

To prove (**), write

$$A_{\hat{0}}^{[s]} = \bigcup (B_{j_1} \times \cdots \times B_{j_s}),$$

where $\{B_j\}$ is a countable set of measurable disjoint sets whose union is A, and the union on the right-hand side ranges over all sets of indices such that $j_1, \ldots, j_s$ are all distinct. Then

$$\begin{aligned}
\|\tilde{\Psi}^{[s]}(A_{\hat{0}}^{[s]})\|^2 &= \sum_{j_1,\ldots,j_s} \sum_{k_1,\ldots,k_s} \langle \tilde{\Delta}_{m_1}(B_{j_1}) \cdots \tilde{\Delta}_{m_s}(B_{j_s}) \tilde{\Delta}_{m_1}(B_{k_1}) \cdots \tilde{\Delta}_{m_s}(B_{k_s})\rangle \\
&= s! \sum_{j_1,\ldots,j_s} \langle \tilde{\Delta}_{m_1}^2(B_{j_1})\rangle \cdots \langle \tilde{\Delta}_{m_s}^2(B_{j_s})\rangle \\
&\le s!\, C_n^s \sum_{j_1,\ldots,j_s} \sigma(B_{j_1}) \cdots \sigma(B_{j_s}) \\
&\le s!\, C_n^s \sigma(A)^s.
\end{aligned}$$

Returning to (*), we note that

$$\|x_1 + \cdots + x_n\|^2 \le n(\|x_1\|^2 + \cdots + \|x_n\|^2).$$

Thus,

$$\begin{aligned}
\sum_{i=1}^{N} \|\mathrm{St}_{<\hat{1}}^{[n]}(A_{iN}^{[n]})\|^2 &\le N(B_n - 1) \sum_{\hat{0}\le\pi<\hat{1}} \|\mathrm{St}_{\pi}^{[n]}(A_{iN}^{[n]})\|^2, \\
&\le N(B_n - 1)^2 K_n \sigma(A_{iN})^2 \\
&= \frac{1}{N}(B_n - 1)^2 K_n \sigma(A)^2,
\end{aligned}$$

where B_n, Bell's number, is the number of partitions of the set $[n]$ (see [29]).

We now bound the cross terms. If A and A' are disjoint, then

$$|\langle \mathrm{St}_{\pi}^{[n]}(A^{[n]}) \mathrm{St}_{\pi'}^{[n]}(A'^{[n]})\rangle| \le C_n^{r+s} \sigma(A)^{|\pi|} \sigma(A')^{|\pi'|},$$

where $r = |\pi|$ and $s = |\sigma|$. Decomposing both $A^{[n]}$ and $A'^{[n]}$ in terms of triangular sets, as above, we get

$$\begin{aligned}
&\langle \mathrm{St}_{\pi}^{[n]}(A^{[n]}) \mathrm{St}_{\pi'}^{[n]}(A'^{[n]})\rangle \\
&\quad = \sum_{j_1,\ldots,j_r} \sum_{k_1,\ldots,k_s} \langle \Delta_{m_1}(B_{j_1}) \cdots \Delta_{m_r}(B_{j_r}) \Delta_{m_1'}(B_{k_1}') \cdots \Delta_{m_s'}(B_{k_s}')\rangle.
\end{aligned}$$

Since the $r+s$ sets of each term are pairwise disjoint, we obtain

$$\begin{aligned}
&|\langle \mathrm{St}_{\pi}^{[n]}(A^{[n]})\mathrm{St}_{\pi'}^{[n]}(A'^{[n]})\rangle| \\
&\quad \le C_n^{r+s} \sum_{j_1,\dots,j_r} \sum_{k_1,\dots,k_s} \sigma(B_{j_1})\cdots\sigma(B_{j_r})\sigma(B'_{k_1})\cdots\sigma(B'_{k_s}) \\
&\quad \le C_n^{2n}\sigma(A)^{|\pi|}\sigma(A')^{|\pi'|}.
\end{aligned}$$

Thus,

$$|\langle \mathrm{St}_{\pi}^{[n]}(A_{iN}^{[n]})\mathrm{St}_{\pi'}^{[n]}(A_{jN}^{[n]})\rangle| \le \frac{1}{N^4} C_n^{2n}\sigma(A)^4,$$

and

$$\sum_{\substack{i,j=1\\ i\ne j}}^{N} |\langle \mathrm{St}_{<\hat{1}}^{[n]}(A_{iN}^{[n]})\mathrm{St}_{<\hat{1}}^{[n]}(A_{jN}^{[n]})\rangle| \le \frac{(B_n-1)^2}{N^2} C_n^{2n}\sigma(A)^4. \qquad \square$$

REFERENCES

[1] BJORKEN, J. D. and DRELL, S. D. (1964). *Relativistic Quantum Mechanics*. McGraw-Hill, New York.

[2] BOGOLIUBOV, N. N. and SHIRKOV, D. V. (1980). *Introduction to the Theory of Quantized Fields*. Wiley, New York.

[3] CAMERON, R. H. and MARTIN, W. T. (1947). The orthogonal development of nonlinear functionals in series of Fourier–Hermite functionals. *Ann. Math.* **48** 385–392.

[4] CARLEN, E. A. and KRÉE, P. (1991). L^p estimates on iterated stochastic integrals. *Ann. Probab.* **19** 354–368.

[5] CHIHARA, T. S. (1978). *An Introduction to Orthogonal Polynomials*. Gordon and Breach, New York.

[6] COX, D. R. and ISHAM, V. (1980). *Point Processes*. Chapman and Hall, London.

[7] DOLÉANS-DADE, C. (1970). Quelques applications de la formule de changement de variables pour les semimartingales. *Z. Wahrsch. Verw. Gebiete* **16** 181–194.

[8] DOUBILET, P. (1972). On the foundations of combinatorial theory. VII: symmetric functions through the theory of distribution and occupancy. *Stud. Appl. Math.* **51** 377–396.

[9] ENGEL, D. D. (1982). The multiple stochastic integral. *Mem. Am. Math. Soc.* **38** 1–82.

[10] FLOREANINI, R., LETOURNEUX, J. and VINET, L. (1993). Quantum mechanics and polynomials of a discrete variable. *Ann. Physics* **226** 331–349.

[11] HIDA, T. (1970). *Stationary Stochastic Processes*. Princeton Univ. Press.

[12] HIDA, T. and IKEDA, N. (1965). Analysis on Hilbert space with reproducing kernel arising from multiple Wiener integral. *Proc. Fifth Berkeley Symp. Math. Statist. Probab.* **2** 117–143. Univ. California Press, Berkeley.

[13] ITÔ, K. (1951). Multiple Wiener integral. *J. Math. Soc. Japan* **3** 157–169.

[14] ITÔ, K. (1956). Spectral type of the shift transformation of differential processes with stationary increments. *Trans. Amer. Math. Soc.* **81** 253–263.

[15] KAKUTANI, S. (1950). Determination of the spectrum of the flow of Brownian motion. *Proc. Nat. Acad. Sci. U.S.A.* **36** 319–323.

[16] KAKUTANI, S. (1961). Spectral analysis of stationary Gaussian processes. *Proc. Fourth Berkeley Symp. Math. Statist. Probab.* **2** 239–247. Univ. California Press, Berkeley.

[17] KALLENBERG, O. (1983). *Random Measures*. Academic Press, New York.

[18] KALLENBERG, O. and SZULGA, J. (1989). Multiple integration with respect to Poisson and Lévy processes. *Probab. Theory Related Fields* **83** 101–134.

[19] LEBEDEV, N. N. (1972). *Special Functions and Their Applications*. Dover, New York.

[20] MAJOR, P. (1981). *Multiple Wiener–Itô Integrals. Lecture Notes in Math.* **849**. Springer, New York.
[21] MCKEAN, JR., H. P. (1955). Sample functions of stable processes. *Ann. Math.* **61** 564–579.
[22] MCKEAN, JR., H. P. (1969). *Stochastic Integrals*. Academic Press, New York.
[23] METIVIER, M. and PELLAUMAIL, J. (1980). *Stochastic Integration*. Academic Press, New York.
[24] MEYER, P.-A. (1976). Un cours sur les integrales stochastiques. *Seminaire de Probabilités X: Université de Strasbourg. Lecture Notes in Math.* **511** 245–400. Springer, Berlin.
[25] MEYER, P.-A. (1993). *Quantum Probability for Probabilists. Lecture Notes in Math.* **1538**. Springer, New York.
[26] NIKIFOROV, A. F., SUSLOV, S. K. and UVAROV, V. B. (1991). *Classical Orthogonal Polynomials of a Discrete Variable*. Springer, Berlin.
[27] OGURA, H. (1972). Orthogonal functionals of the Poisson process. *IEEE Trans. Inform. Theory* **IT-18** 473–481.
[28] PROTTER, P. (1990). *Stochastic Integration and Differential Equations: A New Approach*. Springer, Berlin.
[29] ROTA, G.-C. (1964). The number of partitions of a set. *Amer. Math. Monthly* **71** 498–504.
[30] ROTA, G.-C. (1964). On the foundations of combinatorial theory I. Theory of Möbius functions. *Z. Wahrsch. Verw. Gebiete* **2** 340–368.
[31] ROTA, G.-C. (1975). *Finite Operator Calculus*. Academic Press, New York.
[32] ROTA, G.-C., KAHANER, D. and ODLYZKO, A. (1973). On the foundations of combinatorial theory. VIII. Finite operator calculus. *J. Math. Anal. Appl.* **42** 684–760.
[33] ROTA, G.-C. and TAYLOR, B. D. (1994). The classical umbral calculus. *SIAM J. Math. Anal.* **25** 694–711.
[34] SEGAL, I. E. (1956). Tensor algebras over Hilbert spaces, I. *Trans. Amer. Math. Soc.* **81** 106–134.
[35] SEGALL, A. and KAILATH, T. (1976). Orthogonal functionals of independent increment processes. *IEEE Trans. Inform. Theory* **IT-22** 287–298.
[36] STREATER, R. F. and WIGHTMAN, A. S. (1964). *PCT, Spin and Statistics, and All That*. Benjamin/Cummings, Reading, MA.
[37] WICK, G.-C. (1950). *Phys. Rev.* **80** 268–272.
[38] WIENER, N. (1938). The homogeneous chaos. *Amer. J. Math.* **55** 897–936.
[39] WIENER, N. and WINTNER, A. (1941). The discrete chaos. *Amer. J. Math.* **65** 279–298.
[40] ZINN-JUSTIN, J. (1989). *Quantum Field Theory and Critical Phenomena*. Clarendon Press, Oxford.

DEPARTMENT OF MATHEMATICS
MASSACHUSETTS INSTITUTE
OF TECHNOLOGY
CAMBRIDGE, MASSACHUSETTS 02139
E-MAIL: rota@math.mit.edu

THEORETICAL DIVISION
LOS ALAMOS NATIONAL LABORATORY
LOS ALAMOS, NEW MEXICO 87545
E-MAIL: tcw@lanl.gov

Annals of Combinatorics 3 (1999) 13-25

Annals of Combinatorics

Probability Set Functions*

William J. Bruno[1], Gian-Carlo Rota, and David C. Torney[1]

[1]Theoretical Division, Mail Stop K710, Los Alamos National Laboratory, Los Alamos, New Mexico 87545, USA

{billb, dct}@lanl.gov

Received January 6, 1999

AMS Subject Classification: 60C05, 60G99

Abstract. A probability set function is interpretable as a probability distribution on binary sequences of fixed length. Cumulants of probability set functions enjoy particularly simple properties which make them more manageable than cumulants of general random variables. We derive some identities satisfied by cumulants of probability set functions which we believe to be new. Probability set functions may be expanded in terms of their cumulants. We derive an expansion which allows the construction of examples of probability set functions whose cumulants are arbitrary, restricted only by their absolute values. It is known that this phenomenon cannot occur for continuous probability distributions. Some particular examples of probability set functions are considered, and their cumulants are computed, leading to a conjecture on the upper bound of the values of cumulants. Moments of probability set functions determined by arithmetical conditions are computed in a final example.

Keywords: algebraic enumeration, binary sequences, set partitions into even-size blocks, restricted set partitions, finite residue classes modulo three, moment, centered moment, cumulant, bounds, Fourier analysis, Möbius inversion.

1. Introduction

This work is motivated by the current interest in DNA sequences. Much work has been done on this subject and some joint probability distributions have been proposed as stochastic models of DNA sequences. As will be seen, the four letters of DNA sequences may be encoded profitably as sequences of $+1$'s and -1's as follows: $A = 1\ \ 1$, $C = 1\ \ -1$, $G = -1\ \ 1$, and $T = -1\ \ -1$. Then the problem is that of defining stochastic processes on finite sequences of minus ones and plus ones which will model statistical properties of DNA.

Most models proposed so far rely upon Markovian assumptions, whether as Markov chains or by hidden Markovian behavior [19]. Markovian models, however, cannot

* Dedicated to our friend, W.A. Beyer. Financial support for this work was derived from the U.S.D.O.E. Human Genome Project, through the Center for Human Genome Studies at Los Alamos National Laboratory, and also through the Center for Nonlinear Studies, Los Alamos National Laboratory, LANL report LAUR-97-323.

easily take into account correlations among entries of a sequence which are distant from each other, and which may not be related by any type of Markovianness.

It has long been known that long-range statistical correlations can be accounted for by cumulants. Nevertheless, the project of bringing cumulants to bear in accounting for such long-range correlations by explicit stochastic processes defined in terms of cumulants has remained largely programmatic. Cumulants, despite their striking elegance in characterizing all classical probability distributions, present to this day deep theoretical problems which remain to be solved; paramount among these is the problem of giving a simple set of inequalities which characterize the cumulants of probability distributions, a problem which has been open for over a century but which has hardly been worked upon [6, 7].

The purpose of the present paper is to take a modest step forward in this program, for the special case of stochastic sequences of -1's and $+1$'s. Our objective is twofold. In the first place, we obtain a simplified expression for a probability distribution on a finite sequence of $+1$'s and -1's in terms of its cumulants (Theorem 4.1 below), which, we would like to believe, displays the correlational roles of successive cumulants. As an application of our main result, we show that probability distributions of finite sequences exhibit a behavior which stands in stark contrast to the behavior of the "classical" stochastic processes. It turns out (Theorem 5.1) that the cumulants of probability distributions on finite sequences of $+1$'s and -1's can be pretty much arbitrarily prescribed, subject only to a restriction on their maxima. In particular, probability distributions on finite sequences of $+1$'s and -1's exist for which the first k cumulants are nonzero and all successive cumulants equal zero. Compare this with the case of continuous probability distributions, where this phenomenon can happen only for $k = 2$, namely, for the normal distribution.

We have chosen the language of set theory rather than the language of sequences, both because computations are simplified in this language and because of a remarkable coincidence. It turns out that the moments of a probability set function coincide with a common type of transform which has been used under different guises (Hadamard transform, Walsh transform, etc.).

Torney [20] introduced a new type of cumulants: binary cumulants. Binary cumulants might appear to be more suited to the present setup. However, their use will be limited until a more explicit formula is discovered, which expresses a probability set function in terms of its binary cumulants. Nevertheless, binary cumulants offer some empirical advantages over ordinary cumulants.

Some significant examples appear throughout the paper, which we hope will be found to be of independent interest.

2. Probability Set Functions and Their Moments

A probability set function P is a probability defined on the family of all subsets of a set U, that is, for all $S \subseteq U$, $0 \leq P(S) \leq 1$, and $\sum_{S \subseteq U} P(S) = 1$.

If the set U is the set $\{1, 2, \ldots, n\}$, then a probability set function may be interpreted as a probability distribution on the set of all sequences $\{a_1, a_2, \ldots, a_n\}$, where each of the entries a_i equals $+1$ or -1. Some of the definitions given below may become clearer when interpreted in terms of sequences of $+1$'s and -1's. For example, $P(S)$ may be interpreted as the probability of the sequence having $+1$'s at the positions indexed by S, and having -1's at the complementary positions.

The set-moments $\mu(S)$ are defined for $S \subseteq U$ as follows.

$$\mu(S) = \sum_{T \subseteq U} (-1)^{|S \setminus T|} P(T). \tag{2.1}$$

Here, $S \setminus T$ denotes the relative complement: $S - S \cap T$.

The probabilities are expressed in terms of the moments as follows.

$$P(S) = 2^{-|U|} \sum_{T \subseteq U} (-1)^{|T \setminus S|} \mu(T). \tag{2.2}$$

A short proof of this pair of inverse formulas goes as follows. Substituting (2.2) into (2.1) we obtain

$$\mu(S) = 2^{-|U|} \sum_{T \subseteq U} (-1)^{|S \setminus T|} \sum_{R \subseteq U} (-1)^{|R \setminus T|} \mu(R).$$

Observe that $\sum_{T \subseteq U} (-1)^{|S \setminus T|} (-1)^{|R \setminus T|} = 2^{|U|} \delta_{S,R}$, hence the conclusion.

The preceding pair of inversion formulas of a probability set function and its moments is equivalent to other pairs in the literature [2,9,11]. It is an example of a Fourier inversion because the coefficients of the $P(T)$'s in (2.1) and of the $\mu(T)$'s in (2.2) are essentially the group characters of the $|U|$-fold Cartesian product of the group Z_2 [21].

The $2^{|U|}$ equations of (2.2), satisfied by the moments, are linear in the μ's. These equations embody the necessary and sufficient conditions for a sequence of numbers to be the moments of a probability set function. These conditions are simpler than the classical conditions for a sequence of real numbers to be moments of a probability distribution [8, 16]. Such a simplification will result in further simplification of cumulant computations, as we will shortly see.

3. Cumulants

The set-cumulants $\kappa(S)$ are defined for $S \subseteq U$ as follows:

$$\kappa(S) = \sum_{\pi} (-1)^{|\pi|-1} (|\pi| - 1)! \prod_{B \in \pi} \mu(B). \tag{3.1}$$

Here, the variable π ranges over all partitions of the set S [14, 15, 17]. (Recall that a partition of a set is the set of equivalence classes, called *blocks*, of an equivalence relation.) We denote by $|\pi|$ the number of blocks of the partition π. The variable B ranges over all blocks of π.

The following expression of moments in terms of cumulants by the Möbius inversion of (3.1) is known [14, 15]:

$$\mu(S) = \sum_{\pi} \prod_{B \in \pi} \kappa(B). \tag{3.2}$$

The following technical remark will be used in later computations:

Proposition 3.1. *Let a probability set function P have* $\kappa(U) = a$. *Then there exists another probability set function with* $\kappa(U) = -a$.

Proof. Consider the mapping ρ of a probability set function to another probability set function as follows: $P(S) \mapsto P(S\Delta u)$, with $u \in U$ and $S \subseteq U$, and Δ denoting the symmetric difference. This changes the sign of $\mu(S)$, with $S \subseteq U$, if and only if $u \in S$. From the definition (3.1), it follows that this changes the sign of $\kappa(U)$. ∎

It follows from Proposition 3.1 that the upper and lower bounds for $\kappa(U)$ over all probability set functions have the same magnitude.

Example 3.2. It is easily computed that $|\kappa(U)| \leq 1$ when $|U| \leq 3$ (viz. (6.3)).

Example 3.3. Let $|U| = 2n$, and consider the probability set function

$$\begin{aligned} P(U) &= \tfrac{1}{2}, \\ P(\emptyset) &= \tfrac{1}{2} \end{aligned}$$

and

$$P(S) = 0,$$

otherwise $0 \subset S \subset U$. The moments $\tilde{\mu}$ of this probability set function are as follows.

$$\tilde{\mu}(S) = 1 \text{ if } |S| = 2k,$$

and

$$\tilde{\mu}(S) = 0 \text{ if } |S| = 2k+1,$$

for all $S \subseteq U$. Therefore, from (3.1), the cumulants $\tilde{\kappa}$ of this probabioity set function are as follows:

$$\tilde{\kappa}(U) = \sum_{j=1}^{n} (-1)^{j-1}(j-1)!\mathcal{E}_j^{(2n)}.$$

Here, $\mathcal{E}_j^{(2n)}$ denotes the number of partitions of a set of size $2n$ into j even-size blocks (See Section 7 of [13]). Its exponential generating function, (7.1), gives

$$\tilde{\kappa}(U) = (-1)^{n-1} T_{2n-1};\ 0 < n.$$

Here, T_{2n-1} is the tangent coefficient defined as $\frac{d^{2n-1}}{dx^{2n-1}} \tan(x)|_{x=0}$. From [1], we obtain

$$|\tilde{\kappa}(U)| \sim 2(2/\pi)^{2n}(2n-1)!.$$

We conjecture that $\tilde{\kappa}(U)$ bounds all cumulants of all probability set functions on sets U with $2n$ elements.

4. Probability Set Functions and Their Cumulants

We now come to the main result of this paper. If P is a probability set function on U, if u is an element of the set U and if S is a subset of U, we denote by $P_u(S)$ the following probability:

$$P_u(S) = \begin{cases} \sum_{R\subseteq\{U-u\}} P(R\cup u), & \text{if } u \in S, \\ \sum_{R\subseteq\{U-u\}} P(R), & \text{if } u \notin S. \end{cases}$$

The $P_u(S)$ will be called the *marginals* of the probability set function P. From this definition and (2.2), it follows that

$$P_u(S) = \frac{1}{2}\{1 + (2|u\cap S| - 1)\mu(u)\}, \tag{4.1}$$

which is intuitive from the interpretation of probability set functions in terms of probability distributions on sequences.

The following theorem expresses the values of a probability set function in terms of its set-cumulants and its marginals.

Theorem 4.1. *Let π range over partitions of U and let $S \subseteq U$. Let π' be the set of blocks of π containing more than one element, and π'' the set of blocks of π containing one element, so that $\pi = \pi' \cup \pi''$. Then*

$$P(S) = \sum_{\pi} \prod_{B\in\pi'} (-1)^{|B\setminus S|} 2^{-|B|} \kappa(B) \prod_{u\in\pi''} P_u(S). \tag{4.2}$$

Proof. We first rewrite (3.2) for $T \subseteq U$:

$$\mu(T) = \sum_{R\subseteq T} \prod_{u\in T-R} \mu(u) \sum_{\pi'} \prod_{B\in\pi'} \kappa(B), \tag{4.3}$$

with $\sum_{\pi'}$ ranging over the partitions of R with blocksize greater than unity. This identity results from a transformation of the right-hand side of (3.2) according to a trivial bijection from all partitions of a set onto the partitions with blocksize greater than unity of the elements of its power set. Substituting (4.3) in (2.2) yields an expression for the probability set functions, $P(S)$, in terms of their cumulants. In the resulting expression, for each $R \subseteq U$, the coefficient of $\sum_{\pi'} \prod_{B\in\pi'} \kappa(B)$ (with the summation ranging over the partitions π' of R) is readily found to be $2^{-|U|} \sum_{R\subseteq T\subseteq U} (-1)^{|T\setminus S|} \prod_{u\in T-R} \mu(u)$. Using the definition of the marginals (4.1), this coefficient is also seen to equal

$$(-1)^{|R\setminus S|} 2^{-|R|} \prod_{u\in U-R} P_u(S).$$

Distributing the factors involving R and using the aforementioned bijection establishes (4.2) and the theorem. ∎

An alternative version of the identity (4.2) is described in Section 6.

Example 4.2. If all cumulants of $\kappa(B)$ of a probability set function vanish for $1 < |B|$, then

$$P(S) = \prod_{u \in U} P_u(S).$$

In other words, if all the specified cumulants of a probability set function vanish, then such a probability set function is the product of its marginals. In view of Theorem 4.1, it stands to reason that cumulants $\kappa(B)$, with $1 < |B|$, contain all information relating to dependencies of the probability set function.

Example 4.3. If $\kappa(S) = v$ with $2 = |S|$ and if $\kappa(S) = 0$ with $2 < |S|$ vanish, and if $0 < P_u(S) < 1$ for all $u \in U$, then

$$P(S) = \left(\prod_{u \in U} P_u(S) \right) \sum_{j=0}^{\lfloor \frac{|U|}{2} \rfloor} (2j-1)!!v^j e_{2j}. \tag{4.4}$$

Here, e_{2j} denotes the elementary symmetric polynomial of degree $2j$ of the $|U|$ quantities $(2|u \cap S| - 1)/(2P_u(S))$, $u \in U$, with $e_0 \stackrel{\text{def}}{=} 1$. Observe that $(2j-1)!! \stackrel{\text{def}}{=} (2j)!/(2^j j!)$ and that $(-1)!! \stackrel{\text{def}}{=} 1$. We do not know, in general, under which conditions on the $P_u(S)$'s and v the right-hand side of (4.4) defines a probability set function.

Example 4.4. If $\kappa(S) = w$ with $2 \leq |S|$, and if $0 < P_u(S) < 1$ for all $u \in U$, then

$$P(S) = \left(\prod_{u \in U} P_u(S) \right) \left\{ 1 + \sum_{j=2}^{|U|} e_j \sum_{k=1}^{\lfloor \frac{j}{2} \rfloor} \mathcal{D}_k^{(j)} w^k \right\}. \tag{4.5}$$

Here, e_j denotes the elementary symmetric polynomial of degree j of the $|U|$ quantities $(2|u \cap S| - 1)/(2P_u(S))$, $u \in U$, and $\mathcal{D}_k^{(j)}$ denotes the number of partitions of the elements of a j-set into k blocks of size greater than unity. This example is reminiscent of the Poisson distribution for which all cumulants take the same value. Again, we do not know under which conditions on the $P_u(S)$'s and w the right-hand side of (4.5) defines a probability set function.

5. Probability Set Functions with Given Cumulants

Recall that, for continuous probability distributions of one real variable, Marcinkiewicz's Theorem states that either the first or the second cumulant is the only nonvanishing cumulants or there is an infinite number of nonvanishing cumulants [15]. The following theorem shows that the set cumulants are strikingly less restricted.

Theorem 5.1. *Let $\nu(S)$ be a real set function defined for $S \subseteq U$. Then, provided $|\varepsilon|$ is sufficiently small, there exists a probability set function, $P(S)$, having $\kappa(S) = \varepsilon\nu(S)$ for its cumulants for $S \subseteq U$.*

Proof. Using (4.1), substitute the $(1 \pm \varepsilon\nu(u))/2$'s for the respective marginal probabilities appearing in (4.2). Substitute $\varepsilon\nu(B)$ for the remaining cumulants of (4.2). Then

$P(S)$ is found to be a polynomial in ε, of degree $|U|$. For the purposes of this proof, the polynomial can be represented by $2^{-|U|}+\varepsilon O(1)$, where $O(1)$ denotes a constant and the monomials containing higher powers of ε are suppressed. In general, a different constant is obtained for each $S \subseteq U$. However, we can clearly let $|\varepsilon|$ be sufficiently small so that, for all S, the terms of the form $\varepsilon O(1)$ are less than $2^{-|U|}$ in magnitude. This suffices to ensure that $P(S)$ is a probability set function. ∎

The same argument yields that a probability set function results if the cumulants for single elements are fixed (in the interval $[-1,+1]$), and the remaining $\kappa(B)$ are taken to be equal to $\varepsilon\nu(B)$, provided ε is sufficiently small.

6. Centered Moments

As in the case of probability distributions, one can introduce set-centered moments and obtain simplified expressions of the cumulants in terms of set-centered moments. We include a brief discussion of set centered moments for the sake of completeness and because they yield a pre-existing version of of our identity (4.2) [10].

Set-centered moments $\xi(S)$ of a probability set function are defined for $S \subseteq U$ as follows:

$$\xi(S)=\sum_{R\subseteq S}\mu(R)(-1)^{|S\setminus R|}\prod_{s\in S\setminus R}\mu(s). \tag{6.1}$$

Equation (6.1) is readily inverted to yield

$$\mu(S)=\sum_{R\subseteq S}\xi(R)\prod_{s\in S\setminus R}\mu(s). \tag{6.2}$$

Substituting (3.2) into (6.1) yields

$$\xi(S)=\sum_{\pi}{}'\prod_{B\in\pi}\kappa(B);\ 2\leq|S|. \tag{6.3}$$

The sum $\sum'_{\pi}$ ranges over all partitions of S with all blocks containing two or more elements. Möbius inversion [14] of (6.3) on the poset of partitions with blocks of size at least two yields the following expression for $\xi(S)$ with $2\leq|S|$:

$$\kappa(S)=\sum_{\pi}(-1)^{|\pi|-1}(|\pi|-1)!\prod_{B\in\pi}\xi(B);\ \ 2\leq|S|. \tag{6.4}$$

Here, the summation ranges over all partitions of S, since $\xi(B)$ vanishes when $|B|=1$.

Collecting terms in (4.2) of Theorem 4.1 and using (6.3), we obtain an expression of a probability set function in terms of its set-centered moments and its marginals.

Corollary 6.1.[10]

$$P(S)=\sum_{T\subseteq U}(-1)^{|T\setminus S|}2^{-|T|}\xi(T)\prod_{u\in U\setminus T}P_u(S). \tag{6.5}$$

Note that although this expression for the probability set function involves fewer summands than the expression of Theorem 4.1, the former could be easier to evaluate for many set functions. For instance, many of those easily expressed in terms of their cumulants.

7. Partitions with Even-Sized Blocks

This section describes the combinatorial properties of the numbers of partitions of an even-sized set into all even-sized blocks [13, 18]. These numbers have been used in an example above and will be used in the next example. Denote the number of partitions of a set of size $2n$ into j even-sized blocks by $\mathcal{E}_j^{(2n)}$.

Theorem 7.1. *Clearly,* $\mathcal{E}_1^{(2n)} = 1$ *and* $\mathcal{E}_j^{(2n)} = 0$ *if* $n < j$. *For* $1 < j \leq n$, *the following recurrence is obtained*:

$$\mathcal{E}_j^{(2n)} = j^2 \mathcal{E}_j^{(2n-2)} + (2j-1)\mathcal{E}_{j-1}^{(2n-2)}.$$

Proof. Begin with the evident recurrence for $\mathcal{N}_{j,l}^{(m)}$ – the number of partitions of an m-set into j even-sized and l odd-sized blocks: $\mathcal{N}_{j,l}^{(m)} = (j+1)\mathcal{N}_{j+1,l-1}^{(m-1)} + (l+1)\mathcal{N}_{j-1,l+1}^{(m-1)} + \mathcal{N}_{j,l-1}^{(m-1)}$, for $1 < m$ with $\mathcal{N}_{0,1}^{(1)} = 1$ and $\mathcal{N}_{j,-1}^{(m)} = \mathcal{N}_{-1,l}^{(m)} = 0$. By definition, $\mathcal{E}_j^{(2n)}$ equals $\mathcal{N}_{j,0}^{(2n)}$. The following identity is then used to transform the previous recurrence into one involving only $\mathcal{E}_j^{(2n)}$'s:

$$\mathcal{N}_{j-2,2}^{(2n)} = \binom{j}{2}\mathcal{N}_{j,0}^{(2n)} + (j-1)\mathcal{N}_{j-1,0}^{(2n)}.$$

The identity is established from the following bijection. Start with $\binom{j}{2}$ copies of all partitions of $2n$ elements into j even-sized blocks. In each copy, fuse a different one of the $\binom{j}{2}$ unordered pairs of blocks into one block, yielding a partition with $j-1$ even-sized blocks. Then split the new block, resulting from the fusion, into two odd-sized blocks. There is one more way to split into odd-sized blocks than there is for even-sized blocks, from which the term $(j-1)\mathcal{N}_{j-1,0}^{(2n)}$ originates. (This can be seen by summing "even" and "odd" Binomial coefficients). It is easily verified that every resulting partition is distinct, and that all partitions with two odd-size blocks are created in this procedure. ∎

See [1,3] for analogous results for the Stirling numbers of the second kind. A closed form for $\mathcal{E}_j^{(2n)}$ follows.

Corollary 7.2.

$$\mathcal{E}_j^{(2n)} = \frac{2^{1-j}}{j!}\sum_{k=1}^{j}(-1)^{n-k}\binom{2j}{j-k}k^{2n}.$$

Here, $1 \leq j \leq n$ and $1 \leq n$. The following identity was employed in our derivation: $(2n)! = 2\sum_{k=1}^{n}(-1)^{n-k}k^{2n}\binom{2n}{n-k}$. The exponential generating function of $\mathcal{E}_j^{(2n)}$ is a straightforward consequence of Theorem 7.1 (or of Corollary 7.2).

$$\sum_{n=j}^{\infty}\mathcal{E}_j^{(2n)}\frac{x^{2n}}{(2n)!} = \frac{(\cosh(x)-1)^j}{j!}. \tag{7.1}$$

Let $\mathcal{E}_{2n}$ denote the number of partitions of a set of size $2n$ into any number of even-sized blocks: $\mathcal{E}_{2n} \stackrel{\text{def}}{=} \sum_{j=1}^{n} \mathcal{E}_j^{(2n)}$, and let $\mathcal{E}_0 = 1$. Then [13, 18]

$$\sum_{n=0}^{\infty} \mathcal{E}_{2n} x^{2n}/(2n)! = \exp(\cosh(x) - 1).$$

8. Residual Sequences Modulo Three

We close this paper with examples of probability set functions with remarkable sets of moments. (The cumulants in these examples were also computed but they turned out to be of less interest.) To fix our ideas, we describe the typical example. Consider the set of all m-sequences $\{b_0, b_1, \ldots, b_{m-1}\}$, where each b_j equals 0 or 1, which have the additional property that the integer $b_0 + 2b_1 + 2^2 b_2 + \cdots + 2^{m-1} b_{m-1}$ is a multiple of three. To each such sequence, associate the subset of the set $\{0, 1, \ldots, m-1\}$ containing all integers j for which $b_j = 0$. We thereby obtain a family of subsets of the set $\{0, 1, \ldots, m-1\}$ which we call $Z_0(m)$.

Now define a probability set function on the set $V = \{0, 1, \ldots, m-1\}$ as follows: Assign the same probability to every subset of V which belongs to the family $Z_0(m)$, and assign zero probability to every other subset of V.

Example 8.1. For $Z_0(2)$, the probability set function is defined on the set $V = \{0, 1\}$ by taking $P(\emptyset) = P(\{0, 1\}) = 1/2$.

Instead of choosing integers which are multiples of three, we can repeat the same construction with integers which are congruent to one modulo three or congruent to two modulo three. The families of subsets $Z_1(m)$ and $Z_2(m)$ are defined accordingly. We thereby obtain three families of probability set functions, whose moments are computed below. We refer to these probability set functions as $(3, i, m)$-residual probability set functions with $i \in \{0, 1, 2\}$.

We now restate our probability set functions in more precise terms. Let $N_i(m) = \{n : 0 \le n < 2^m;\ n \equiv i \pmod 3\}$. Let $\Omega_i(m)$ denote $|N_i(m)|$. For any integer, construct the set of indices of its binary digits that are equal to zero. Let $Z_i(m)$ denote the set, for all $n \in N_i(m)$, of the corresponding sets of indices $i \in \{0, 1, 2\}$. Consider the probability set functions defined on the subsets of $V = \{0, 1, \ldots, m-1\}$:

$$P(S) = \frac{1}{\Omega_i(m)},$$

if $S \in Z_i(m)$, and

$$P(S) = 0,$$

otherwise. These are the $(3, i, m)$-residual probability set functions. The moments of these probability set functions will be derived next.

Theorem 8.2. *For each $(3, i, m)$-residual probability set function, with $i \in \{0, 1, 2\}$ and $1 \le m < \infty$, the moments $\mu_i(S)$ are the following:*

(a) *m and |S| even,*

$$\begin{aligned}
\mu_0(S) &= 2\cdot 3^{|S|/2}\cdot(-1)^{t-t_0}/(2^m+2);\\
\mu_1(S) &= -3^{|S|/2}\cdot(-1)^{t-t_0}/(2^m-1);\\
\mu_2(S) &= -3^{|S|/2}\cdot(-1)^{t-t_0}/(2^m-1);
\end{aligned}$$

(b) *m even and |S| odd,*

$$\begin{aligned}
\mu_0(S) &= 0;\\
\mu_1(S) &= -3^{(|S|+1)/2}\cdot(-1)^{t-t_0}/(2^m-1);\\
\mu_2(S) &= 3^{(|S|+1)/2}\cdot(-1)^{t-t_0}/(2^m-1);
\end{aligned}$$

(c) *m odd and |S| even,*

$$\begin{aligned}
\mu_0(S) &= 3^{|S|/2}\cdot(-1)^{t-t_0}/(2^m+1);\\
\mu_1(S) &= 3^{|S|/2}\cdot(-1)^{t-t_0}/(2^m+1);\\
\mu_2(S) &= -2\cdot 3^{|S|/2}\cdot(-1)^{t-t_0}/(2^m-2);
\end{aligned}$$

(d) *m and |S| odd,*

$$\begin{aligned}
\mu_0(S) &= 3^{(|S|+1)/2}\cdot(-1)^{t-t_0}/(2^m+1);\\
\mu_1(S) &= -3^{(|S|+1)/2}\cdot(-1)^{t-t_0}/(2^m+1);\\
\mu_2(S) &= 0.
\end{aligned}$$

Here, $t \stackrel{def}{=} \sum_{h\in S} h$ *and* $t_0 \stackrel{def}{=} |S|(|S|+1)/2$.

Proof. There is a simple rule for congruences modulo three of binary numbers analogous to well-known rules [5]. Because $2^{2j} \equiv 1 \pmod 3$ and $2^{2j+1} \equiv -1 \pmod 3$, the following formula, involving the foregoing binary representation of n, is congruent to n modulo three.

$$\sum_{j=0,2,4,\ldots} b_j - \sum_{j=1,3,5,\ldots} b_j. \tag{8.1}$$

To facilitate the proof, we now employ the bijection of the elements of the $Z_i(m)$'s onto binary m-sequences introduced in Section 1. That is, each $S \in Z_i(m)$, a subset of $V = \{0, 1, \ldots, m-1\}$, maps to the m-sequence with $+1$'s at the positions indexed by the elements of S and with -1's at the positions indexed by the elements of $V - S$. Thus, to evaluate all the moments for $(3, i, m)$-residual probability set functions, we begin by counting the number of times, for all $S \in Z_i(m)$, the product of the m corresponding digits ($\in \{-1, +1\}$) equals $+1$ and equals -1. Denote these numbers $C_i^+(m)$ and $C_i^-(m)$, respectively. From (8.1), the following linear recurrences apply.

$$\begin{aligned}
C_0^\pm(m+2) &= C_1^\mp(m) + C_2^\mp(m) + 2C_0^\pm(m);\\
C_1^\pm(m+2) &= C_0^\mp(m) + C_2^\mp(m) + 2C_1^\pm(m);\\
C_2^\pm(m+2) &= C_0^\mp(m) + C_1^\mp(m) + 2C_2^\pm(m).
\end{aligned}$$

(These formulas constitute six recurrences: the upper and lower superscripts each hold in three). Subtracting one set of recurrences from the other leads to three recurrences for the differences: $D_i(m) \stackrel{\text{def}}{=} C_i^+(m) - C_i^-(m)$. Only these differences are required for determining the moments. Thus, by induction, using the values of $D_i(1)$ and $D_i(2)$, it follows, for larger values of the arguments, that

$$D_0(2n) = 2\cdot 3^{n-1};\quad D_0(2n+1) = 3^n;$$
$$D_1(2n) = -3^{n-1};\quad D_1(2n+1) = -3^n;$$
$$D_2(2n) = -3^{n-1};\quad D_2(2n+1) = 0.$$

We now evaluate $\Delta_i(l;m)$: the number of times the product of the rightmost l of the m digits, a_1 through a_l, equals $+1$ minus the number of times the product equals -1 for each $N_i(m)$. We note the values of $2^m \pmod{3\cdot 2^l}$ because, for all $N_i(m)$, within blocks of consecutive sequences of length $3\cdot 2^l$, the number of such products equal to $+1$ equals the number of products equal to -1.

$$2^{2n} \pmod{3\cdot 2^l} \equiv 4^{\lfloor\frac{l-1}{2}\rfloor+1};$$
$$2^{2n+1} \pmod{3\cdot 2^l} \equiv 2\cdot 4^{\lfloor\frac{l}{2}\rfloor}.$$

Considering only the residual, final block of sequences of the specified length, the $\Delta_i(l;m)$'s are seen to involve either one or two of the $D_j(l)$, and it follows that

$$\begin{aligned}
\Delta_0(2k;2n) &= 2\cdot 3^{k-1}; & \Delta_0(2k;2n+1) &= 3^{k-1};\\
\Delta_1(2k;2n) &= -3^{k-1}; & \Delta_1(2k;2n+1) &= 3^{k-1};\\
\Delta_2(2k;2n) &= -3^{k-1}; & \Delta_2(2k;2n+1) &= -2\cdot 3^{k-1};\\
\Delta_0(2k+1;2n) &= 0; & \Delta_0(2k+1;2n+1) &= 3^k;\\
\Delta_1(2k+1;2n) &= -3^k; & \Delta_1(2k+1;2n+1) &= -3^k;\\
\Delta_2(2k+1;2n) &= 3^k; & \Delta_2(2k+1;2n+1) &= 0.
\end{aligned}$$

Recall the definition of the $\Delta_i(l;m)$'s. Within the subset of m-sequences corresponding to $N_i(m)$, the magnitude of the corresponding quantity for an arbitrary l-subset of digit positions equals $\Delta_i(l;m)$, for the following reasons. From the divisibility rule, (8.1), these subsets of sequences are invariant to interchanges of pairs of columns of digits, provided all digits in both columns are multiplied by -1 whenever columns of even and odd index are interchanged. Therefore, $\Delta_i(l;m)$ applies to the l-subsets of positions with a congruent number of even indices, modulo two, to the number in the right-most l indices $\{0, 1, \ldots, l-1\}$, and $-\Delta_i(l;m)$ obtains for the l-subsets with a non-congruent number. To convert the Δ's into the desired moments, from (2.1) it suffices to divide by the respective size of the set of integers, $\Omega_i(m)$. One computes the following: (a) for m even,

$$\Omega_0(m) = (2^m+2)/3;$$
$$\Omega_1(m) = (2^m-1)/3;$$
$$\Omega_2(m) = (2^m-1)/3;$$

(b) for m odd,

$$\Omega_0(m) = (2^m + 1)/3;$$
$$\Omega_1(m) = (2^m + 1)/3;$$
$$\Omega_2(m) = (2^m - 2)/3.$$

∎

Example 8.3. The integer eleven has $b_0 = b_1 = b_3 = 1$ and all other $b_j = 0$, and (8.1) correctly places eleven in the residue class of integers congruent to two modulo three.

In view of the preceding result, we see that, for all $(3, i, m)$-residual probability set functions, the absolute value of $\mu_i(S)$ depends only upon $|S|$. Furthermore, the appropriate sign of the moments is readily determined as it only involves $t - t_0$. Thus, at most $2m$ distinct moments suffice to characterize the $(3, i, m)$-residual probability set functions. The magnitudes of moments increase, essentially and geometrically with $|S|$. Related results are found in [4, 12].

The cumulants of the $(3, i, m)$-residual probability set functions are evaluated using Theorem 8.2 and (3.1).

Example 8.4. For $(3, 0, 2n)$-residual probability set functions, if $|S| = 2k - 1$, then $\kappa_0(S) = 0$, and if $|S| = 2k$, then

$$\kappa_0(S) = 3^k \sum_{j=1}^{k} (-1)^{(j-1)} (j-1)! \mathcal{E}_j^{(2k)} \left\{ \frac{2}{2^m + 2} \right\}^j .$$

Acknowledgments. This work benefited from discussions with our colleagues at Los Alamos National Laboratory, in particular, Drs. G. Bell, C. Doering (now University of Michigan), C. Macken, I. Saias, T. Warnock, and G. Xie (now Merck, Inc.). In addition, we thank Drs. D.J. Balding of the University of Reading, R. Karp of the University of Washington, and L. Székely of the University of South Carolina for sharing their insights.

References

1. M. Abramowitz and I.A. Stegun, Handbook of Mathematical Functions, U.S. Government Printing Office, Washington, 1964, pp. 75 & 805.
2. R.R. Bahadur, A representation of the joint distribution of responses to n dichotomous items, In: Studies in Item Analysis and Prediction, H. Solomon, Ed., Stanford University Press, Stanford, 1961, pp. 158–168.
3. C. Berge, Principles of Combinatorics, J. Sheehan, Translator, Academic Press, New York, 1971, pp. 37–44.
4. J. Coquet, A summation formula related to the binary digits, Invent. Math. **73** (1983) 107–115.
5. C.F. Gauss, Disquisitiones Arithmeticae, A. A. Clarke, Translator, Springer-Verlag, New York, 1966, p. 4, in English.
6. A.B. Gorchakov, Upper estimates for semi-invariants of the sum of multi-indexed random variables, Discrete Appl. Math. **5** (1995) 317–331.

7. L. Heinrich, Some bounds of cumulants of m-dependent random fields, Math. Nachr. **149** (1990) 303–317.
8. T.H. Hildebrandt and I.J. Schoenberg, On linear functional operations and the moment problem for a finite interval on one or several dimensions, Ann. Math. **34** (1933) 317–328.
9. M. Iwasaki, Spectral analysis of multivariate binary data, J. Japan Statist. Soc. **22** (1992) 45–65.
10. P.F. Lazarsfeld, The algebra of dichotomous systems, In: Studies in Item Analysis and Prediction, H. Solomon, Ed., Stanford University Press, Stanford, 1961, pp. 111–157.
11. N. Linial, Y. Mansour, and N. Nisan, Constant depth circuits, Fourier transform, and learnability, J. Assoc. Comp. Mach. **40** (1993) 607–620.
12. D.J. Newman, On the number of binary digits in a multiple of three, Proc. Amer. Math. Soc. **21** (1969) 719–721.
13. A. Rényi, Új módszerek és eredmények a kombinatorikus analízisben, I. Magyar Tudomanos Akademia III, Osztály Közlményei **16** (1966) 77–105.
14. G.-C. Rota, On the foundations of combinatorial theory I. Theory of Möbius functions, Z. Wahrsch. Verw. Gebiete **2** (1964) 340–368.
15. A.N. Shiryaev, Probability, 2nd Ed., R.P. Boas, Translator, Springer-Verlag, New York, 1996, 290–292.
16. J.A. Shohat and J.D. Tamarkin, (1963). The Problem of Moments, Mathematical Surveys, Vol. 1, American Mathematical Society, Providence, 1963.
17. T.P. Speed, Cumulants and parition lattices, Austral. J. Statist. **25** (1983) 378–388.
18. R. P. Stanley, Exponential structures, Stud. Appl. Math. **59** (1978) 73–82.
19. A. Thomas and M.H. Skolnick, A probabilistic model for detecting coding regions in DNA sequences, IMA J. Math. Appl. Med. Biol. **11** (1994) 149–160.
20. D.C. Torney, Binary cumulants. Adv. Appl. Math., submitted.
21. B.L. Van der Waerden, Algebra, Vol. 1, Springer-Verlag, New York, 1991, pp. 156–160.

Journal of Combinatorial Theory, Series A 91, 283–304 (2000)
doi:10.1006/jcta.1999.3017, available online at http://www.idealibrary.com on IDEAL®

On the Combinatorics of Cumulants

Gian-Carlo Rota

Department of Mathematics, Massachusetts Institute of Technology, Cambridge, Massachusetts 02139

and

Jianhong Shen

School of Mathematics, University of Minnesota, Minneapolis, Minnesota 55455
E-mail: jhshen@math.umn.edu

Communicated by the Managing Editors

Received July 26, 1999

DEDICATED TO THE MEMORY OF GIAN-CARLO ROTA

We study cumulants by Umbral Calculus. Various formulae expressing cumulants by umbral functions are established. Links to invariant theory, symmetric functions, and binomial sequences are made.

1. INTRODUCTION

Cumulants were first defined and studied by Danish scientist T. N. Thiele. He called them *semi-invariants*. The importance of cumulants comes from the observation that many properties of random variables can be better represented by cumulants than by moments. We refer to Brillinger [3] and Gnedenko and Kolmogorov [4] for further detailed probabilistic aspects on this topic.

Given a random variable X with the moment generating function $g(t)$, its nth cumulant K_n is defined as

$$K_n = \left.\frac{d^n}{dt^n}\right|_{t=0} \log g(t).$$

That is,

$$\sum_{n\geqslant 0} \frac{m_n}{n!} t^n = g(t) = \exp\left(\sum_{n\geqslant 1} \frac{K_n}{n!} t^n\right), \tag{1}$$

where, m_n is the nth moment of X.

Generally, if σ denotes the standard deviation, then

$$K_1 = m_1, \qquad K_2 = m_2 - m_1^2 = \sigma^2.$$

Cumulants of some important and familiar random distributions are listed as follows. For the *Poisson* distribution with mean λ,

$$K_n \equiv \lambda, \qquad n \geqslant 1.$$

The *exponential* distribution with mean μ has cumulants

$$K_n = (n-1)!\, \mu^n, \qquad n \geqslant 1.$$

The *Gaussian* distribution $N(\mu, \sigma)$ possesses the simplest list of cumulants:

$$K_1 = \mu, \qquad K_2 = \sigma^2, \qquad K_n = 0, \qquad n \geqslant 3.$$

These classical examples clearly demonstrate the simplicity and efficiency of cumulants for describing random variables. It is apparently not fortuitous for cumulants to encode the most important information of the associated random variables. The underlying reason may well reside in the following two invariant properties (which are in fact related to each other).

- (Translation Invariance) Let $K_n(X)$ denote the nth cumulant of a random variable X. Then, for any constant c,

$$K_1(X+c) = c + K_1(X), \qquad K_n(X+c) = K_n(X), \quad n \geqslant 2.$$

- (Additivity) Let X and Y be any two independent random variables. Then,

$$K_n(X+Y) = K_n(X) + K_n(Y), \qquad n \geqslant 1.$$

Our combinatorial interests in cumulants are very much inspired by these algebraic properties, especially the translation invariance. Notice that these are truly *algebraic* relations, which has very little to do with the *positivity* attribute of random variables (though in probability theory or statistics, positivity is extremely important; see, for instance, the famous Moment Problem [1, 2, 8]). Given any sequence of (real) numbers: $a_0 = 1$, a_1, $a_2, \ldots$, the associated cumulants $K_1, K_2, \ldots$ are always well defined according to Eq. (1). While enjoying this fresh degree of freedom, we have to face the difficulty in interpreting $X + c$, since for an arbitrary sequence, it is unclear whether there exists a random variable X or not (unless we solve the moment problem).

This difficulty is overcome by the notion of umbrae. An umbra is a *formal*, or *dry*, vividly speaking, random variable, which has no probabilistic flesh, but is indeed endowed with the algebraic spirit of a random variable. It is a powerful tool in dealing with a sequence of numbers (such as the Bell numbers; see Rota [10]), and for studying combinatorial algebraic objects like binomial sequences and algebraic invariants of polynomial systems (Hilbert [5], Kung and Rota [7]). We refer to Roman and Rota [9] and Rota, Kahaner, and Odlyzko [11] for the history and development of Umbral Calculus. An algebraic treatment was given in Joni and Rota [6]. New developments can be found in Rota and Taylor [13] and Rota, Shen, and Taylor [12]. Also see Shen [14] for a recent application in wavelet analysis.

In this paper, we shall employ the full freedom of Umbral Calculus to study cumulants. Umbral Calculus leads to various formulae for the cumulant sequence, each of which reveals one portion of the secret encoded in cumulants. Through these formulae, cumulants are connected to familiar combinatorial objects such as binomial sequences and symmetric functions. In return, the study of cumulants has stimulated new extension of the existing theory of Umbral Calculus. For instance, for the first time in this paper, we discuss umbral derivatives (or the *star algebra*).

Our plan is the following. In Section 2, we survey briefly the literature and recent development of Umbral Calculus. By doing so, we make our readers comfortable with the notations and symbols necessary for the rest of the paper. The following five sections study five different ways of understanding cumulants, among which, four are based on Umbral Calculus. The fith employs partition lattices and similar work can be found in Speed [15].

2. UMBRAL CALCULUS

An umbra α is the generalization of the expectation E of a random variable X. It is a powerful tool for dealing with a sequence of numbers $a_0, a_1, \ldots$, whether positive definite or not. (A sequence of real numbers is said to be *positive definite* if it can be realized as the moment sequence of certain random variables.) In some sense, an umbra can be called *a formal random variable*. The formality makes it much easier to understand operators like $\partial/\partial\alpha$.

2.1. *Fundamentals*

Umbral Calculus is axiomatized by the following definition (See Rota and Taylor [13]).

DEFINITION 2.1. An umbral calculus consists of the following data and rules:

(1) An alphabet A, whose elements are called *umbrae* and denoted by Greek letters.

(2) A commutative integral domain D whose quotient field is of characteristic zero.

(3) A D-linear functional **eval**: $D[A] \to D$ such that $\mathbf{eval}(1) = 1$, and

$$\mathbf{eval}(\alpha^i \beta^j \cdots \gamma^k) = \mathbf{eval}(\alpha^i)\, \mathbf{eval}(\beta^j) \cdots \mathbf{eval}(\gamma^k)$$

whenever $\alpha, \beta, ..., \gamma$ are distinct umbrae.

(4) A distinguished element ε of the alphabet A, such that

$$\mathbf{eval}(\varepsilon^n) = \delta_n,$$

where δ is the Kronecker delta.

If two umbral polynomials p and q are evaluated to a same element in D, then they are said to be *umbrally equivalent* and denoted by $p \simeq q$. A sequence $a_0, a_1, ...$ is said to be *umbrally represented* by an umbra α, if

$$\mathbf{eval}(\alpha^n) = a_n, \qquad \text{for} \quad n = 0, 1, ...$$

According to Axiom (3), a_0 must be 1. If two distinct umbrae α and β represent the same sequence, then they are said to be *exchangeable* and we write

$$\alpha \equiv \beta.$$

Notice the difference between $\alpha \simeq \beta$ and $\alpha \equiv \beta$. The notion of exchangeability plays a significant role in the algebraic development of Umbral Calculus. It models the familiar concept of "i.i.d" (independent and identical distributions) in probability theory.

Umbral Calculus is usually assumed to be *saturated*, which means that any sequence in D can be represented by infinitely many umbrae in A (see Rota and Taylor [13]). Especially, A cannot be countable.

A taste of Umbral Calculus can be sought from the famous example of the *Bell Umbra* β. An umbra β is called a *Bell umbra* if it umbrally represents the sequence of Bell numbers B_0, B_1, ... (B_n is the number of distinct partitions of a set with n elements. See Rota [10].) This umbra is completely characterized by

$$\beta^{n+1} \simeq (\beta+1)^n, \qquad n=0, 1, 2, \ldots.$$

All other properties of the Bell numbers are the derivatives of this simple relation.

By the saturation assumption, any formal power series in $D[[t]]$

$$g(t)=\sum_{n\geqslant 0} \frac{a_n}{n!} t^n, \qquad a_n \in D$$

can be umbrally represented by a formal power series in $D[A][[t]]$. In fact, if α umbrally represents the sequence (a_n), then

$$g(t) \simeq e^{\alpha t},$$

assuming that we naturally extend **eval** to be $D[t]$-linear. We say that $g(t)$ is umbrally represented by α. If $g(t)$ and $f(t)$ are umbrally represented by two distinct umbrae α and β, then $g(t)\, f(t)$ is represented by $\alpha+\beta$ (or any one that is exchangeable with it). This is precisely a restatement of the fact that $g(t)\, f(t)$ defines a convolution in terms of their coefficients.

2.2. *Further Developments*

2.2.1. *The Annihilating Umbra and Auxiliary Umbrae*

For a given umbra $\alpha \in A$, we define its *annihilating umbra* $-.\alpha$ to be the unique umbra (up to umbral exchangeability) whose moment generating function is the reciprocal of α's. It is uniquely characterized by the umbral equivalence

$$\alpha+(-.\alpha) \equiv \varepsilon.$$

More generally, for any integer n, we use $n.\alpha$ to denote the unique (up to exchangeability) umbra whose moment generating function is the nth power of α's. When n is positive, then

$$n.\alpha \equiv \alpha_1+\alpha_2+\cdots+\alpha_n$$

for any n distinct umbrae α_i, all exchangeable with α. If n is negative, say $n=-k$, then

$$-k.\alpha \equiv \beta_1+\beta_2+\cdots+\beta_k$$

for any k distinct umbrae β_i exchangeable with the annihilating umbra $-.\alpha$. We call $n.\alpha$ an *auxiliary umbra* of α. More discussion can be found in Rota and Taylor [13].

2.2.2. *Partial Derivatives*

In the algebra $D[A]$, for any $\alpha \in A$, let α^* denote the partial derivative operator ∂_α. It is clear that for any $\alpha, \beta \in A$,

$$\alpha^* \circ \beta^* = \beta^* \circ \alpha^*.$$

Hence, if $D[A^*]$ denotes the subalgebra generated by all star elements α^* in the D-linear operator algebra of $D[A]$, then $D[A^*]$ is isomorphic to the polynomial algebra generated by the star alphabet

$$A^* = \{\alpha^* \mid \alpha \in A\}.$$

The mapping $\alpha \to \alpha^*$ from A to A^* introduces a natural isomorphism between $D[A]$ and $D[A^*]$. That is, for any umbral polynomial $p(\alpha, \beta, ..., \gamma) \in D[A]$,

$$p \to p^* \in D[A^*]: \qquad p^* = p(\alpha^*, \beta^*, ..., \gamma^*).$$

We call it *the star isomorphism.*

Furthermore, $D[A]$ can D-linearly act on itself by multiplication. In this sense, the elements in $D[A]$ are called *multipliers* and $D[A]$ is also treated as a subalgebra of its D-linear operator algebra. Let $D[A \mid A^*]$ denote the subalgebra generated by all multipliers and star elements. Because of the uncertainty principle (Strang [16]),

$$\alpha^*\alpha - \alpha\alpha^* = 1,$$

$D[A \mid A^*]$ cannot be commutative. Yet the following result seems to need no proof.

Proposition 2.1. *Algebra $D[A \mid A^*]$ is isomorphic to $(D[A])[A^*]$, the polynomial algebra generated by the star alphabet A^* on the umbral algebra $D[A]$.*

It is now the time for exploring various umbral approaches to understanding cumulants.

3. FIRST REPRESENTATION OF CUMULANTS: CUMULANT UMBRAE

As for the Bell numbers, it is possible to give a clean umbral recursion formula for cumulants of a given sequence $a_0 = 1$, a_1, a_2, ... The concept of cumulant umbra is coined for this purpose. This first representation is the simplest and most direct, though it may not be optimal for understanding the algebraic properties hidden in cumulants.

DEFINITION 3.1 [The Cumulant Umbra]. For a given umbra α, up to exchangeability, there exists a unique umbra, say κ, such that,

$$\alpha^n \simeq \kappa(\kappa+\alpha)^{n-1} \tag{2}$$

for all $n = 1, 2, \ldots$. We call κ the *cumulant umbra* associated with α.

The existence and uniqueness follow readily from the recursive relation

$$\kappa^n \simeq \alpha^n - \sum_{m=0}^{n-2} \binom{n-1}{m} \kappa^{m+1} \alpha^{n-1-m}. \tag{3}$$

On the other hand, suppose α represents a sequence (a_n). Define

$$K_n = \frac{d^n}{dt^n}\bigg|_0 \ln\left(\sum_{m\geq 0} \frac{a_m}{m!} t^m\right),$$

for $n = 1, 2, \ldots$. As for random variables, we call K_n the nth cumulant of the sequence (and the umbra α).

PROPOSITION 3.1.

$$\kappa^n \simeq K_n, \qquad n = 1, 2, \ldots.$$

Proof. Define the generating functions

$$g(t) = \mathbf{eval}(e^{\alpha t}), \qquad K(t) = \mathbf{eval}(e^{\kappa t}).$$

From Eq. (2),

$$\alpha \sum_{n\geq 0} \frac{\alpha^n}{n!} t^n \simeq \kappa \sum_{n\geq 0} \frac{(\kappa+\alpha)^n}{n!} t^n,$$

$$\alpha e^{\alpha t} \simeq \kappa e^{(\kappa+\alpha)t} = \kappa e^{\kappa t} e^{\alpha t},$$

$$\frac{de^{\alpha t}}{dt} \simeq \frac{de^{\kappa t}}{dt} e^{\alpha t}.$$

Noticing that d/dt commutes with **eval**, we eventually have the ordinary differential equation

$$g'(t) = K'(t)\, g(t),$$

with the initial condition $g(0)=1$. This gives the unique solution

$$g(t)=\exp(K(t)-1),$$

which proves that $K_n=\mathbf{eval}(\kappa^n)$. ▮

Thus, we can derive the classical result.

COROLLARY 3.1. *Suppose that* α *represents a sequence* a_n, $n=0, 1, \ldots$. *Then its* n*th cumulant* K_n *is given by*

$$K_n=(-1)^{n-1}(n-1)!\det\begin{bmatrix} a_1 & 1 & 0 & 0 & \cdots \\ a_2 & a_1 & 1 & 0 & \cdots \\ a_3/2! & a_2/2! & a_1 & 1 & \cdots \\ a_4/3! & a_3/3! & a_2/2! & a_1 & \cdots \\ \cdots & \cdots & \cdots & \cdots & \cdots \end{bmatrix}_{n\times n}. \tag{4}$$

Proof. According to the preceding proposition, the recursion formula (3), after evaluation, is a linear system of equations on cumulants K_m, with coefficients from the sequence $a_0, a_1, \ldots$. We apply the evaluated formula for $m=1, 2, \ldots, n$. Then K_n can be expressed explicitly using Cramer's formula for linear systems. Formula (4) is finally obtained by moving the last column to the first in Cramer's formula for K_n.

Our next representation starts from a further umbralization of this matrix formula.

4. SECOND REPRESENTATION: UMBRAL SYMMETRIC FUNCTIONS

4.1. *A Notation for Generalized Vandemondes*

Let R be an arbitrary ring (commutative or non-commutative). For any n by n matrix $M=[a_{i,j}]$, $a_{i,j}\in R$, if a_{ij} commutes with a_{kl} whenever $k\neq i$, and $l\neq j$, then we can define the determinant of M by the classical expansion formula. Properties such as skew-symmetry and Laplace expansion still hold.

Especially, suppose $x_1, x_2, \ldots, x_n$ are n mutually commutative elements in R. For any partition λ of length n:

$$\lambda=(\lambda_1, \lambda_2, \ldots, \lambda_n)$$

with non-negative integer parts λ_i, we define

$$(x_1, x_2, ..., x_n \mid \lambda) := \det[x_i^{\lambda_j}]_{n\times n}.$$

For example, the classical Vandemonde $V_n(x_1, x_2, ..., x_n)$ is now written as

$$(x_1, x_2, ..., x_n \mid 0, 1, ..., n-1).$$

We shall call $(x_1, x_2, ..., x_n \mid \lambda)$ the Vandemonde of $(x_1, x_2, ..., x_n)$ with respect to λ. Notice that

$$\frac{(x_1, x_2, ..., x_n \mid \lambda)}{(x_1, x_2, ..., x_n \mid 0, 1, ..., n-1)}$$

is the classical Schur function S_λ.

In what follows, we apply this notation to the umbral algebra $D[A]$ and operator algebra $D[A \mid A^*]$. For star elements, we can even allow the partition λ to have negative parts under the following assumption. For any star element $\alpha^* \in A^*$, and any positive integer k, we define $[\alpha^*]^{-k}$ to be the multiplier α^k. That is, "–" seems to annihilate "*."

Notice that for a star element, negative powers do not commute with positive ones. However, for any two *distinct* umbrae $\alpha, \beta \in A$, and any two integers n, m (negative maybe), $[\alpha^*]^n$ commutes with $[\beta^*]^m$. Hence, for any n distinct umbrae $\alpha_1, \alpha_2, ..., \alpha_n$, and partition λ of length n and with arbitrary integer parts,

$$(\alpha_1^*, \alpha_2^*, ..., \alpha_n^* \mid \lambda)$$

is well defined and is an element in $D[A \mid A^*]$.

4.2. *Symmetric Umbral Representation for Cumulants*

Suppose K_n is the nth cumulant of a given umbra α, which represents a sequence a_n, $n = 0, 1, ...$ Let $\alpha_1, \alpha_2, ..., \alpha_n \in A$ be any n distinct umbrae, each exchangeable with α. The main task of this section is to establish the following theorem:

THEOREM 4.1 (Symmetric and Translation Invariant Representation). *For any* $n = 1, 2, ...,$

$$K_n \simeq C_n(\alpha_1^*, \alpha_2^*, ..., \alpha_n^* \mid -1, 0, ..., n-2) \cdot (\alpha_1, \alpha_2, ..., \alpha_n \mid 0, 1, ..., n-1)$$

with

$$C_n = (-1)^{n-1} [((n-2)!)! \, n!]^{-1}.$$

Here $(k!)! = (1!)(2!)\cdots(k!)$ *for any non-negative integer* k *and* $((-1)!)!$ *is assumed to be* 1.

Proof. In Eq. (4), use α_i to represent the ith column of the matrix involved. Since all columns are now umbrally unrelated, we can exchange the order of the det operator and **eval**. Therefore,

$$K_n \simeq (-1)^{n-1}(n-1)! \det \begin{bmatrix} \alpha_1 & 1 & 0 & 0 & \cdots \\ \alpha_1^2 & \alpha_2 & 1 & 0 & \cdots \\ \alpha_1^3/2! & \alpha_2^2/2! & \alpha_3 & 1 & \cdots \\ \alpha_1^4/3! & \alpha_2^3/3! & \alpha_3^2/2! & \alpha_4 & \cdots \\ \cdots & \cdots & \cdots & \cdots & \cdots \end{bmatrix}_{n\times n}$$

$$= (-12)^{n-1}(n-1)!$$

$$\times \det \begin{bmatrix} \alpha_1 \cdot 1 & 1 & \alpha_3^* 1 & [\alpha_4^*]^2 1 & \cdots \\ \alpha_1 \cdot \alpha_1 & \alpha_2 & \alpha_3^* \alpha_3 & [\alpha_4^*]^2 \alpha_4 & \cdots \\ \alpha_1 \cdot \alpha_1^2/2! & \alpha_2^2/2! & \alpha_3^* \alpha_3^2/2! & [\alpha_4^*]^2 \alpha_4^2/2! & \cdots \\ \alpha_1 \cdot \alpha_1^3/3! & \alpha_2^3/3! & \alpha_3^* \alpha_3^3/3! & [\alpha_4^*]^2 \alpha_4^3/3! & \cdots \\ \cdots & \cdots & \cdots & \cdots & \cdots \end{bmatrix}_{n\times n}$$

$$= \frac{(-1)^{n-1}}{((n-2)!)!} [\alpha_1^*]^{-1} [\alpha_2^*]^0 [\alpha_3^*]^1 \cdots [\alpha_n^*]^{n-2} V_n(\alpha_1, \alpha_2, ..., \alpha_n).$$

The linearity of **eval** and exchangeability of α_k's allow the symmetrization:

$$K_n \simeq \frac{(-1)^{n-1}}{((n-2)!)!\, n!} \sum_{\sigma \in S_n} [\alpha_{\sigma[1]}^*]^{-1} [\alpha_{\sigma[2]}^*]^0 [\alpha_{\sigma[3]}^*]^1 \cdots [\alpha_{\sigma[n]}^*]^{n-2}$$

$$\times V_n(\alpha_{\sigma[1]}, \alpha_{\sigma[2]}, ..., \alpha_{\sigma[n]})$$

$$= C_n \sum_{\sigma \in S_n} \operatorname{sign}(\sigma)[\alpha_{\sigma[1]}^*]^{-1}$$

$$\times [\alpha_{\sigma[2]}^*]^0 [\alpha_{\sigma[3]}^*]^1 \cdots [\alpha_{\sigma[n]}^*]^{n-2} V_n(\alpha_1, \alpha_2, ..., \alpha_n)$$

$$= C_n(\alpha_1^*, \alpha_2^*, ..., \alpha_n^* \mid -1, 0, ..., n-2) \cdot (\alpha_1, \alpha_2, ..., \alpha_n \mid 0, 1, ..., n-1).$$

This completes the proof. ∎

Let $p(\alpha_1, \alpha_2, ..., \alpha_n)$ denote the umbral polynomial

$$(\alpha_1^*, \alpha_2^*, ..., \alpha_n^* \mid -1, 0, ..., n-2) \cdot (\alpha_1, \alpha_2, ..., \alpha_n \mid 0, 1, ..., n-1).$$

It is easy to see that p is a symmetric function of $\alpha_1, ..., \alpha_n$. Furthermore, we now show that when $n \geq 2$, p is translation invariant. Namely, for any real constant c,

$$p(\alpha_1 + c, \alpha_2 + c, ..., \alpha_n + c) = p(\alpha_1, \alpha_2, ..., \alpha_n).$$

First,

$$(\alpha_1, \alpha_2, ..., \alpha_n \mid 0, 1, ..., n-1) = \prod_{j>i} (\alpha_j - \alpha_i)$$

is apparently translation invariant. Second, the property of translation invariance is preserved under the action of the star algebra $D[A^*]$. Finally, Laplace expansion with respect to the first and second columns represents

$$(\alpha_1^*, \alpha_2^*, ..., \alpha_n^* \mid -1, 0, ..., n-2)$$

by

$$\sum_{j>i} (-1)^{i+j} (\alpha_i, \alpha_j \mid 0, 1) \times (\alpha_1^*, ..., \alpha_{i-1}^*, \alpha_{i+1}^*, ..., \alpha_{j-1}^*, \alpha_{j+1}^*, ..., \alpha_n^* \mid 1, 2, ..., n-2).$$

For each index pair j, i,

$$(\alpha_1^*, ..., \alpha_{i-1}^*, \alpha_{i+1}^*, ..., \alpha_{j-1}^*, \alpha_{j+1}^*, ..., \alpha_n^* \mid 1, 2, ..., n-2)$$

is in the star algebra $D[A^*]$ and the multiplier $(\alpha_i, \alpha_j \mid 0, 1) = \alpha_j - \alpha_i$ is clearly translation invariant. All together, these facts confirm the translation invariance of p.

Accordingly, we have established umbrally the translation invariance of cumulants addressed in the Introduction.

COROLLARY 4.1. *Let $K_n(\alpha)$ denote the nth cumulant of an umbra α. Then for any real constant c, and $n \geq 2$,*

$$K_n(\alpha + c) = K_n(\alpha).$$

5. THIRD REPRESENTATION: UMBRAL PARAMETRIC FORM

5.1. *The Generalized Umbral Calculus*

The *Generalized Umbral Calculus* (see Rota, Shen, and Taylor [12]) is obtained from the classical one by replacing axiom (3) in Definition 2.1 by

(3′) A D-linear functional **eval**: $D[A] \to D[x]$ (instead of D) such that **eval**$(1)=1$, and

$$\mathbf{eval}(\alpha^i \beta^j \cdots \gamma^k) = \mathbf{eval}(\alpha^i)\, \mathbf{eval}(\beta^j) \cdots \mathbf{eval}(\gamma^k)$$

whenever $\alpha, \beta, ..., \gamma$ are distinct umbrae.

We call α a *scalar* umbra if it represents a sequence of D elements, and χ a *polynomial umbra* if it represents a sequence of $D[x]$ elements. In this paper, we always use Greek letters α, β, γ to represent scalar umbrae, and χ, ϕ, ψ to represent polynomial umbrae.

There is an important operator called *restriction* in the generalized Umbral Calculus that relates polynomial umbrae to scalar ones.

Given an element in D, say b, for any polynomial umbra χ, its *restriction* at b is the unique scalar umbra (up to exchangeability) that represents

$$1, \mathbf{eval}(\chi)(b), \mathbf{eval}(\chi^2)(b), \mathbf{eval}(\chi^3)(b), ...$$

We denote this scalar umbra by $\langle \chi \mid b \rangle$.

The following properties hold apparently.

(1) $\langle \phi \mid b \rangle + \langle \psi \mid b \rangle \equiv \langle \phi + \psi \mid b \rangle$ for any two distinct polynomial umbrae ϕ and ψ.

(2) $c \langle \phi \mid b \rangle \equiv \langle c\phi \mid b \rangle$ for any constant c.

For simplicity, we now assume that D is the real number field.

5.2. *Binomial Sequences and Binomial Umbrae*

There is a close relation between the moment–cumulant pair and the binomial sequence of polynomials. We refer to Rota, Shen, and Taylor [12] for the most recent development in binomial sequences of polynomials.

A sequence of real polynomials $p_n(x)$, $n=0, 1, ...,$ is called a *binomial sequence* if $p_1 \neq 0$, and for any non-negative integer n and real numbers a and b,

$$p_n(a+b) = \sum_{0 \leqslant k \leqslant n} \binom{n}{k} p_k(a)\, p_{n-k}(b).$$

It is easy to see that $p_0 = 1$ and $p_1 = cx$ for some non-zero constant c.

A polynomial umbra χ representing a binomial sequence $p_n(x)$, $n=0, 1, ...,$ is called a *binomial umbra.*

Proposition 5.1. *A polynomial umbra χ is a binomial umbra if and only if* **eval**$(\chi) \neq 0$, *and for any real numbers a and b,*

$$\langle \chi \mid a+b \rangle \equiv \langle \chi \mid a \rangle + \langle \chi \mid b \rangle.$$

DEFINITION 5.1. A binomial umbra χ is called *unital* if

$$\chi \simeq x.$$

Similarly, we call the associated binomial sequence $p_n(x)$ *unital* when $p_1(x) = x$.

A unital binomial sequence is the most direct generalization of the sequence x^n, $n = 0, 1, \ldots$. Another household binomial sequence is the factorial sequence:

$$p_n(x) = x(x-1)\cdots(x-n+1) = (x)_n, \qquad n = 0, 1, \ldots.$$

The preceding proposition together with the properties of the restriction operator brings us to

COROLLARY 5.1 (Linearity). *Suppose χ_1 and χ_2 are two binomial umbrae, and c_1, c_2 two real numbers such that*

$$\mathbf{eval}(c_1\chi_1 + c_2\chi_2) \neq 0.$$

Then, any polynomial umbra exchangeable with $c_1\chi_1 + c_2\chi_2$ must be a binomial umbra.

Rota, Shen, and Taylor proved the following theorem in [12] to uniformly characterize an arbitrary binomial sequence.

THEOREM 5.1. *A polynomial umbra χ is binomial if and only if there exist a scalar umbra β and a non-zero real number c, such that for any $n = 1, 2, \ldots$*

$$\chi^n \simeq cx(cx + n.\beta)^{n-1}.$$

Furthermore, c and β are unique (up to exchangeability).

5.3. *Moment–Cumulant vs Binomial Sequence*

THEOREM 5.2. *Suppose κ is the cumulant umbra of a given scalar umbra α whose "mean" $\mathbf{eval}(\alpha)$ is not zero. Then there exists a unique binomial umbra (up to exchangeability), say χ_α, such that*

$$\alpha \equiv \langle \chi_\alpha \mid 1 \rangle, \qquad \kappa^n \simeq \frac{d}{dx}\bigg|_0 \chi_\alpha^n, \qquad n = 1, 2, \ldots,$$

where $d/dx(\chi_\alpha^n)$ stands for $d/dx(\mathbf{eval}(\chi_\alpha^n))$.

Proof. Uniqueness follows from the fact that any binomial umbra can be reconstructed (up to exchangeability) from its restricted scalar umbra at some non-zero value.

Let $g(t)$ and $K(t)$ denote the generating functions of α and κ. Set $f(t) = K(t) - 1$. Then $g(t) = e^{f(t)}$. Assume that

$$e^{f(t)x} = \sum_{n \geqslant 0} \frac{p_n(x)}{n!} t^n. \tag{5}$$

Then the sequence $p_n(x)$, $n = 0, 1, \ldots$, must be a binomial sequence. Let χ denote the binomial umbra representing (p_n). Then $\exp(f(t)\,x) \simeq \exp(t\chi)$ and

$$g(t) = e^{f(t)} \simeq e^{t\langle\chi|1\rangle}.$$

On the other hand, taking derivative with respect to x on both sides of (5) and setting $x = 0$, we have

$$\kappa^n \simeq p_n'(0) \simeq \frac{d}{dx}\bigg|_0 \chi^n.$$

Hence, χ is the demanded binomial umbra. ▮

As a result, one can easily prove (by noticing that $\chi_{\alpha+\beta} \equiv \chi_\alpha + \chi_\beta$) that

COROLLARY 5.2 (Additivity). *For any two distinct scalar umbra α and β,*

$$K_n(\alpha + \beta) = K_n(\alpha) + K_n(\beta), \qquad n = 1, 2, \ldots.$$

The preceding two theorems establish the umbral parametric representation for cumulants.

COROLLARY 5.3 (Umbral Parametric Form). *For any umbra α with a non-zero mean $c = \mathbf{eval}(\alpha)$, there exists a unique umbra, say β, such that*

$$\alpha^n \simeq c(c + n.\beta)^{n-1} \qquad \textit{and} \qquad \kappa^n \simeq c(n.\beta)^{n-1}, \qquad n = 1, 2, \ldots.$$

In this representation, α and κ are "parameterized" by another scalar umbra β.

EXAMPLE 1. Suppose β is the Bell umbra and $c = 1$. The associated binomial sequence $p_n(x)$ is exactly the factorial sequence $(x)_n$. Hence, $\alpha^n \simeq p_n(1) = 0$ for all $n \geqslant 2$, and $\kappa^n \simeq (-1)^{n-1}(n-1)!$. Since $K_2 = D_2 = -1$, α is cannot be positive, or equivalently, α cannot be the moment umbra of a real random variable. However, if we take $c = -1$, then

$$\kappa^n \simeq (-1)^n (n-1)!,$$

which is exactly the nth cumulant of the exponential distribution with mean -1 (see Section 1).

EXAMPLE 2. Suppose β represents sequence $1, b, 0, 0, \ldots$ and $c=1$. By the preceding corollary,

$$\kappa^n \simeq (n)_{n-1}\, b^{n-1} = n!\, b^{n-1}.$$

Therefore, the moment generating function of the underlying moment umbra α is

$$g(t) = \exp\left(\frac{t}{1-bt}\right).$$

6. FOURTH REPRESENTATION: PARTITION LATTICE AND MÖBIUS INVERSION

This section parallels Speed's work [15].

6.1. *Sequence Partitions and Set Partitions*

A finite sequence of real numbers $\lambda = (\lambda_1, \lambda_2, \ldots, \lambda_k)$ is a *sequence partition* if $\lambda_1 \geqslant \lambda_2 \geqslant \cdots \geqslant \lambda_k$. Another way to represent a sequence partition is the *multiset* form: $(a_1^{m_1}, a_2^{m_2}, \ldots, a_h^{m_h})$ with $a_1 > a_2 > \cdots > a_h$ and $m_i > 0$. λ has exactly m_i a_i's. The following quantities are standard but helpful for the rest of the paper.

(1) $\lambda! = \lambda_1!\, \lambda_2! \cdots \lambda_k!$. $|\lambda| = \lambda_1 + \lambda_2 + \cdots + \lambda_k$. $\prod \lambda = \lambda_1 \lambda_2 \cdots \lambda_k$. $\lambda \vdash n$ means that $|\lambda| = n$.

(2) k is the *length* of λ and is also denoted by l_λ.

(3) Vector $m_\lambda = (m_1, m_2, \ldots, m_h)$ encodes the *multiplicity* of λ. $m_\lambda! = m_1!\, m_2! \cdots m_h!$. Clearly, $l_\lambda = m_1 + m_2 + \cdots + m_h$.

(4) For any two partitions λ and μ, their sum $\lambda + \mu$ is the unique sequence partition whose associated multiset is the union. For example, suppose $\lambda = (322)$, $\mu = (4321)$; then $\lambda + \mu = (4332221)$.

(5) A function g_λ defined for all sequence partitions λ is said to be *decomposable* if

$$g_{\lambda+\mu} = g_\lambda g_\mu, \qquad \text{for any } \lambda, \mu.$$

For example, the product function $\prod \lambda$ is decomposable. Apparently, if g_λ is decomposable, then

$$\text{span}_{\mathbb{Z}}\{g_\lambda \mid \text{all } \lambda\} = \mathbb{Z}[g_1, g_2, g_3, \ldots].$$

Now we review some basic facts about partitions of a *set*. Let $\prod(S)$ denote the partition lattice of a set S. If $S = [n] = \{1, 2, \ldots, n\}$, we simply write $\prod_n$ for $\prod(S)$. A partition is denoted by $\sigma = \sigma_1 \mid \sigma_2 \mid \cdots \mid \sigma_k$. Each σ_i is a block of σ.

For a given partition $\sigma = \sigma_1 \mid \sigma_2 \mid \cdots \mid \sigma_k$, we define

$$\lambda(\sigma) = (\lambda_1, \lambda_2, \ldots, \lambda_k), \qquad l_i = \#\sigma_i, \quad i = 1, 2, \ldots, k.$$

$\lambda(\sigma)$ is called the *type* of σ and k the *size* of σ and sometimes is denoted by b_σ. Hence, $b_\sigma = l_{\lambda(\sigma)}$.

Suppose T and S are two disjoint sets and $\sigma \in \prod(T)$, $\theta \in \prod(S)$. Their sum $\sigma + \theta$ is the unique partition of $T \cup S$ which refines the two-block partition $T \mid S$ and whose restrictions in T and S are σ and t. Then, $\lambda(\sigma + \theta) = \lambda(\sigma) + \lambda(\theta)$.

Let $\prod_\infty = \bigcup_{n \geq 1} \prod_n$ be the poset whose order is defined by $\sigma \leq \theta$ if and only if for some n, $\sigma, \theta \in \prod_n$, and $\sigma \leq \theta$. A function g_σ in $\prod_\infty$ (or $\prod_n$) is *indistinguishable* if

$$g_\sigma = g_\theta, \qquad \text{whenever} \quad \lambda(\sigma) = \lambda(\theta).$$

If g_σ is indistinguishable, then it introduces a type class function g^*_λ such that $g_\sigma = g^*_{\lambda(\sigma)}$. For simplicity, we still denote g^*_λ by g_λ. The size function b_σ is indistinguishable, for instance. A distinguishable function g_σ is said to be *decomposable* if its associated type class function g_λ is so.

6.2. *Möbius Inversion*

The integral f_σ of a given function g_σ on $\prod_\infty$ is

$$f_\sigma = \sum_{\theta \leq \sigma} g_\theta.$$

According to the Möbius inversion formula,

$$g_\sigma = \sum_{\theta \leq \sigma} \mu(\theta, \sigma) f_\theta,$$

if $\mu(\theta, \sigma)$ denotes the Möbius function of the lattice.

By the order structure of $\prod_\infty$, if $\sigma \in \prod_n$, we can replace $\sum_{\theta \leqslant \sigma}$ by $\sum_{\theta \leqslant \sigma, \theta \in \prod_n}$ in the above equations. The following proposition can be checked easily.

PROPOSITION 6.1. *A function g_σ is indistinguishable (or decomposable) if and only if its integral f_σ is.*

For any $\lambda \vdash n$, let $\binom{n}{\lambda}^*$ denote the number of partitions σ in $\prod_n$ whose type is λ. It is not difficult to see that

$$\binom{n}{\lambda}^* = \frac{n!}{\lambda!\, m_\lambda!} = \binom{n}{\lambda} \frac{1}{m_\lambda!}.$$

Denote $[n] \in \prod_n$ by $\mathbf{1}_n$ and $g_{\mathbf{1}_n}$ by g_n for any function g_σ in $\prod_\infty$.

THEOREM 6.1 (Inversion Formula). *Suppose that g_σ is a function on $\prod_\infty$ with integral f_σ.*

(a) *If g is indistinguishable, then*

$$f_n = \sum_{\lambda \vdash n} \binom{n}{\lambda}^* g_\lambda, \qquad \textit{and} \qquad g_n = \sum_{\lambda \vdash n} \binom{n}{\lambda}^* f_\lambda (-1)^{l_\lambda - 1} (l_\lambda - 1)!.$$

(b) *If g is decomposable and $\lambda = (\lambda_1 \lambda_2 \cdots)$, then*

$$f_\lambda = f_{\lambda_1} f_{\lambda_2} \cdots \qquad \textit{and} \qquad g_\lambda = g_{\lambda_1} g_{\lambda_2} \cdots.$$

Especially, $\{f_\lambda \mid \lambda\}$ is a $\mathbb{Z}$ basis for $\mathbb{Z}[g_1, g_2, \ldots]$.

Proof. (a) Notice that in the lattice $\prod_n$, for any σ,

$$\mu(\sigma, \mathbf{1}_n) = (-1)^{b_\sigma - 1} (b_\sigma - 1)!.$$

(b) Notice that as sequence partitions,

$$\lambda = (\lambda_1 \lambda_2 \cdots \lambda_k) = (\lambda_1) + (\lambda_2) + \cdots + (\lambda_k). \quad \blacksquare$$

Therefore, if g_σ on $\prod_\infty$ is decomposable, then both g_σ and its integral f_σ are uniquely determined by two sequences:

$$g_1, g_2, \ldots \qquad \text{and} \qquad f_1, f_2, \ldots.$$

The relation between their exponential generating functions is revealed by this simple lemma.

LEMMA 6.1.

$$\exp\left(\sum_{n\geq 1} \frac{g_n}{n!} t^n\right) = \sum_{\lambda} \frac{g_\lambda t^{|\lambda|}}{\lambda!\, m_\lambda!}.$$

It can be checked directly by expansion. Comparing it to the preceding theorem, we have thus established

COROLLARY 6.1. *Suppose function g_σ on $\prod_\infty$ is decomposable. Then the exponential generating functions of g_n and its integral f_n satisfy*

$$\sum_{n\geq 0} \frac{f_n}{n!} t^n = \exp\left(\sum_{n\geq 1} \frac{g_n}{n!} t^n\right).$$

Therefore, f_n's and g_n's in fact embody the exact relation between moments and cumulants, which eventually leads to the fourth formula for cumulants.

THEOREM 6.2. *Let K_n's be the cumulants of a moment sequence (a_n). Then*

$$a_n = \sum_{\lambda \vdash n} \binom{n}{\lambda}^* K_\lambda, \qquad K_n = \sum_{\lambda \vdash n} \binom{n}{\lambda}^* (-1)^{l_\lambda - 1} (l_\lambda - 1)!\, a_\lambda, \tag{6}$$

where $K_\lambda = K_{\lambda_1} K_{\lambda_2} \ldots$, $a_\lambda = a_{\lambda_1} a_{\lambda_2} \ldots$, if $\lambda = (\lambda_1 \lambda_2 \cdots)$.

7. FIFTH REPRESENTATION: UMBRAL "FOURIER" TRANSFORM

In this section, we assume that the integral domain D in umbral calculus is the complex number field.

Suppose that umbra α represents a sequence $a_0, a_1, \ldots$ and K_n is its nth cumulant. Let

$$f(t) = \sum_{n\geq 1} \frac{K_n}{n!} t^n$$

be the exponential generating function of K_n. Then

$$e^{\alpha t} \simeq e^{f(t)}.$$

THEOREM 7.1 ("Fourier" Transform). *For any integer n, let $\omega = \exp(2\pi i/n)$ be the nth unit root. Then,*

$$K_n \simeq \frac{1}{n}\left(\sum_{i=0}^{n-1} \alpha_i \omega^i\right)^n \tag{7}$$

for any n distinct umbrae $\alpha_0, \alpha_1, ..., \alpha_{n-1}$ exchangeable with α.

Proof. For each $i = 0, 1, ..., n-1$,

$$e^{\alpha_i \omega^i t} \simeq e^{f(\omega^i t)}.$$

Since $e^{\alpha_i \omega^i t}$, $i = 0, 1, ..., n-1$, are unrelated, we have

$$\exp\left(\sum_{i=0}^{n-1} \alpha_i \omega^i t\right) \simeq \exp\left(\sum_{i=0}^{n-1} f(\omega^i t)\right) = \exp\left(n\left(\frac{K_n}{n!} t^n + \frac{K_{2n}}{(2n)!} t^{2n} + \cdots\right)\right). \tag{8}$$

Comparing the nth power coefficients of both sides yields Eq. (7). ∎

Let $\mathbb{N}$ denote the set of all non-negative integers. For any fixed positive integer n and an n-tuple $I = (I_1, ..., I_n) \in \mathbb{N}^n$, define $\int_n I$ to be the weighted sum

$$\int_n I = \sum_{i=0}^{n-1} i I_i.$$

For any sequence partition $\lambda = (\lambda_1, \lambda_2, ..., \lambda_k) \vdash n$, and $I = (I_1, I_2, ..., I_n) \in \mathbb{N}^n$, we write $I \vdash \lambda$ if as multisets,

$$\{I_1, I_2, ..., I_n\} = \{\lambda_1, \lambda_2, ..., \lambda_k, 0^{n-k}\}.$$

COROLLARY 7.1. *For any sequence partition $\lambda \vdash n$,*

$$\frac{1}{n} \sum_{I \in \mathbb{N}^n, I \vdash \lambda} \omega^{\int_n I} = \frac{(-1)^{l_\lambda - 1}(l_\lambda - 1)!}{m_\lambda!}. \tag{9}$$

Proof. By Eq. (7),

$$K_n \simeq \frac{1}{n}(\alpha_0\omega^0+\alpha_1\omega^1+\cdots+\alpha_{n-1}\omega^{n-1})^n$$

$$=\frac{1}{n}\sum_{I\in\mathbb{N}^n,\,|I|=n}\binom{n}{I}\alpha_0^{I_0}\alpha_1^{I_1}\cdots\alpha_{n-1}^{I_{n-1}}\omega^{\int_n I}$$

$$\simeq\sum_{\lambda\vdash n}\binom{n}{\lambda}\left(\frac{1}{n}\sum_{I\in\mathbb{N}^n,\,I\vdash\lambda}\omega^{\int_n I}\right)a_\lambda.$$

Comparing the last expression with Eq. (6), we obtain the identity (9). ∎

For any given positive integer d, let ω denote the dth unit root and θ_d any umbra exchangeable with

$$\alpha_0\omega^0+\alpha_1\omega^1+\cdots+\alpha_{d-1}\omega^{d-1}.$$

Equation (8) provides us with the following information:

- $K_n(\theta_d)=0$ unless when $d\,|\,n$, $K_n=dK_n(\alpha)$.
- $\theta_d^n\simeq 0$ unless $d\,|\,n$.

Therefore, Theorem 6.2 generalizes to

THEOREM 7.2. *For any non-negative integer m,*

$$\theta_d^{dm}\simeq\sum_{\lambda\vdash m}d^{l_\lambda}\binom{dm}{d\lambda}^* K_{d\lambda},\tag{10}$$

where $d\lambda=(d\lambda_1, d\lambda_2, ..., d\lambda_k)$ *if* $\lambda=(\lambda_1, ..., \lambda_k)$; *and the inversion is given by*

$$K_{dm}\simeq\frac{1}{d}\sum_{\lambda\vdash m}\binom{d_{\prime}n}{d\lambda}^*\theta_d^{d\lambda}(-1)^{l_\lambda-1}(l_\lambda-1)!.\tag{11}$$

EXAMPLE 3. Suppose we want to estimate the dth cumulant K_d of a real random variable X. Let ω_d be the dth unit root. Assume that X_0,

$X_1, ..., X_{d-1}$ are available d independent occurrences of X. To estimate K_d, we define our statistic T_d to be

$$T_d = \frac{1}{d}(X_0 + X_1\omega + \cdots + X_{d-1}\omega^{d-1})^d.$$

By Theorem 7.1, T_d is an unbiased estimator of K_d. According to the last theorem, the standard error SE is

$$\begin{aligned} \text{SE} &= \sigma(T_d) = (ET_d^2 - K_d^2)^{1/2} \\ &= \left(\frac{1}{d^2}\left\{d\binom{2d}{2d}K_{2d} + d^2\binom{2d}{d}\frac{1}{2!}K_d^2\right\} - K_d^2\right)^{1/2} \\ &= \left(\left(\frac{1}{2}\binom{2d}{d} - 1\right)K_d^2 + \frac{K_{2d}}{d}\right)^{1/2}. \end{aligned}$$

Therefore, the standard error contains the information of K_{2d}.

8. A FINAL NOTE FROM THE SECOND AUTHOR

The paper is "distilled" from a longer draft by the authors and was rewritten by the second author after Gian-Carlo's unexpected departure from his beloved world, and his little umbral pets: α, β, ... Gian-Carlo's initial intention was to understand the positivity property of the moments and cumulants of a real random variable by Umbral Calculus. His major conjecture is

Conjecture. Any positive quantity (i.e., polynomial functions of the moments) of a real random variable can be expressed as a sum of squares of umbral polynomials.

This is different from Hilbert's problem. Umbrae certainly provide more flexibility. The conjecture was very much inspired by the following observation. A real unital sequence (a_n) is non-negative (i.e., can be realized as the moment sequence of a real random variable) if and only if all the Hankel determinants of (a_n) are non-negative. It is not difficult to show that the nth order Hankel determinant is precisely (up to a multiplicative constant) the evaluation of the squared Vandemonde

$$[V(\alpha_1, \alpha_2, ..., \alpha_n)]^2,$$

where $\alpha_1, ..., \alpha_n$ are distinct umbrae representing the sequence. Expressing cumulants (and even orthogonal polynomials) by umbrae is believed to be a first step. The present paper partially fulfills this goal.

Mixed into the Chinglish writing of the present paper is my evergreen memory of Gian-Carlo, my dear mentor and friend.

REFERENCES

1. N. I. Akhiezer, "The Classical Moment Problem" (N. Kemme, Transl.), Hafner, New York, 1965.
2. C. Berg, J. P. R. Christensen, and C. U. Jensen, A remark on the multidimensional moment problem, *Math. Ann.* **223** (1979), 163–169.
3. D. R. Brillinger, "Time Series, Data Analysis and Theory," Holt, Rinehart & Winston, New York, 1975.
4. B. V. Gnedenko and A. N. Kolmogorov, "Limit Distributions for Sums of Independent Random Variables," Addison–Wesley, Reading, MA, 1954.
5. D. Hilbert, "Theory of Algebraic Invariants," Cambridge Univ. Press, Cambridge, UK, 1993.
6. S. A. Joni and G.-C. Rota, Coalgebras and bialgebras in combinatorics, *Stud. Appl. Math.* **61** (1979), 93–139.
7. J. P. S. Kung and G.-C. Rota, The invariant theory of binary forms, *Bull. Amer. Math. Soc.* **10** (1984), 27–85.
8. H. J. Landau (Ed.), "Moments in Mathematics," Proceedings of Symposia in Applied Mathematics, Vol. 37, Amer. Math. Soc., Providence, RI, 1987.
9. S. M. Roman and G.-C. Rota, The umbral calculus, *Adv. in Math.* **27**, No. 2 (1978), 95–188.
10. G.-C. Rota, The number of partitions of a set, *Amer. Math. Monthly* **71**, No. 5 (1964), 498–504.
11. G.-C. Rota, D. Kahaner, and A. Odlyzko, On the foundations of combinatorial theory. VIII. Finite operator calculus, *J. Math. Anal. Appl.* **42** (1973), 684–760.
12. G.-C. Rota, J. Shen, and B. Taylor, All polynomials of binomial type are represented by Abel polynomials, *Ann. Scuola Norm. Sup. Pisa Cl. Sci. (4)* **25** (1997).
13. G.-C. Rota and B. D. Taylor, The classical umbral calculus, *SIAM J. Math. Anal.* **25**, No. 2 (1994), 694–711.
14. J. Shen, Combinatorics for wavelets: The umbral refinement equation, *Stud. Appl. Math.* **103**, No. 2 (1999), 121–147.
15. T. P. Speed, Cumulants and partition lattices, *Austral. J. Statist.* **25**, No. 2 (1983), 378–388.
16. G. Strang, "Introduction to Applied Mathematics," Wellesley–Cambridge Press, Wellesley, MA, 1986.

Printed by Catherine Press, Ltd., Tempelhof 41, B-8000 Brugge, Belgium

Rota's Foundations of Probability Theory

Joseph P. S. Kung

The mathematical career of Gian-Carlo Rota, like that of Beethoven, can be divided into three periods. In the early period, Rota's interest centered mainly on continuous mathematics, around the areas of differential equations, operator theory, ergodic theory, and probability theory. His papers from this period are collected in the first part of this volume. In the early 1960's, Rota's focus shifted to combinatorics, the main subject of his middle period. There must have been many motivations for this change. An important one was Rota's desire to uncover the mathematical structure behind identities used by Glen Baxter in fluctuation theory (see, for example, [7]). Rota found combinatorics in a primordial and chaotic state of having "too many theorems and too little theory" and proceeded to lay out a program to give the subject its "foundations".

It is likely that at the same time, Rota also found that probability theory itself was in need of a similar foundational program. However, unlike combinatorics, probability theory was (and is) a more developed subject and a similar *á outrance* attack on its foundations would have met with much resistance. Thus, Rota's "On the foundations of probability theory" took some time to appear. I first became aware of his interest in this area when I took a course from him at the Massachusetts Institute of Technology in 1977. Later, in the fall of 1978, Rota asked me to be the joint author of an expository article on "Probabilità" for the *Enciclopedia del Novecento.* The *Enciclopedia* was a project of the Istituto dell'Enciclopedia Italiana, an agency associated with the Italian government, and had reached the letter "P" around 1978. So it was a matter of some urgency to get the project finished before the twentieth century ran out. As was normal, Rota left the writing until the last minute, but he had obviously been thinking about it. His ideal of an encyclopedia article was very far from the dry recounting of conventional wisdom many would expect. Instead, he proposed to imitate M. W. Crofton, who set down the basic ideas of geometric probability in a famous article in the ninth edition of the *Encyclopædia Britannica.* I am not certain whether Rota achieved his aim (since we finished the article in four intensive weeks), but the final product [9], which appeared between "Preistoria" and "Produttivitá", had a decidedly unorthodox point of view and touched on many uncharted and controversal topics. It included, in particular, a slightly misnamed section on the "Fondamenti filosofici della probabiltà".

Rota's intense interest in probability theory began early in his career. Indeed, his conception of probability theory was formed by Willy Feller, from whom he took his first probability course as an undergraduate at Princeton (see [6]). Feller's classic two-volume textbook [2] was, to him, the "old testament". Rota agreed with Feller that probability theory is not a branch of measure theory. In particular, Kolmogorov's axiomatization of probability theory using measure theory and random variables is a distraction from the main issue. It is not that Kolmogorov did not have the truth, but that he did not have the whole truth. Consonant with his interest in combinatorics, Rota agreed with Feller that the study of discrete random variables is more – not less – difficult than the study of continuous random variables. He also held the opinion that there is a probabilistic way of thinking and it was characteristic of him that he tried to convey this way of thought in "his" undergraduate probability course, 18.313 at M. I. T., which he taught almost every year. One mathematician whose way of thought was essentially probabilistic was Alfréd Rényi (see, for example, [5]). Rota always thought that Rényi's work was grossly unappreciated in comparison with one of his coauthors, who enjoyed much better publicity.

Rota began writing seriously on the foundations of probability theory in the 1990's. Were it not for his early death in 1999, we would have seen the full flowering of his "third period". It is fortunate, therefore, that we have a detailed description of Rota's foundational program in probability in the text of his Fubini Lectures [8], lovingly brought to publication in the memorial volume [1]. It is perhaps not a vain hope that just as Rota's ideas in combinatorics were spread by his Hedrick Lectures and his summer course at Bowdoin College, his ideas in probability will receive a similar dissemination through his Fubini Lectures.

Four papers of Rota are reprinted in this chapter. The first, "A measure theoretic approach to logical quantification", was written with David Ellerman in the late 1970's. It explores the idea that "logic is, in some sense, a limiting case of probability theory". Valuation and generator rings of Boolean algebras are defined and used to overcome the technical difficulty that there does not exist a positive finitely-additive real-valued probability measure on an uncountable set. Because every finite Boolean algebra is atomic, the generator ring is "locally" a polynomial ring, in the sense that every finitely generated subalgebra is a polynomial ring. Hence, generator rings over a base field are integral domains and one can take the field of fractions to obtain a generator field for a Boolean algebra (over a given base field). Taking the base field to be the rationals (or the integers modulo a prime), the generator field gives a universal "field of values" for finitely-additive measures on the Boolean algebra. More details can be found in Section 6 of [4]. Another reconsideration of the foundations of logic by Rota can be found in [3].

The other three papers were written in the 1990's, at about the same time as [8]. The paper "Stochastic integrals: a combinatorial approach", written with Timothy Wallstrom, gives a combinatorial definition of multiple stochastic integrals. This paper gives connections between stochastic integrals, Möbius inversion over partition lattices, and sequences of polynomials of binomial type.

The other two papers are about cumulants. In the paper "On the combinatorics of cumulants", Rota and Jianhong Shen use umbral calculus to develop formulas and identities for cumulants. Combinatorialists may enjoy pondering the observation (in Section 2) that an umbra (or umbral symbol) can be considered a formal random variable and that evaluating the umbra can be viewed as formally taking the expectation. Finally, in the paper "Probability set functions", written with William Bruno and David Torney, cumulants of probability distributions on binary sequences of a fixed length are studied. This special case was motivated by DNA sequences.

References

[1] H. Crapo and D. Senato, eds., *Algebraic Combinatorics and Computer Science. A Tribute to Gian-Carlo Rota.* Springer-Verlag Italia, Milano, 2001.

[2] W. Feller, *An Introduction to Probability Theory and its Applications.* Vol. I, 3rd edition, Wiley, New York, 1968; Vol. II, 2nd edition, Wiley, New York, 1971.

[3] D. Finberg, M. Mainetti, and G.-C. Rota, The logic of commuting equivalence relations, In A. Ursini and P. Aglianò, eds., *Logic and Algebra (Pontignano, 1994),* Marcel Dekker, New York, 1996, pp. 69–96.

[4] J. P. S. Kung and C. H. Yan, Six problems of Gian-Carlo Rota in universal algebra and lattice theory. *Algebra Universalis,* to appear.

[5] A. Rényi, *Probability Theory.* North-Holland, Amsterdam, 1970.

[6] G.-C. Rota, Fine Hall in its golden age: remembrances of Princeton in the early fifties. In *A Century of Mathematics in America, Part II,* P. Duren, ed., Amer. Math. Soc., Providence RI, 1989, pp. 223–236.

[7] G.-C. Rota, Baxter operators, an introduction. In *Gian-Carlo Rota on Combinatorics,* J. P. S. Kung, ed., Birkhäuser, Boston and Basel, 1995, pp. 504–512.

[8] G.-C. Rota, Twelve problems in probability no one likes to bring up (The Fubini Lectures, Torino, Italy, June 3–5, 1998). In [1], pp. 57–93.

[9] G.-C. Rota and J. P. S. Kung, Probabilità. In *Enciclopedia del Novecento,* Vol. 5, Istituto dell'Enciclopedia Italiana, Rome, 1981, pp. 552–571.

Cumulants and Probability Set Functions

David C. Torney

Rota was keen to derive insights into what he called the cumulant problem: the discovery of elemental necessary and sufficient conditions for a collection of numbers to be the cumulants of a probability distribution. His preference was for these conditions to be expressed in terms of inequalities involving small numbers of cumulants. He believed that the umbral calculus had much potential in this regard, and strong corroboration is at hand [5].

In the paper "Probability Set Functions", reprinted in this chapter, by expressing distributions on binary n-sequences in terms of cumulants, the first steps were taken toward resolving the cumulant problem in this special case. The conciseness of our formulas is due principally to Rota's insistence upon simplification. Our results were applied to biological sequences in [7], and cumulants have also been applied recently to genotypes [1]. Such applications will, no doubt, be greatly facilitated by the determinantal formula for cumulants [5, Corollary 3.1].

Our combinatorial analysis of distributions on binary n-sequences provided the foundation for a subsequent extension into a general statistical method for classification of binary n-sequences, based upon examples [6]. This extension turned out to be, in part, an exercise in geometric probability [3]. I selected the moment parameterization of distributions to facilitate exploration of such extensions. Rota, however, believed that cumulants often provide an uncannily concise parameterization, and he urged the substitution of cumulants for moments in my method. Because the full implementation involving moments was sufficiently trying, I have yet to pursue this advice.

Shortly after I first described the subject of "Probability Set Functions" to Rota, he introduced me to much of the wonderland of the cumulant problem: for instance, moment problems, Edgeworth expansions, Clifford algebras, multivariate Hermite polynomials, and Hilbert's 17th problem (for the latter, see [2, p. 57], [4], and [5, p. 303].

Rota encouraged his collaborators to explore new ideas, which nucleated in his presence. With his encyclopædic knowledge, he could attest their novelty. We increased our appreciation for the importance of reformulation of established results in facilitating research. Many investigations germinated from one or two of his lucid statements; he could keep us busy, all right. Thus, we were profoundly uplifted by these collaborations.

References

[1] N. H. Barton, Estimating multilocus linkage disequilibria. *Heredity*, **84** (2000), 373–389.

[2] G. H. Hardy, J. E. Littlewood, and G. Pólya, *Inequalities.* Second Edition, Cambridge University Press, Cambridge, 1952.

[3] D. A. Klain and G.-C. Rota, *Introduction to geometric probability.* Lezioni Lincee, Cambridge Univ. Press, Cambridge, 1997.

[4] A. Pfister, Hilbert's seventeenth problem and related problems on definite forms. In F. E. Browder, ed., *Mathematical developments arising from Hilbert problems, Part 2,* Proceedings of Symposia in Pure Mathematics, Vol. XXVIII, Amer. Math. Soc., Providence RI, 1976, pp. 483–489.

[5] G.-C. Rota and J. Shen, On the combinatorics of cumulants. *J. Combin. Theory Ser. A*, **91** (2000), 283–304. Reprinted in this chapter.

[6] D. C. Torney, Bayesian analysis of binary sequences. *Adv. Appl. Math.*, submitted.

[7] D. C. Torney, C. C. Whittaker, and G. Xie, The stationary statistical properties of human coding sequences. *J. Mol. Biol.*, **286** (1999), 1461–1469.

GPSR Compliance

The European Union's (EU) General Product Safety Regulation (GPSR) is a set of rules that requires consumer products to be safe and our obligations to ensure this.

If you have any concerns about our products, you can contact us on ProductSafety@springernature.com

In case Publisher is established outside the EU, the EU authorized representative is:

Springer Nature Customer Service Center GmbH
Europaplatz 3
69115 Heidelberg, Germany

Batch number: 10578424

Printed by Printforce, the Netherlands